Foreword

Deep Foundations Institute is an international non-profit association of engineers, contractors, manufacturers, equipment suppliers in the deep foundations and excavations operated from the United States of America. DFI brings together members through networking, education, communication and collaboration. DFI promotes the advancement of the deep foundation industry through technical committees, educational programs, seminars, webinars, workshops and conferences, publication of guides and specifications, a peer-reviewed journal, a flagship magazine, research, government relations, and outreach.

Deep Foundations Institute of India (DFI of India), a non-profit organisation, was registered in Chennai with the Ministry of Company Affairs as a regional chapter of DFI USA in 2013. DFI of India's mission is to support the Indian foundation industry on a continuous and sustained basis in measured and measurable steps, become very professional, and embrace new technologies for faster development of India. The chapter looks to provide a platform for continuous interaction with the Indian foundation industry stakeholders, including those international agencies with an India focus. Conferences, workshops, seminars, and webinars are conducted regularly to highlight best practices and expertise worldwide. Interesting case studies and panel discussions are presented to raise technical awareness and propel advanced foundation design, construction, testing, contracting methods, and risk-sharing in the Indian foundation industry.

Focussing on the key mission of technological enhancement of the deep foundation industry, DFI of India organises annual conferences with the common theme 'Deep Foundation Technologies for Infrastructure Development in India'. These yearly conferences showcase advanced deep foundation technologies and their application in Indian infrastructure projects through expert keynote presentations and contributory papers from industry and academia. The first conference was held in 2011 in Hyderabad, followed by the conferences at Chennai (2012), Mumbai (2013), New Delhi (2014), Bangalore (2015), Kolkata (2016), Chennai (2017), Gandhinagar (2018), Hyderabad (2019) two virtual conferences in 2020 and 2021, and Tirupati (2022).

The contributory papers are invited from the professional and academia through a call for abstracts. The abstracts are reviewed, and the authors of the selected abstracts are asked to develop the full-length paper. A pool of reviewers in the respective fields thoroughly examines the writings. At least two reviewers review and re-review each paper before accepting it for publication.

DFI of India considered it appropriate to publish the selected papers that were presented live during the conference as a conference proceedings volume under the Springer series 'Lecture Notes in Civil Engineering.' Publishing the conference proceedings through a prestigious publisher, Springer, would enhance the visibility of the commendable works done in deep foundation technologies in India. DFI of India places on record its gratitude to Springer for bringing out this conference volume to benefit the geotechnical engineering community.

Er. I. V. Anirudhan
Chairman, DFI of India
Chennai, Tamil Nadu, India

Prof. A. Murali Krishna
IIT Tirupati, DFI India 2022
Conference Chairmen
Tirupati, Andhra Pradesh, India

Preface

Deep Foundations Institute of India conducted its 11th annual conference on 'Deep Foundation Technologies for Infrastructure Development in India': DFI India 2022, during 15–17 September 2022 at Tirupati. The conference was organised under different subthemes as listed below:

1. Deep Foundation and Deep Excavation Technologies for Metro and High Speed Rail,
2. Innovative Techniques and Testing for Geotechnical Investigations Monitoring, Performance, and QA-QC of Deep Foundations,
3. Challenges to Working and the Impact of Existing Codes with Respect to Technical Innovations in Deep Foundations,
4. Ground Improvement Techniques for Mass Stabilization,
5. Geotechnics for Marine, Near-Shore, and Coastal Construction,
6. Research, Experimental and Numerical Methods in Deep Foundations and Deep Excavation Technologies,
7. Mono-piles and Other Advanced Deep Foundation Systems,
8. Deep Foundations in Coralline and Carbonate Rocks, and Any Other Unique and Interesting Papers Related to Deep Foundation Themes.

A total of 117 abstracts were received under different sub-themes listed above. After peer review, finally 50 full papers got accepted for publication of the conference. Each article was peer-reviewed by two reviewers, each reviewer not reviewing more than six papers. We sincerely thank all the reviewers for their valuable time and effort in reviewing the papers. Out of the accepted papers, 24 papers that are presented live in the conference are getting published through Springer Nature Publication.

Most of these articles were developed based on the field experiences requiring the theories put into practice. Under the theme 1, Deep Foundation and Deep Excavation Technologies for Metro and High Speed Rail, six articles are featured. Five articles are featured in themes two and three. Further, five articles are featured under the theme of Ground Improvement Techniques for Mass Stabilisation covering effective use of stone columns, Controlled Modulus Columns, Deep Mixing technique, and Micro Piles. Two articles on application of recycled polymer and Stabilisation of

Saline Soil with fly ash are discussed in the theme Geotechnics for Marine, Near-shore, and Coastal Construction. One article is featured in theme 6. Three interesting articles on monopiles, pedestal piles and Trench cutter technology are featured in the Seventh theme. Two interesting articles are featured in theme 8. We thank all the authors for their contributions to the conference.

We, the editors, sincerely hope that the articles covered in this volume would immensely help the professionals and academicians alike to improve the work practice.

Tirupati, India — Prof. A. Murali Krishna
Chennai, India — Prof. Subhadeep Banerjee
Chennai, India — Dr. N. Kumar Pitchumani
Technical Editors

Contents

Deep Foundation and Deep Excavation Technologies for Metro and High Speed Rail

Lecture Notes in Civil Engineering

Volume 373

Lecture Notes in Civil Engineering (LNCE) publishes the latest developments in Civil Engineering—quickly, informally and in top quality. Though original research reported in proceedings and post-proceedings represents the core of LNCE, edited volumes of exceptionally high quality and interest may also be considered for publication. Volumes published in LNCE embrace all aspects and subfields of, as well as new challenges in, Civil Engineering. Topics in the series include:

- Construction and Structural Mechanics
- Building Materials
- Concrete, Steel and Timber Structures
- Geotechnical Engineering
- Earthquake Engineering
- Coastal Engineering
- Ocean and Offshore Engineering; Ships and Floating Structures
- Hydraulics, Hydrology and Water Resources Engineering
- Environmental Engineering and Sustainability
- Structural Health and Monitoring
- Surveying and Geographical Information Systems
- Indoor Environments
- Transportation and Traffic
- Risk Analysis
- Safety and Security

To submit a proposal or request further information, please contact the appropriate Springer Editor:

- Pierpaolo Riva at pierpaolo.riva@springer.com (Europe and Americas);
- Swati Meherishi at swati.meherishi@springer.com (Asia—except China, Australia, and New Zealand);
- Wayne Hu at wayne.hu@springer.com (China).

All books in the series now indexed by Scopus and EI Compendex database!

A. Murali Krishna · Subhadeep Banerjee ·
N. Kumar Pitchumani
Editors

Deep Foundations for Infrastructure Development in India

Proceedings of DFI-India 2022 Annual Conference

Springer

Editors
A. Murali Krishna
Department of Civil and Environmental Engineering
Indian Institute of Technology Tirupati
Tirupati, Andhra Pradesh, India

Subhadeep Banerjee
Department of Civil Engineering
Indian Institute of Technology Madras
Chennai, Tamil Nadu, India

N. Kumar Pitchumani
AECOM India Pvt. Ltd.
Chennai, Tamil Nadu, India

ISSN 2366-2557 ISSN 2366-2565 (electronic)
Lecture Notes in Civil Engineering
ISBN 978-3-031-37255-1 ISBN 978-3-031-37256-8 (eBook)
https://doi.org/10.1007/978-3-031-37256-8

This Springer imprint is published by the registered company Springer Nature Switzerland AG
The registered company address is: Gewerbestrasse 11, 6330 Cham, Switzerland

Foreword

Deep Foundations Institute is an international non-profit association of engineers, contractors, manufacturers, equipment suppliers in the deep foundations and excavations operated from the United States of America. DFI brings together members through networking, education, communication and collaboration. DFI promotes the advancement of the deep foundation industry through technical committees, educational programs, seminars, webinars, workshops and conferences, publication of guides and specifications, a peer-reviewed journal, a flagship magazine, research, government relations, and outreach.

Deep Foundations Institute of India (DFI of India), a non-profit organisation, was registered in Chennai with the Ministry of Company Affairs as a regional chapter of DFI USA in 2013. DFI of India's mission is to support the Indian foundation industry on a continuous and sustained basis in measured and measurable steps, become very professional, and embrace new technologies for faster development of India. The chapter looks to provide a platform for continuous interaction with the Indian foundation industry stakeholders, including those international agencies with an India focus. Conferences, workshops, seminars, and webinars are conducted regularly to highlight best practices and expertise worldwide. Interesting case studies and panel discussions are presented to raise technical awareness and propel advanced foundation design, construction, testing, contracting methods, and risk-sharing in the Indian foundation industry.

Focussing on the key mission of technological enhancement of the deep foundation industry, DFI of India organises annual conferences with the common theme 'Deep Foundation Technologies for Infrastructure Development in India'. These yearly conferences showcase advanced deep foundation technologies and their application in Indian infrastructure projects through expert keynote presentations and contributory papers from industry and academia. The first conference was held in 2011 in Hyderabad, followed by the conferences at Chennai (2012), Mumbai (2013), New Delhi (2014), Bangalore (2015), Kolkata (2016), Chennai (2017), Gandhinagar (2018), Hyderabad (2019) two virtual conferences in 2020 and 2021, and Tirupati (2022).

The contributory papers are invited from the professional and academia through a call for abstracts. The abstracts are reviewed, and the authors of the selected abstracts are asked to develop the full-length paper. A pool of reviewers in the respective fields thoroughly examines the writings. At least two reviewers review and re-review each paper before accepting it for publication.

DFI of India considered it appropriate to publish the selected papers that were presented live during the conference as a conference proceedings volume under the Springer series 'Lecture Notes in Civil Engineering.' Publishing the conference proceedings through a prestigious publisher, Springer, would enhance the visibility of the commendable works done in deep foundation technologies in India. DFI of India places on record its gratitude to Springer for bringing out this conference volume to benefit the geotechnical engineering community.

Er. I. V. Anirudhan
Chairman, DFI of India
Chennai, Tamil Nadu, India

Prof. A. Murali Krishna
IIT Tirupati, DFI India 2022
Conference Chairmen
Tirupati, Andhra Pradesh, India

Preface

Deep Foundations Institute of India conducted its 11th annual conference on 'Deep Foundation Technologies for Infrastructure Development in India': DFI India 2022, during 15–17 September 2022 at Tirupati. The conference was organised under different subthemes as listed below:

1. Deep Foundation and Deep Excavation Technologies for Metro and High Speed Rail,
2. Innovative Techniques and Testing for Geotechnical Investigations Monitoring, Performance, and QA-QC of Deep Foundations,
3. Challenges to Working and the Impact of Existing Codes with Respect to Technical Innovations in Deep Foundations,
4. Ground Improvement Techniques for Mass Stabilization,
5. Geotechnics for Marine, Near-Shore, and Coastal Construction,
6. Research, Experimental and Numerical Methods in Deep Foundations and Deep Excavation Technologies,
7. Mono-piles and Other Advanced Deep Foundation Systems,
8. Deep Foundations in Coralline and Carbonate Rocks, and Any Other Unique and Interesting Papers Related to Deep Foundation Themes.

A total of 117 abstracts were received under different sub-themes listed above. After peer review, finally 50 full papers got accepted for publication of the conference. Each article was peer-reviewed by two reviewers, each reviewer not reviewing more than six papers. We sincerely thank all the reviewers for their valuable time and effort in reviewing the papers. Out of the accepted papers, 24 papers that are presented live in the conference are getting published through Springer Nature Publication.

Most of these articles were developed based on the field experiences requiring the theories put into practice. Under the theme 1, Deep Foundation and Deep Excavation Technologies for Metro and High Speed Rail, six articles are featured. Five articles are featured in themes two and three. Further, five articles are featured under the theme of Ground Improvement Techniques for Mass Stabilisation covering effective use of stone columns, Controlled Modulus Columns, Deep Mixing technique, and Micro Piles. Two articles on application of recycled polymer and Stabilisation of

Saline Soil with fly ash are discussed in the theme Geotechnics for Marine, Near-shore, and Coastal Construction. One article is featured in theme 6. Three interesting articles on monopiles, pedestal piles and Trench cutter technology are featured in the Seventh theme. Two interesting articles are featured in theme 8. We thank all the authors for their contributions to the conference.

We, the editors, sincerely hope that the articles covered in this volume would immensely help the professionals and academicians alike to improve the work practice.

Tirupati, India
Chennai, India
Chennai, India

Prof. A. Murali Krishna
Prof. Subhadeep Banerjee
Dr. N. Kumar Pitchumani
Technical Editors

Contents

Deep Foundation and Deep Excavation Technologies for Metro and High Speed Rail

A Case Study of Deep Shaft Excavation in River Channel Soil Deposit in Kolkata

Kaustuv Bhattacharya, Manish Kumar, Praveen Soni, Nirbhay Singh, and Prodyot Kr. Ray

Abstract This paper highlights salient features of an excavation 38 m (Length) by 10 m (Width) by 24 m (Depth) carried out in Central Kolkata. Secant pile wall (SPW) formed using a combination of alternating 1200 mm and 800 mm diameter piles was used. Six layers of struts and waler beams including bottom slab were utilized as lateral restraints to piles. The geotechnical profile for excavation zone was predominantly silty sand to sandy silt. The profile is characteristic of river channel deposit usually encountered in Kolkata. Temporary stage analysis (including seepage analysis) and design were carried out in FEM software. Design checks for clay bursting, piping, bottom heave, etc. were also carried out. Extensive instrumentation and monitoring were carried out during the excavation period to check parameters such as ground movement, pile movement, piezometric heads, etc. as per review levels. During excavation, ground improvement using Tube-a-manchette grouting technique was also required to be carried out as per field observations. The paper discusses some of the major challenges faced as well as measures taken to counter them, during deep excavation carried out in silty soils in a very congested urban setting in the heart of a Megacity.

Keywords Kolkata river channel deposit · Braced excavation · Secant pile wall

K. Bhattacharya (✉) · M. Kumar · P. Soni · N. Singh · P. Kr. Ray
ITD Cementation India Ltd., Chennai, India
e-mail: kaustuv.bhattacharya@itdcem.co.in

M. Kumar
e-mail: manish.kumar@itdcem.co.in

P. Soni
e-mail: praveen.soni@itdcem.co.in

N. Singh
e-mail: nirbhay.singh@itdcem.co.in

P. Kr. Ray
e-mail: prodyot.ray@itdcem.co.in

A. M. Krishna et al. (eds.), *Deep Foundations for Infrastructure Development in India*, Lecture Notes in Civil Engineering 373, https://doi.org/10.1007/978-3-031-37256-8_1

1 Introduction

In a fast developing India, where urban public transport sector is blooming, all major cities are seeing a development of Metro-railway corridors. In cities, where availability of land is a major concern, the authorities and civil contractors look for underground corridors, which would typically involve underground metro stations and Bored/NATM tunnels. The retaining system for deep excavations is generally of several types—Soldier pile with lagging, contiguous bored pile walls, secant pile walls, diaphragm walls, etc. Soil/rock anchors, strut/waler arrangements usually form a part of the retaining system. The selection of the type depends on several factors like depth of excavation, geotechnical profile, position of groundwater table, local conditions, construction timeline planned, etc. Based on boreholes conducted in the previous construction works of the North–South line of the existing Kolkata Metro railway, extensive geotechnical characterization was carried out by Ghosh and Gupta (1972). Current paper has utilized some of their findings in current project work.

2 Brief Summary of the Case Study

A cut and cover box tunnel was envisaged to be constructed at a depth of approximately 24 m below the existing ground level. A shaft with plan dimensions of 38 m by 10 m was designed for the same. Secant Pile wall comprising 1200 mm dia Hard Pile and 800 mm Soft Pile, with overlap of 200 mm, was envisaged as the retaining structure for the deep excavation. The details are shown in Fig. 1.

The selection of the system was based on analysis of soil conditions, the existing groundwater table and existing site conditions. The depth of the excavation was 24 m approximately. The pile embedment ratio (Pile Depth/Depth of Excavation)

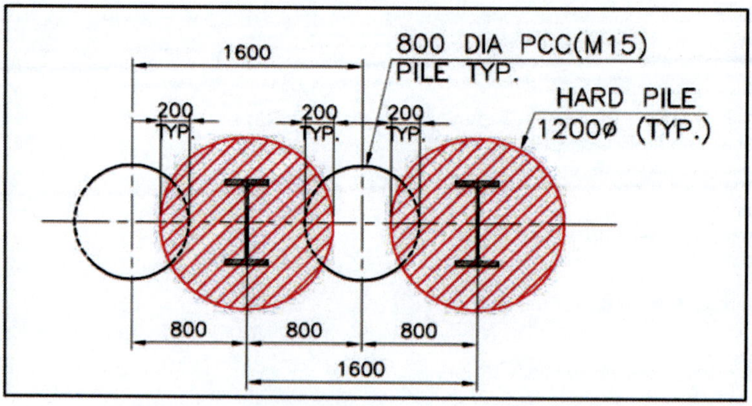

Fig. 1 Secant pile arrangement showing hard and soft piles

was found to be approximately 1.58. The support system comprised five layers of strut/waler system and final layer of base slab for tunnel box. The first and fifth level had steel strut/waler arrangement, whereas second, third and fourth levels had reinforced concrete strut/waler arrangement.

3 Geological and Geotechnical Discussion

The site geomorphology is characterized as Kolkata River Channel Deposit. As per Ghosh and Gupta (1972), the River Channel deposits are superimposed features over the typical subsoil deposits of the Kolkata region and comprise silty sand to sandy silt of considerable thickness, with clayey silt to silty clay in the upper portion of the deposit, at certain locations. The authors had reported sand deposits to be fine to medium coarse. A typical profile of Kolkata River Channel deposit as presented by Ghosh and Gupta (1972) is shown in Fig. 2. The Geomorphology in several other areas of Kolkata Municipal Corporation was discussed by Alpana (2004).

Six number of boreholes were conducted in the shaft vicinity. These were conducted by Shell and Auger methodology. Standard penetration test was conducted at 1.5 m depth interval and undisturbed samples collected at 2.5–3 m depth intervals. The frequency was modified as per site condition during change of soil layer. The borelog profile from the conducted boreholes is shown in Fig. 3. The description of the soil strata is given in Table 1. The variation of SPT results is plotted in Fig. 4. The water table was found to be approximately 4.0 m below existing ground level, but for design purpose considered 1.2 m below ground level.

Based on the site investigation data, the Geotechnical Interpretative Report was prepared, which was utilized in FEM analysis. The summary of geotechnical parameters used in the analysis and design is shown in Table 2. The brief categorization of soil strata for design purpose is given below.

The top soil/filled up soil had a varied thickness of 3–6 m. Soil type was found to be silty sand/sandy silt of varying plasticity. Traces of mica were observed. The average SPT value was 8, with a wide variation within short depth. The average Atterberg limits are LL = 36%, PL = 20%. The average undrained shear strength was estimated to be 6N = 48 kPa.

The second layer had an average thickness of 4 m. This is dominant silt layer, with local silty clay layer in some boreholes. The average SPT is around 8. The average Atterberg limits are LL = 40%, PL = 22%, for silt layer and LL = 44%, PL = 21%, for silty clay layer. The average undrained shear strength was estimated to be 5N = 40 kPa.

The third layer is the predominant layer with an average thickness of 13.5 m. This layer has laminations of silty sand/sand. Traces of mica were observed. The average SPT is around 16. The average Atterberg limits for sandy silt are LL = 34%, PL = 22%. Low plastic sandy silt is treated as c-phi soil. The undrained shear strength was estimated to be 6N = 96 kPa.

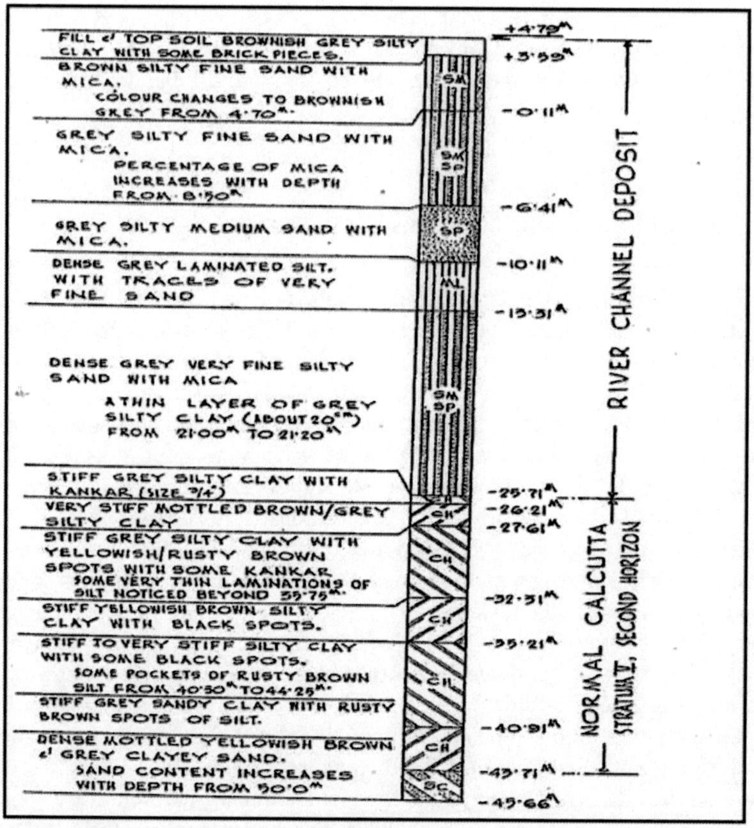

Fig. 2 Typical Kolkata river channel deposit (Ghosh and Gupta 1972)

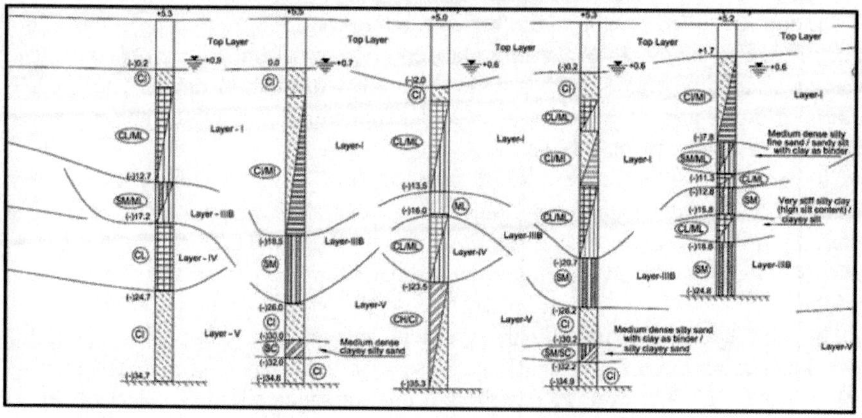

Fig. 3 Borelog profile at the Shaft zone

Table 1 Description of soil layers encountered in borelog

Strata	Soil description
Top layer	Road material followed by clayey silt/clayey silty sand/clayey sandy silt/kankars/ brick bats
I	Firm to stiff/ hard silty clay (high silt content)/clayey silt with occ. laminations of sand
IIIB	Medium dense sandy silt with clay as binder/silty fine sand with traces of mica
IV	Very stiff/hard silty clay (high silt content)/clayey silt with sand/laminations of sand
V	Stiff/very stiff to hard silty clay with yellow/brown spots

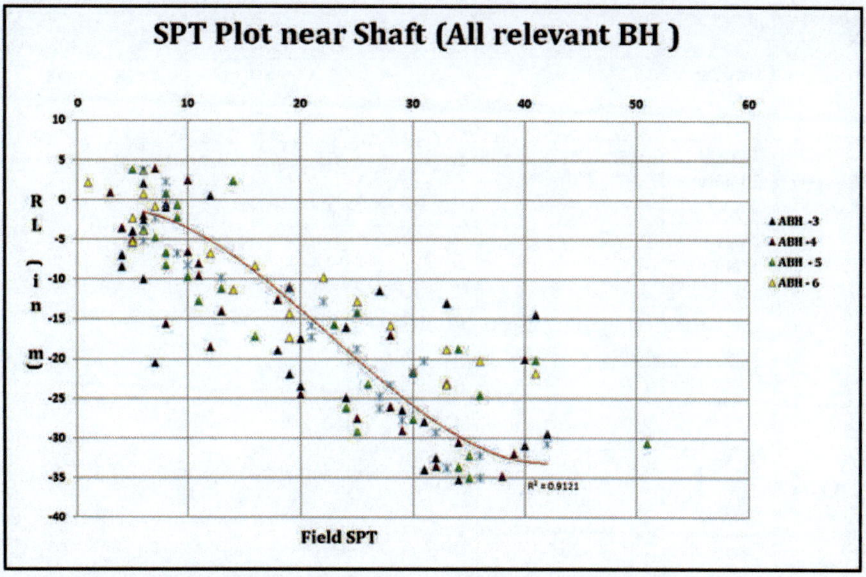

Fig. 4 SPT variation along depth (with SPT best-fit line), in shaft vicinity

The fourth layer is silty sand/sand layer and has a thickness of 7–8 m below the third layer. There is intermixing of layers with bands of sandy silt and clayey silt. The average recorded SPT is 25. These are generally non-plastic in nature. These exhibit drained behavior.

Underlying the sand layer is clayey silt with silty clay with subordinate laminations of sand. The average SPT is around 27. The average Atterberg limits for clayey silt indicate LL is 40%, PL is 21% and for silty clay is LL is 45%, PL is 20%. The undrained shear strength was estimated to be 4.5N = 122 kPa.

The correlation of SPT with shear strength was based on Stroud (1974).

Table 2 Summary of geotechnical parameters used in FEM analysis and design

Elevation	Soil description	Unit weight (kN/m^3)	Avg. SPT	Strength parameters			Ko	Eu (Mpa)	E' (MPa)
				Total stress cu (Kpa)	Effective c′ (Kpa)	Stress phi′ (°)			
(+) 5.50–(−) 0.5	Top soil	17.3	8	48	0	30	1.2	20	14
(−) 0.50–(−) 4.0	Silt/silty clay	18.5	8	40	50	28	1	25	18
(−) 4.0–(−) 18.5	Sandy silt	19	16	96	3	30	0.5	22	16
(−) 18.50–(−) 25.0	Silty sand/ sand	19.5	25	–	3	34	0.4	28	25
(−) 25.0–(−) 35.0	Clayey silt/silty clay	19.5	27	122	65	23	0.6	30	45
(−) 35.0–(−) 50.0	Clayey silt/silty clay	19.5	27	122	65	23	0.6	30	45

		Poison ratio	E_{oed} *(Mpa)*	*Permeability, k(cm/s)*
(+) 5.50–(−) 0.5	Top soil	0.3	–	10^{-4}
(−) 0.50–(−) 4.0	Silt/silty clay	0.3	24	10^{-4}
(−) 4.0–(−) 18.5	Sandy silt	0.35	26	10^{-4}
(−) 18.50–(−) 25.0	Silty sand/ sand	0.35	40	5×10^{-4}
(−) 25.0–(−) 35.0	Clayey silt/silty clay	0.25	54	10^{-5}
(−) 35.0–(−) 50.0	Clayey silt/silty clay	0.25	54	10^{-6}

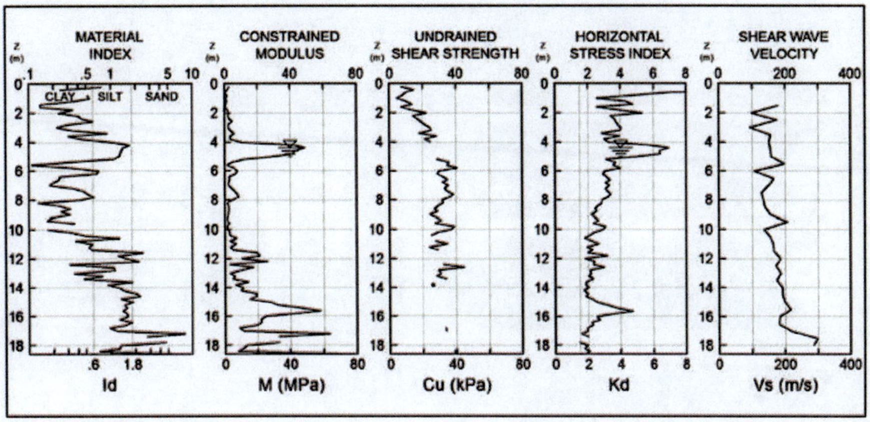

Fig. 5 Sample results from the dilatometer test

Additionally, flat dilatometer tests were carried out in the vicinity of the shaft, to determine the in-situ strength and deformation properties, at intervals of 500 mm. The basis of the test consists of inserting vertically into the soil a blade-shaped steel probe with a thin expandable circular steel membrane mounted flush on one face and determining, at selected depths, the contact pressure exerted by the soil against the membrane when the membrane is flush with the blade and subsequently the pressure exerted when the central displacement of the membrane reaches 1.10 mm. The shear wave velocity was measured in every 0.50 m of depth interval. A sample result from the test is shown in Fig. 5.

4 Finite Element Analysis and Results

The analysis for shaft excavation was carried out in FEM software, Plaxis 2D. The geotechnical parameters as summarized in Table-2 were provided as inputs. Plain strain modeling was done. Hardening Soil Model (Schanz and Vermeer 1999) was selected. The model used a non-linear elastic behavior that relates the elastic modulus to the stress level.

The construction sequence flowchart is as follows.

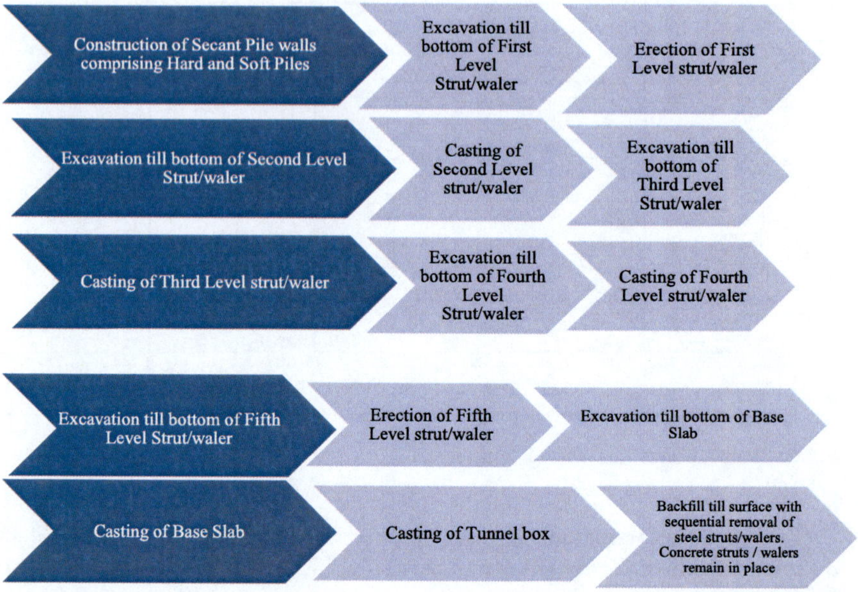

The finite element model is shown in Fig. 6. The general arrangement of shaft is shown in Fig. 7.

The reinforced hard piles were designed to take earth and water pressure, whereas soft piles were only for water tightness. Thus the hard piles were to be designed for Bending Moment, Shear and Axial forces. Hence, they were modeled as a plate element. Soil structure interaction was considered in model. Equivalent section based on pile spacing was considered to compute normal and bending stiffness. The struts (both steel and concrete) were modeled as node to node anchors, as they were predominantly members taking axial load. A typical strut spacing of 6 m c/c was maintained for construction activity.

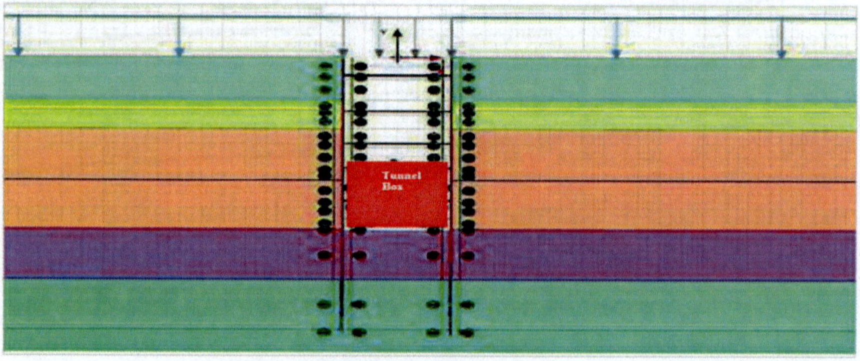

Fig. 6 Finite element model with tunnel box imprinted

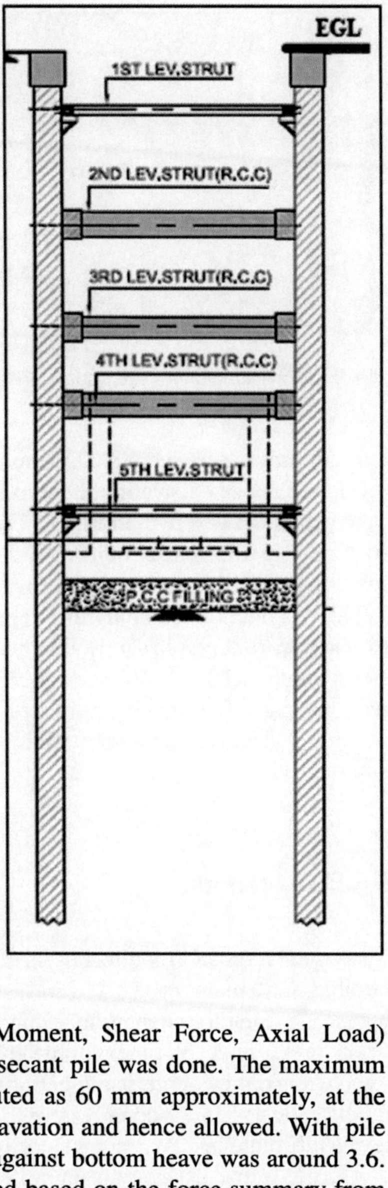

Fig. 7 General arrangement of shaft

Based on the force summary (Bending Moment, Shear Force, Axial Load) obtained from the FEM model, the design of secant pile was done. The maximum deflection of the retaining system was computed as 60 mm approximately, at the fifth strut level and was 0.25% of depth of excavation and hence allowed. With pile embedment ratio of 1.58, the factor of safety against bottom heave was around 3.6. The struts and waler beams were also designed based on the force summary from the FEM analysis. Surface settlement contours were developed based on the FEM analysis. These contours were later utilized in the damage assessment of adjoining civil structures (mainly buildings) and further planning for monitoring the same.

A seepage model was separately carried out, in SEEP/W software, to check the hydraulic cutoff of the retaining wall system. The secant pile wall was modeled as a plate element with very low permeability ($k = 5 \times 10^{-13}$ m/s). The boundary

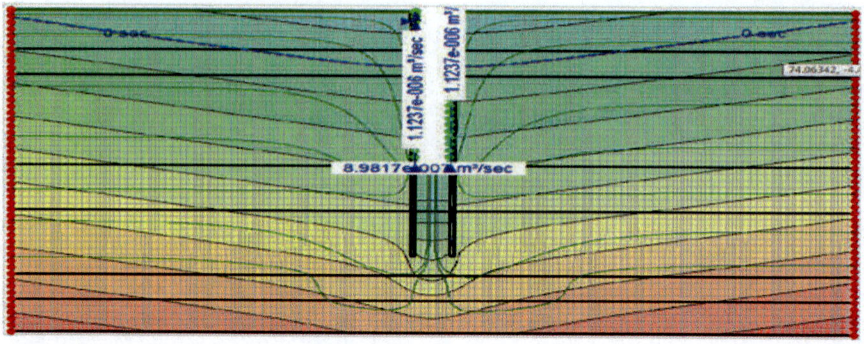

Fig. 8 Seepage model for excavation of shaft

conditions were set such that location of groundwater table was 1.2 m below ground level, and inside excavation the maximum depth was considered. A snapshot of the seepage model is shown in Fig. 8. The exit gradient was checked against the critical hydraulic gradient and found okay. Checks for piping effects were also made, at maximum excavation level and found within limits, with FOS achieved = 2.08 > 1.30. The effects of clay bursting were checked and found negligible as the maximum excavation level was in sandy silt/silty sand strata and was underlain by clayey strata. Based on the analysis, only dewatering inside shaft area, during construction period was envisaged for working purpose.

Based on above, the braced retaining system was found to be adequate for current site scenario.

5 Field Results

Instrumentation and monitoring form an integral part of any deep excavation work. Ground settlement markers were established in array pattern. Stand pipes were installed to monitor changes in groundwater table. Piezometers were also installed to check pressure development at different levels. The deflections of piles were monitored using optical targets and inclinometers. Strain gauges were installed in struts to monitor the developed compressive forces. A representative plot showing piezometer levels (absolute levels) around shaft is shown in Fig. 9.

During initial excavation, pump out test was carried out inside the shaft, with pumping wells and monitoring wells, to check the efficacy of the retaining system.

Surface Settlement markers were provided to monitor the ground settlement, whereas the deflection of secant pile wall was measured by sample inclinometers. The ratio between actual deflection and predicted deflection was less than 0.8 till excavation up to third level strut and reached close to (but less than) unity till final level of excavation. The strut loads were monitored by strain gauges, to check that the loads coming onto the struts were within design limits.

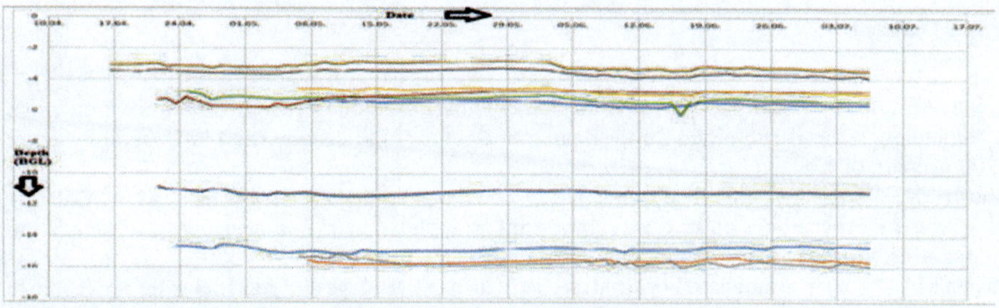

Fig. 9 Plot showing piezometer levels (absolute levels) around shaft, wrt time

Ground improvement was carried out by Tube-a-manchette grouting technique, as a precautionary measure in the exact vicinity of the shaft, to minimize any surface settlement. In this method, the grout was pumped into the ground through a perforated pipe with sleeve valves, which act like no return valves. The hole that is used is controlled by the usage of a double packer.

Adjoining buildings situated within the zone of influence were carefully monitored, based on the Alert-Action-Alarm indicators. Foundation grouting was carried out and temporary supports were provided for critical structures.

6 Discussion

Some of the major challenges of this project were:

a. Due to the geomorphology of the project site, i.e. Kolkata River Channel deposit, parts of the ground had higher permeability, whereas others had very low permeability, leading to confined zones with higher piezometric pressures. It was hard to detect these small bands during boring for soil investigation. The Dilatometer test conducted at 0.5 m intervals had shown these bands, however, with the available equipment, the data were limited till 18 m below ground level.
b. Due to the depth of excavation of 24 m, with pile embedment ratio as 1.58 and pile diameter of 1200 mm, stabilizing the soil during boring activity was a challenge, especially with the presence of the sandy silt/silty sand strata, where chances of collapse are very high. Casing had to be used. This is also the reason why diaphragm wall was not used here.
c. Due to the high head of water and non-plastic nature of the soil, at certain depths, there were risks of water and soil ingress through the joints of the retaining wall. Hence, additional grouting had to be carried out to stop the water as well as stabilize the disturbed soil, and reduce impact at the surface. The cutoff pressures and grout volumes had to be ascertained.
d. Civil structures adjacent to the structure had to be protected. Hence, close monitoring had to be done to check for surface settlement, pile deflection, building settlement. Mitigation measures were adopted accordingly.

References

Alpana M (2004) Geomorphology of Kolkata municipal corporation (KMC)—its impact on urban functions. A thesis submitted for the degree of Doctor of Philosophy (Sc.) in Geography under University of Kolkata

Ghosh PK, Gupta S (1972) The cementation company limited. Subsoil character of Calcutta region. In: Proceedings of the third symposium on application of soil mechanics and foundation engineering in Eastern India, February, pp 49–61

Stroud MA (1974) The standard penetration test in insensitive clays and soft rocks. In: Proceedings of the European symposium on penetration testing, Stockholm, June 5–7, 1974, pp 367–375

Schanz T, Vermeer PA, Bonnier PG (1999) The hardening soil model: formulation and verification. In: Proceedings of the international symposium beyond 2000 in computational geotechnics, Amsterdam, The Netherlands, 18–20 March 1999, pp 281–296

Water Powered Percussive Downhole Hammers—Energy Savings and Productivity

Shyam Chengalath, Johan Jeansson, and Magnus Hörman

Abstract Down-The-Hole (DTH) drilling usually uses compressed air to power the hammer. However, there are now DTH hammers on the market which instead of compressed air use high pressure water to power the hammer. There are several benefits with using water DTH compared to air DTH, which are discussed in more detail in this paper. From a physics point of view, many of the benefits arise from the fact that water is an incompressible media and air is not. Energy consumption is roughly a quarter compared to air, resulting in lower carbon emission and a lower energy bill. In the case of water DTH, the water leaving the drill bit in the hole is at a low velocity and ambient pressure. However, with air DTH, the air leaving the drill bit in the hole is at a high velocity and high pressure. This means that the risk of damaging surrounding structures in sensitive areas such as urban environments and dams is greatly reduced with water DTH. In this paper, two cases where water DTH is used in jet-grouting applications are discussed in detail. As these cases show there is a clear improvement in productivity when using water DTH hammers.

Keywords Water DTH hammer · Low carbon emission · Jet-grouting

1 Short Introduction to Different Drilling Technologies

In principle, there are two major types of drilling, rotary drilling and percussion drilling. Rotary drilling is mainly used for drilling in softer rock and non-consolidated ground since the mechanism of rock destruction is grinding by applying high torque

S. Chengalath (✉)
Panasia Technical Services Ltd., Gurgaon, Haryana, India
e-mail: shyam@panasia.biz

J. Jeansson · M. Hörman
LKAB Wassara AB, Stockholm, Sweden
e-mail: johan.jeansson.wassara@lkab.com

M. Hörman
e-mail: magnus.horman.wassara@lkab.com

A. M. Krishna et al. (eds.), *Deep Foundations for Infrastructure Development in India*, Lecture Notes in Civil Engineering 373, https://doi.org/10.1007/978-3-031-37256-8_2

rotation and high weight on the drill bit. It should, however, be noted that there is one type of rotary drilling, diamond drilling, which is used for taking core samples in rock. In this case, the drill string is rotated at high speed but with low torque and low weight on bit. The rock-breaking mechanism in percussion drilling is crushing of the rock by continuous blows under continuous low torque rotation of the drill string. Weight on bit is low. Percussion drilling can be subdivided into Top hammer drilling and Down-The-Hole drilling (DTH), see Fig. 1.

In Top hammer drilling, the percussion unit is outside the hole hammering on the drill string whereas in DTH drilling, the percussion unit is inside the hole hammering directly on the drill bit. This avoids losses in drill rod connections and makes DTH more suitable for drilling deeper holes than top hammer. In DTH, drilling water or air is used to both power the hammer and to flush drill cuttings out of the hole. For more details, see Bruce et al. (2013).

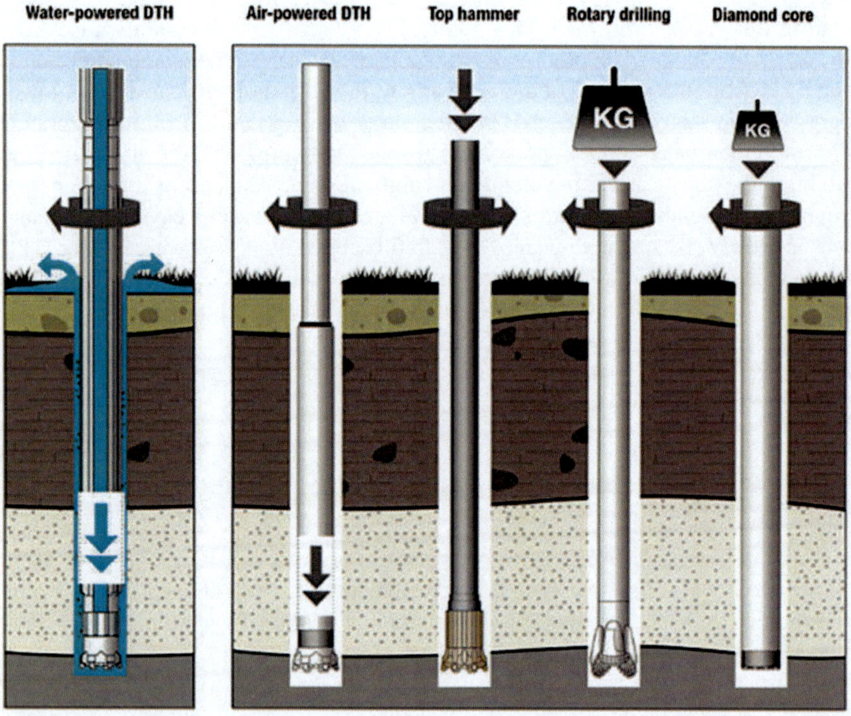

Fig. 1 Rotary drilling and percussion drilling. Percussion drilling can be subdivided into top hammer and down-the-hole (DTH) drilling

2 Water DTH Versus Air DTH

The fact that water is an incompressible liquid and air a compressible gas leads to some interesting differences between these two drilling methods. To compress air consumes more energy than pressurizing water. The needed installed power for the high-pressure water pump is only a quarter of the power needed for the air compressor. During a comparison test with a 4″ Wassara water DTH hammer and a 4″ 32 bar air DTH hammer, the diesel consumption for the water DTH was measured at 25% of the air DTH diesel consumption (Lindholm 2011). This of course translates to a similar reduction in the carbon footprint. As compressed air leaves the drill bit it will do so at full pressure and high air speed. For water this will not happen due to its incompressibility. The principle of the water DTH hammer is that it is powered by the pressure difference between the high pressure from the pump and the ambient pressure at the bottom of the hole. Compressed air leaving the drill bit at full pressure can cause changes in pore pressures and lead to ground settlements. With water DTH this risk is minimized (see Ahlund and Ögren 2016; Asplind 2017). For a 25-bar air system, air speeds can reach 30 m/s whereas the speed of the outgoing water is roughly 1 m/s. Air and water are mixed with drill cuttings on its way out of the hole. The kinetic energy of the cuttings in the air case is thus more than a factor 900 times higher than in the water case. These high-speed cuttings can wear out the bit, drill rods and stabilizers in the drill string. Air DTH can also cause caverns and voids in the ground, which can cause settlements, see Fig. 2.

A result of the high wear in the air case results in more hole deviation than in the water case since stabilizers will not wear out as fast with the low speed of the

Fig. 2 Air DTH to the left and water DTH to the right

Fig. 3 Air DTH holes in yellow, water DTH holes in blue and planned holes in grey. To the right, the stabilized water DTH is shown

return water. In a real case from Stockholm 3, holes were drilled with air DTH and three holes with water DTH. Total hole depth was 220 m. Hole deviation was then measured. The average hole deviation was 19.2% for the air DTH holes and 1.3% for the water DTH holes, see Fig. 3.

3 The Venturi Effect

Bernoulli's equation:

$$\frac{p}{\rho * g} + \frac{u^2}{2 * g} + z = constant \tag{1}$$

here p is the static pressure, ρ is the density of the fluid, g the gravitational acceleration, u is the speed of the fluid and z is the height over a horizontal reference plane.

The speed of the outgoing drilling fluid will cause a pressure decrease in front of the drill bit, see Fig. 4. This can cause fine material from the surrounding formation to be sucked out of the hole. More material than what is defined by the volume of the drilled hole can thus be removed from the ground. This is called over-drilling

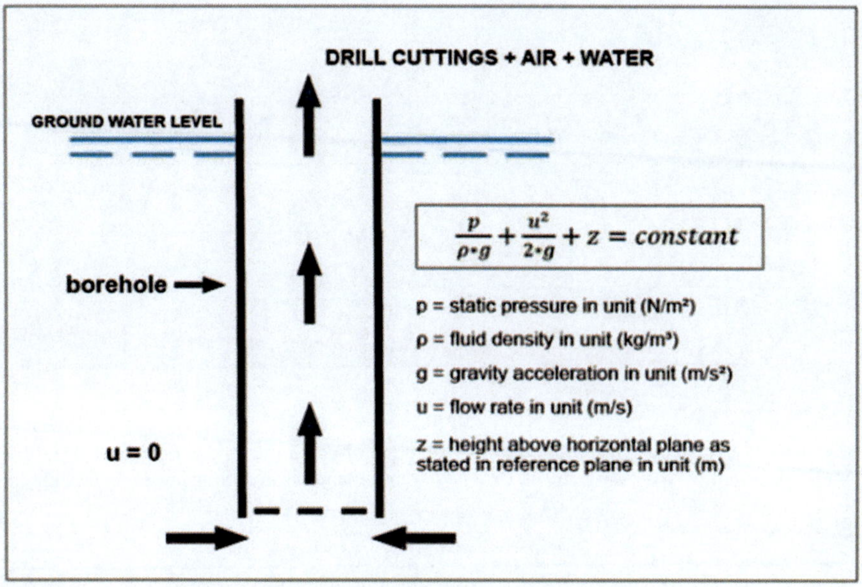

Fig. 4 The Venturi effect (see Asplind 2017, p. 35)

and can be dangerous as it may cause ground settlements. Since the speed of the outgoing air is much higher than the outgoing water speed the pressure decrease in front of the drill bit in the case of air DTH is significantly higher than in the case with water DTH. Thus, the risk of over-drilling is greatly reduced with water DTH.

If you are drilling below the groundwater level, the Venturi effect will effectively turn the air DTH into a water pump. Large amounts of water will be pumped out of the ground at the expense of Rate Of Penetration (ROP), see Fig. 5.

4 Water Hammers in Ground Treatment—Application Jet-Grouting

One particularly interesting technology in ground engineering is jet-grouting, which can be combined with water DTH technology to improve productivity under difficult ground conditions such as the presence of hard layers and boulders. Jet grouting is becoming more popular as a method to reduce the Lugeon values in sandy formations such as riverbeds in order to create positive cut off walls in coffer dam and barrage construction. Drilling has to progress through RBM (River Bed Material), which consists of boulders and soft strata with water bearing zones. Drilling through these strata is not possible with simple rotary bits, which is the most common set-up for a jet-grouting system with a jet-grouting monitor above, see Fig. 6.

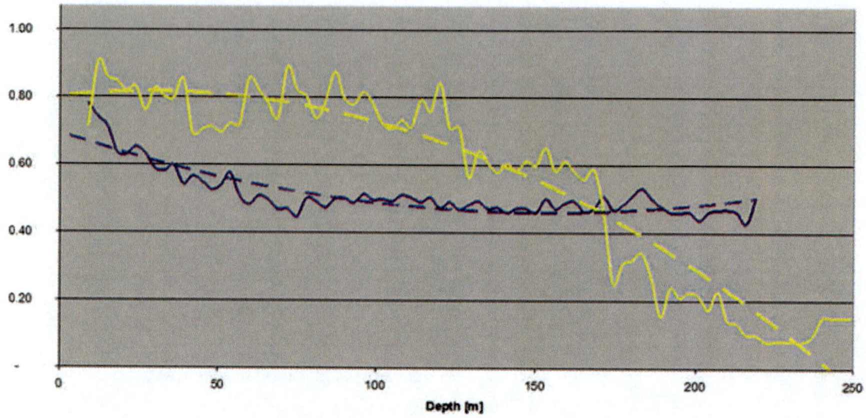

Fig. 5 Rate of penetration (meter/minute) as a function of depth (meter). Yellow Air DTH, blue water DTH. Groundwater level at 160 m. Real data from a comparison drilling project in Stockholm

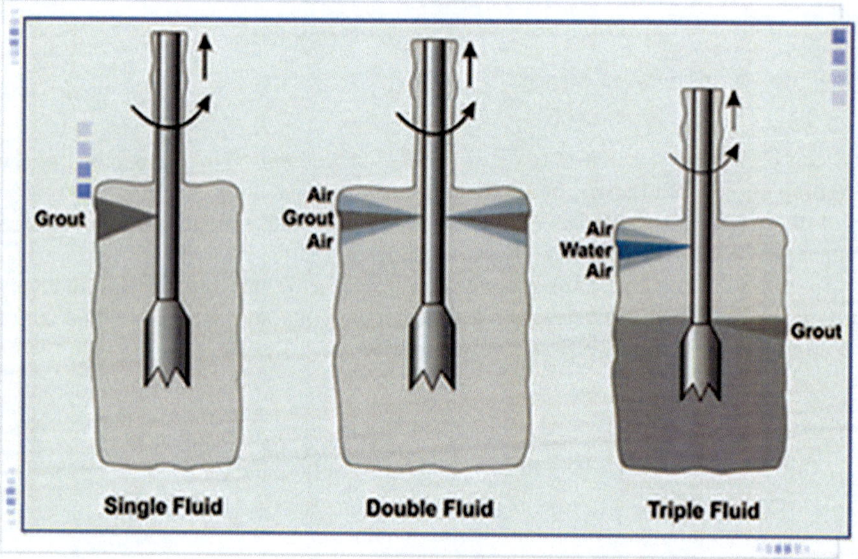

Fig. 6 Conventional Jet-Grouting with rotary+monitor

There are two ways to accomplish jet-grouting with water DTH hammers. The first way is to replace the rotary drill bit with the water DTH hammer, see Fig. 7.

The other option is to integrate the water DTH hammer and jet-grouting monitor into one unit, a Jet-Grouting Hammer.

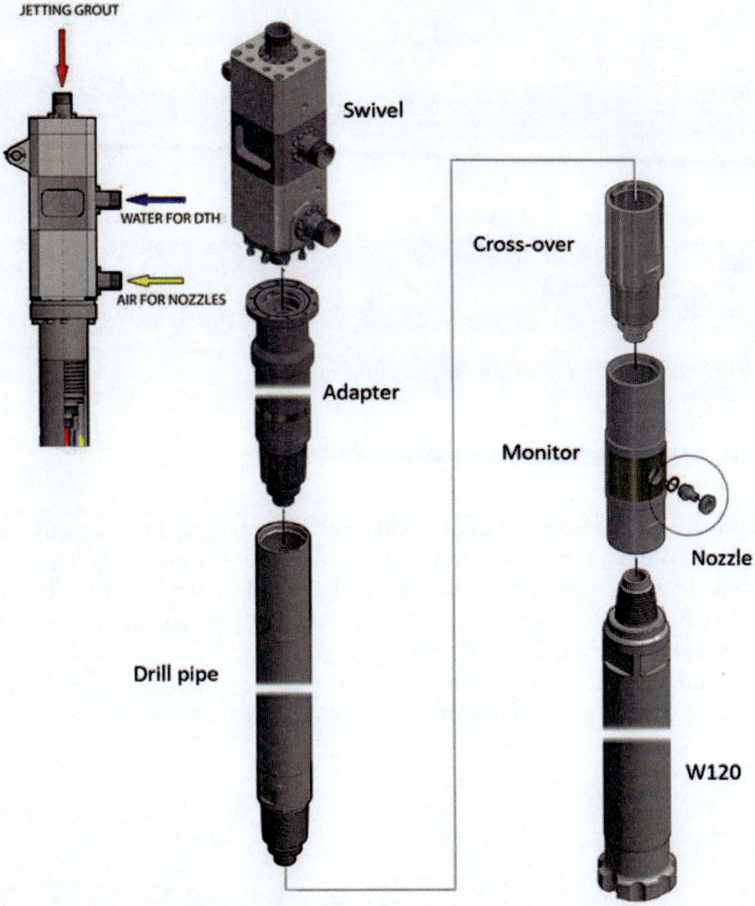

Fig. 7 The jet-grouting system with the CRM jet-grouting system in combination with Wassara W120 Hammer. Bit size is 152–165 mm

5 Jet-Grouting Hammer

During the drilling phase, high pressure water is supplied to the hammer through the outer channel in the double-walled drill pipe. A small waterflow can be applied through the nozzle to prevent clogging. When the desired depth is reached, the grouting can commence while pulling up and rotating the drill string. High pressure grout, up to 450 bar, is supplied through the inner channel in the double-walled drill pipe. The advantage with a Jet-Grouting hammer compared with a hammer+jet-grout monitor system is that the Jet-Grouting nozzle is much closer to the drill bit, which reduces the total length of the drilled hole (Fig. 8).

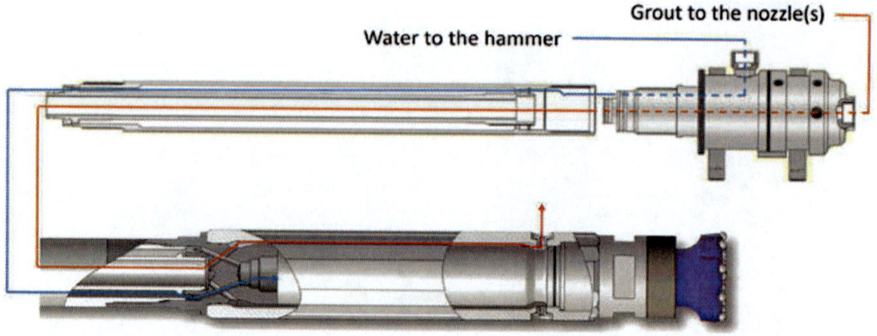

Fig. 8 Jet-grouting hammer principle. Bit size is 165 mm

6 Reference Case Brenner Base Tunnel

The Brenner Base Tunnel is the main segment of a new railway line that will connect Munich with Verona. It runs between Innsbruck and Fortezza, making it the longest underground railway stretch in the world. The BBT ends in Innsbruck in the existing railway bypass tunnel, which ends in Tulfes. The two-tube tunnel system between Innsbruck and Fortezza will be 55 km long (Fig. 9).

The contract for the construction work for the underpass of the Isarco River was awarded in October 2014 to a consortium led by Webuild (formerly Salini-Impegilo).

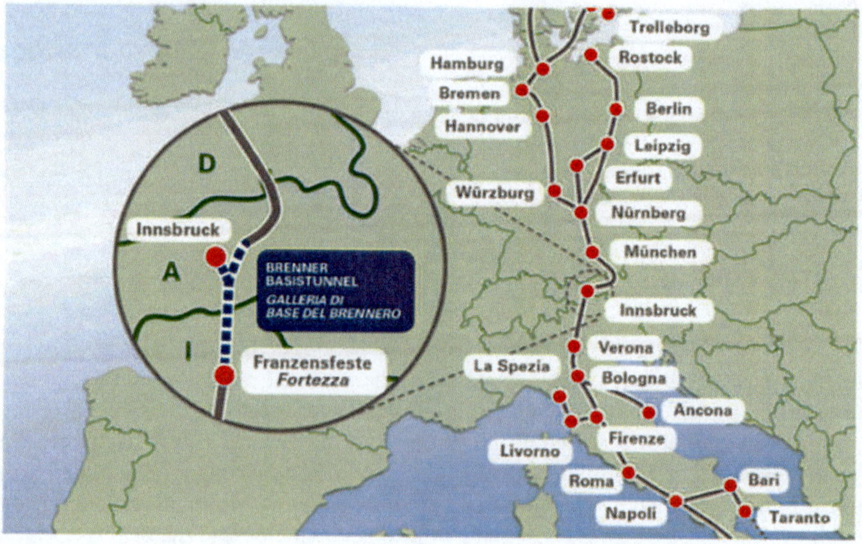

Fig. 9 Near Fortezza in Italy, the Brenner base tunnel underpass the Isarco River

Fig. 10 Worksite during jet-grouting for the four shafts for the river pass. Photo Webuild

The joint venture consists of Webuild, Strabag AG, Strabag S.p.A., Collini Lavori S.p.A. and Consorzio Integra (Lunardi et al. 2018).

The excavation of the tunnel started from four shafts on the riverbanks. Before the excavation of the shafts ground improvement by jet-grouting was performed. The jet-grouting was performed both to stabilize and seal of the water in the shaft as well as improving the grout from the surface before excavating the tunnels (Fig. 10).

The columns are made up to 40 m in one single pass. Components used are W120 Hammer with a Ø152 mm Bit. Above the hammer, a monitor is mounted. The monitor works both as a nozzle holder for the jet-grouting and a flow diverter that diverts water from the water channel in the drill pipe to the hammer. The monitor is of a double-flow type where compressed air is used to improve the performance of the jet-grouting. The water, air, and jet grout are supplied through a 127 mm triple wall drill tube securing that the fluids are not mixed. The most time-consuming part is the jet-grouting, which takes about twice the time as drilling down to target depth.

When producing 31 m grout columns, the drilling process took about 45 min to drill and about 1.5 h to grout. With this, each rig produced seven to eight columns per day during 24-h operation. Drilling was performed with a water pressure of 110–120 Bar, water flow depending on hammer wear from 240 l/min up to 350 l/min.

The drilling was normally relatively easy with a few boulders that was drilled through in a short time. The biggest obstacle was loss of circulation in the hole, which led to 30% of the drilling time was spent flushing to retrieve circulation in the hole. With the double fluid system, columns up to 2 m in diameter were achieved.

Fig. 11 Left: W120 hammer and monitor: middle monitor and triple fluid rods: Right Soilmec SM 28 in action. *Photo* LKAB Wassara

Geology

The Isarco River Underpass passes through loose deposits, heterogeneous both in composition and in granulometry, consisting of gravel and rounded sand, with frequent boulders and some large granite blocks.

Choice of Drilling Method

The Wassara W120 hammer was used because of the challenging geology where traditional rotary drilling was discarded due to the soil conditions with frequent boulders and large blocks.

In the beginning of the project, a comparing test was performed with Wassara hammer and an air DTH hammer. A 36-m-deep hole was drilled with the two different techniques. The Wassara hammer drilled the 36 m in 40 min and the air DTH hammer used 1 1/2 hour for the same depth. The low performance of the air hammer was most likely caused by the high water table (Fig. 11).

7 Reference Case Rio De Janeiro Metro—Jet-Grouted Columns for Station Box Construction

Station box construction for urban metro generally requires the construction of a diaphragm wall that is socketed into the bedrock if applicable. This requires the use of expensive and large equipment deployed in compact construction sites. In the case of secant pile walls, large piling rigs are used but there is still a problem of water leakage in the pile interface especially if the water table is high. Jet-grouted columns offer a cost-effective solution and will also assist the safe launching of the TBM without the need for a special "soft eye" with synthetic reinforcement.

The Wassara Jet-Grouting hammer was used for constructing two entrance and exit shaft at the Gavea metro station. The Gavea station is part of the new metro Line 4 in Rio de Janeiro. The construction phase included jet-grouting 200 columns placed in two rings for each shaft. The jet-grouted columns reached down to 15–20 m and 20–25 m depth, respectively (Fig. 12).

The formation presented a drilling challenge as it had layers of sand and stone combined with layers of large boulders and finally hard rock. That's why the contractor decided to test the Wassara Jet Grouting hammer and compare traditional rotary technology. A comparison between the rotary system and drilling with the Wassara Jet-Grouting hammer showed a clear productivity improvement with the Jet-Grouting hammer, see Table 1.

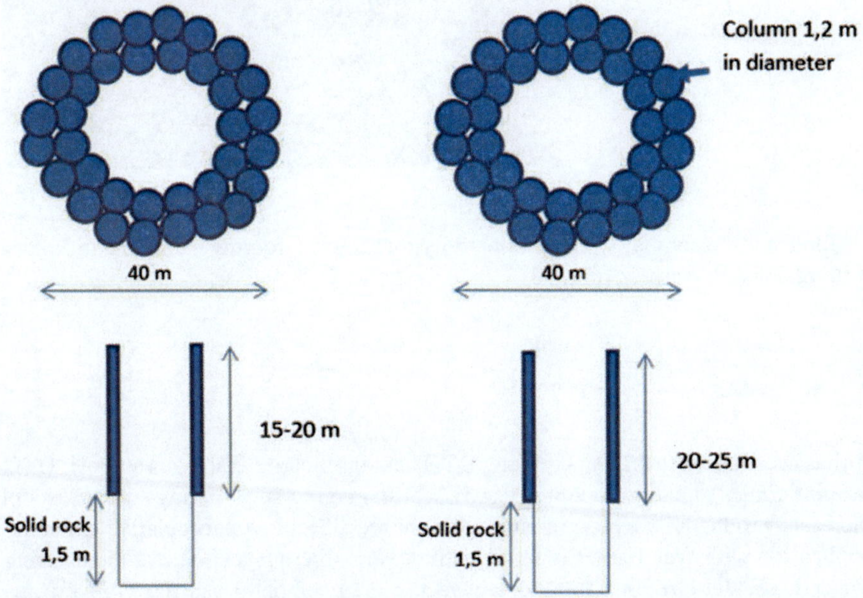

Fig. 12 Drill pattern layout (http://www.wassara.com)

Table 1 Wassara jet-grouting hammer performance versus rotary

Formation	Jet-grouting hammer	Rotary
Soft soil, 19 m	20 min	20 min
Rock (granite/quartz)	3 min	1 h
Total time	23 min	1 h 20 min

Fig. 13 Jet-grouting hammer at the drill site in Rio de Janeiro. *Photo* LKAB Wassara

Table 1 is the comparison of productivity for the Jet-Grouting hammer and Rotary drilling. LKAB Wassara (Fig. 13).

8 Summary

As has been shown in this paper, Water DTH has significant advantages over Air DTH when it comes to energy savings, less risk of damaging surrounding structures and increased productivity when ground conditions are difficult. A particularly interesting application with Water DTH is Jet-Grouting. Here, there is a choice of either using Water DTH+Jet-Grouting monitor above the hammer or to use the Wassara Jet-Grouting hammer where the monitor is integrated into the hammer. Two reference projects were described, which highlighted the advantages of using Water DTH technology in terms of increased productivity.

Acknowledgements Antonio Celot/Project Manager at Webuild, for providing information and photos for the Brenner Base reference case.

Valerio Salvi, Research and Development, CRM, for providing information about the CRM jet-grouting monitor.

References

Ahlund R, Ögren O (2016) Pore pressures and settlements generated from two different pile drilling method. Master of Science thesis, Royal Institute of Technology, Stockholm

Asplind M (2017) Pore water pressure and settlements generated from water driven DTH drilling. Master of Science thesis, Royal Institute of Technology, Stockholm

Bruce DA, Lyon R, Swartling S (2013) The history of down-the.hole drilling and the use of water-powered hammers. Providence, RI, Association of State Dam Safety Officials Annual Conference

http://www.wassara.com

Lindholm J (2011) Cost calculation and market analysis of geothermal drilling methods: a study of the competitiveness of Wassara water powered hammer drilling equipment. Master Thesis Luleå Technical University

Lunardi G, Cassani G, Gatti M, Bellardo L, Palomba A (2018) Brenner Base Tunnel & Isarco River underpass Section: several technical and operational solutions. Gallerie e grandi opera sotterranee n. 125-mars

Excavation Techniques for Underground Drinking Water Pipeline with Micropiles as a Special Application

B. V. Nagarajkumar

Abstract As a part of Mumbai City Mega Drinking Water Project at Amar Mahal Tunnel-1 Ghatkopar East Mumbai, drinking water tunnels were proposed by MCGM from Amar Mahal to Trombay reservoirs, Amar Mahal to Wadala and further up to Parel. The pipeline has to be underground and laid at a depth of 5.5 m. The initial proposal was to provide sheet piling to enable the safe excavation and installation of underground pipeline. However, installation of sheetpiles was practically not feasible due to the existing soil strata for driving the sheet piles. Hence, it was proposed to install micropiles instead of sheet piles. In this paper, we discuss about the purpose, basic design, detailed design approaches and actual work implementation of micropile with temporary struts for shoring protection works in this project.

Keywords Micropile · Grout · Casing pipe · Waler beam · Strut

1 Micropiles

Micropiles as well known for their application of substructure stabilization are utilized for the special application. In this project, micropiles are formed as a long wall with steel waler and capping beams to provide free and secured space for the purpose of facilitating safe installation of drinking water pipeline and relation construction activities. The use of micropiles to stiffen and support an excavation face by soldier piles with structural steel offers a technically sound method to construct excavations. Micropiles at the face can be used with soil nails or waler beam and strut system to construct an excavation, which has to be adapted to difficult site conditions. This can provide additional lateral and vertical deflection control, particularly when a depth of excavation is high and/or has surcharge behind shoring wall.

B. V. Nagarajkumar (✉)
Stresza Systems Global Tech Pvt Ltd., Visakhapatnam, Andhra Pradesh, India
e-mail: nagaraj@streszasystems.in

A. M. Krishna et al. (eds.), *Deep Foundations for Infrastructure Development in India*, Lecture Notes in Civil Engineering 373, https://doi.org/10.1007/978-3-031-37256-8_3

2 Stratigraphy and Subsurface Units

The geotechnical design parameters are determined based on the results of field and laboratory tests conducted on samples retrieved from the respective boreholes as per geotechnical subsurface investigation report. The geotechnical subsurface investigation report of project site (for interconnection works at Hedgewar Udyan Shaft location) comprises three borehole logs (AMT 1-14, AMT 1-15 and AMT 1-16) and all laboratory test results.

Based on the information presented in geotechnical subsurface investigation report, the subsurface stratigraphy along the project stretch covered by these boreholes can be described as three main subsurface units, like a soil overburden comprising Sand and Clay deposits, followed by a layer of highly weathered Basalt rock layer which in turn is underlain by slightly weathered to fresh Basalt rock layer. As the soil strata are configured in layers, i.e., a layered geologic structure exists; the soil is classified on the basis of the soil classification of the weakest soil layer. Each layer is classified individually as a more stable layer lies below a less stable layer, i.e., a Type C soil rests on top of rock formation.

Figure 1 presents the indicative subsurface profile considering maximum thickness of soil overburden encountered in soil investigation to represent critical soil profile for excavation.

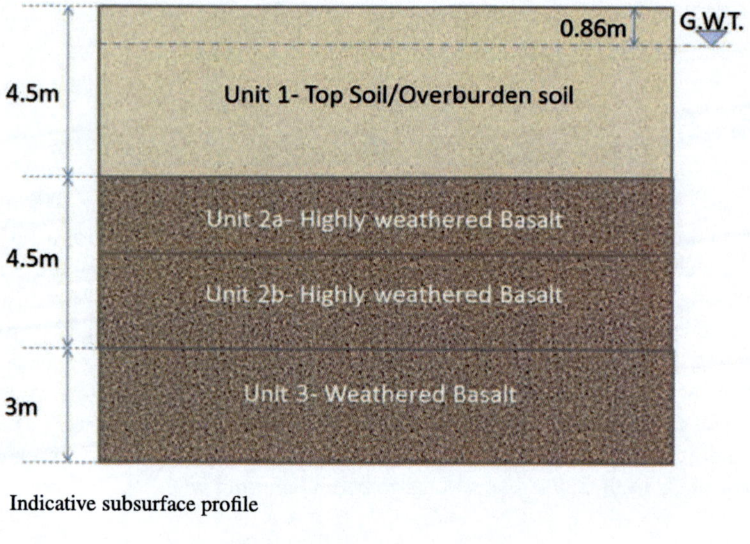

Fig. 1 Indicative subsurface profile

3 Geotechnical Design Parameters and Their Basis

The analysis of the in-situ and laboratory test results presented in subsurface investigation report allowed determining the main design parameters of the encountered soil layers at the location of current project site. The subsurface stratigraphy is defined in three main subsurface units. The thicknesses and the strength parameters of all these layers are varying in each borehole. Hence, the geotechnical design parameters considered for design of shoring system of planned trench excavation are based on the weakest strength parameters to have safe excavation.

The in-situ and laboratory test results (Table 1) are interpreted to derive the main design parameters of these rock layers, such as the Rock Mass Rating (RMR), the Geological Strength Index (GSI), the Rock Modulus (E) and the characteristic strength parameters (UCS, C and ϕ). The selected design parameters for three rock formations are summarized in Table 2.

Design parameters of rock mass strength presented in Table 2 are determined by using Rocklab software, which is based on generalized Hoek-Brown failure criteria. As can be noted, the upper rock layer Unit 2a presents lower strength and deformation parameters with respect to the other two formations. Thus, this formation will be considered as poor rock in this analysis.

Core box photographs of borehole AMT 1-14 are shown in Fig. 2, which presents the nature of soil/rock formation encountered (weakest rock) at project site.

Table 1 Summary of test results

Layer description	Thickness	Grade	SPT-N*	RQD	RMR	UCS
Unit 1—top soil/overburden soil	4.5	–	24			
Unit 2a—highly weathered basalt	2	V to II		12	32	
Unit 2b—highly weathered basalt	2.5	V to I		31	50	66,49,91,81
Unit 3—weathered basalt	3	I, II		43	55	

SPT-N* = Average-1/2 STD deviation

Table 2 Summary of design parameters

Layer description	Thickness (m)	Bulk density (kN/m^3)	C (kPa)	Phi (°)	E mass (MPa)
Unit 1—top soil/overburden soil	4.5	18	0	27	8–12
Unit 2a—highly weathered basalt	2	26	50	33	84
Unit 2b—highly weathered basalt	2.5	26	60	36	2000
Unit 3—weathered basalt	3	26	600	38	5200

AMT 1-14 Box 1

Fig. 2 Core box photographs

4 Design of Micropile Wall (Shoring System)

With the trend towards micropiles that can support higher loads at low displacements and for the requirement to sustain lateral loads, steel-pipe reinforcement has become more common. Pipe reinforcement can provide significant steel area to support high loading and contribution to the micropile stiffness, while providing high shear and reasonable bending capacity to resist the lateral loads. Pipe reinforcement is placed by either using the drill casing as permanent reinforcement, or by placing a smaller diameter permanent pipe inside the drill casing. The use of the drill casing for full-length reinforcement is typical only for micropiles founded in rock, where extraction of the casing for pressure grouting is not necessary. Based on the preliminary design calculations, a tentative layout of the micropile wall is presented in Figs. 3 and 4. Nagarajkumar B. V., DFI 2020 Conference, Special Application of Micro-Piles for the construction of a bridge well foundation, may be referred for the application of Micropile with soil nail and waler beam system.

Design Parameters/Assumptions

1. Excavation details
   - Width: 5.1 m, Depth: 5.5 m
2. Soil stratigraphy
   - As presented in Table 2
3. Design water levels
   - Design water level @ 0.86 m from NGL (as per SI report)
4. Micropile wall

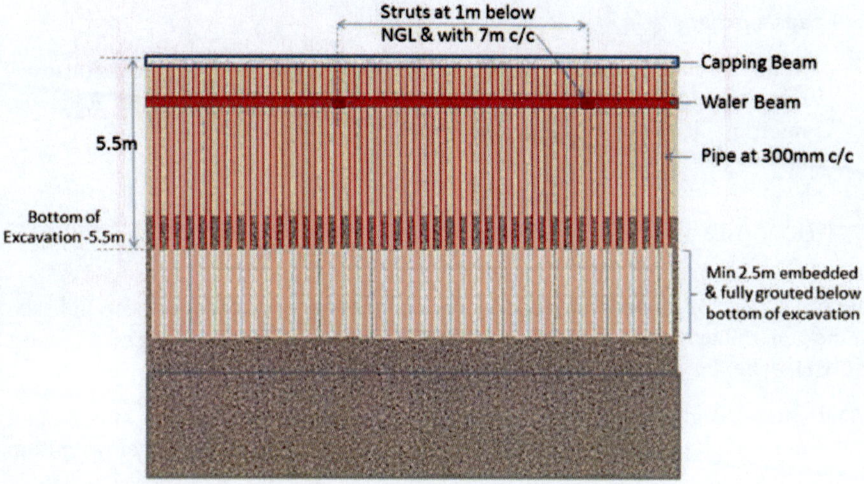

Fig. 3 Micropile wall

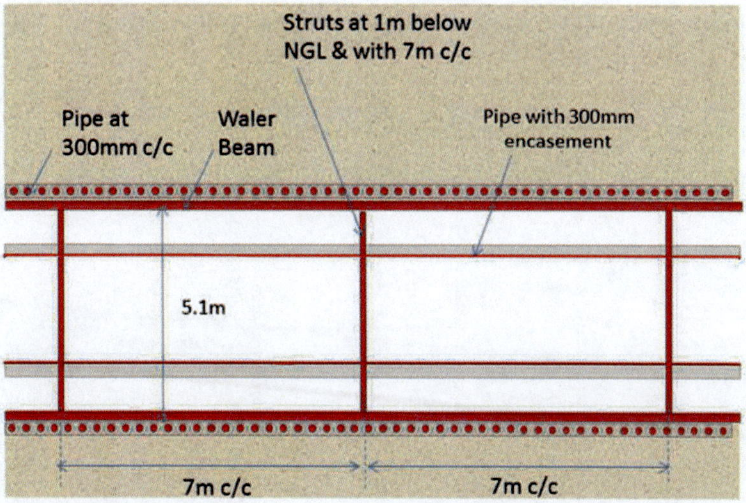

Fig. 4 Micropile wall plan view

- Unsupported length: Steel pipe of OD 168.3 mm and 6.3 mm thickness
- Rock embedded length: hole of 300mm diameter with steel pipe and grouting

5. Struts

- Single level strut at 1.0 m from Top/NGL with c/c spacing @ 7.0 m (considering 6.0 m of pipe length intended to be installed at bottom of excavation)

6. Loads/surcharge
   - Vehicular load or during construction phase to replicate loads from construction machineries, a uniform surcharge of 25 kPa is considered along the micropile wall at 1.0 m distance from wall/excavation face

5 Micropile Wall Stability Analysis

The full stretch of micropile wall is preliminary designed as a single section based on the generalized stratigraphy along the full excavation stretch (based on SI report of 3 BH). The design is based on the following parameters:

- Minimum length of micropile wall designed for excavation of 5.5 m is around 8.0 m from NGL. This will insure the embedment depth of 2.5 m below bottom of excavation.
- Steel pipe for Micropile—NB 150; OD 168.3 mm; class—Heavy; thickness of 6.3 mm
- Moment of Inertia—1053 cm^4.
- Micropile wall is propped against each other by strut at level of 1.0 m below NGL to restrict its excessive deflection as presented in Fig. 5.

WALLAP software is used for the analysis of Micropile wall system. Staged excavation is modeled to replicate the construction sequence.

Stage 1 Apply surcharge number 1 at elevation 0.00m

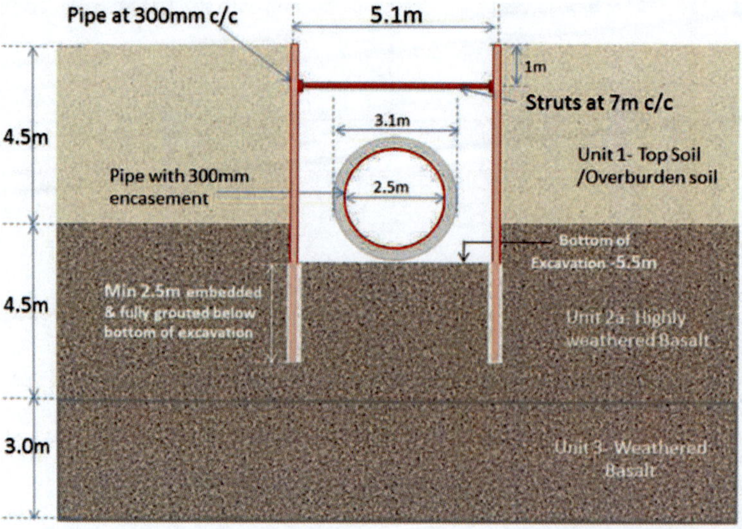

Fig. 5 MicroPile wall section

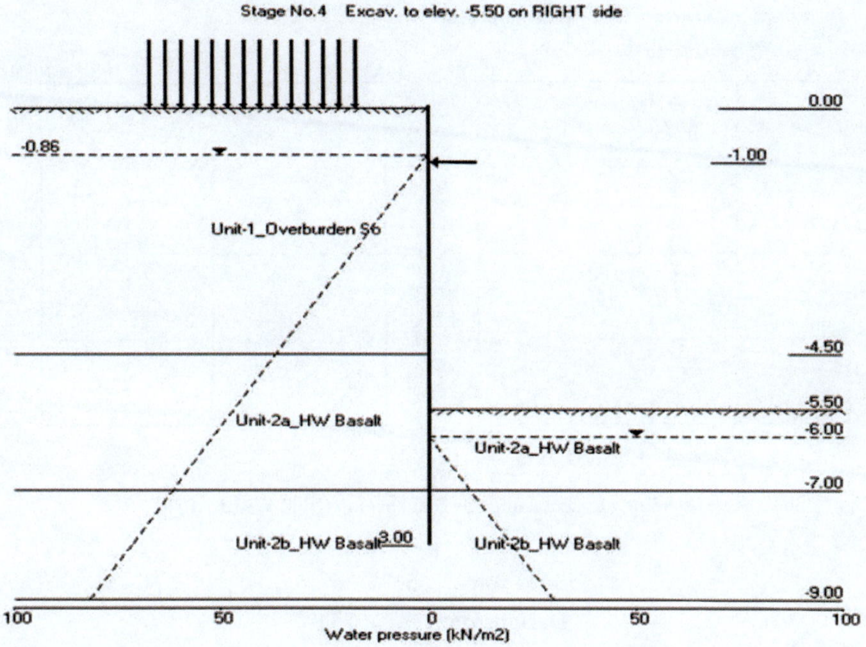

Fig. 6 Micropile wall detailing (model from WALLAP)

Stage 2 Excavate to elevation −1.50m on RIGHT side
Stage 3 Install strut or anchor number 1 at elevation −1.00m
Stage 4 Excavate to elevation −5.50m on RIGHT side

Detailed arrangement of micropile wall, surcharge and strut are shown in Fig. 6. Figure 7 shows the Bending Moment, Shear Force and Deflection Envelope from the WALLAP Analysis.

6 Strut and Waler Beam Design

In the above micropile wall analysis, the wall is propped against each other by Waler and strut (spaced at 7 m c/c) arrangement (as presented in Fig. 5). Based on the analysis of micropile wall in WALLAP, the maximum (working case) axial force of 530 kN per strut is observed.

Strut: Length 5.0 m, Max. Axial force (working) 530 kN.

Waler Beam: Span/Length: 7 m, with UDL of 76kN/m (530 kN/7 m).

WALLAP Design Summary for the Shoring System:

I. Strut Design Data:

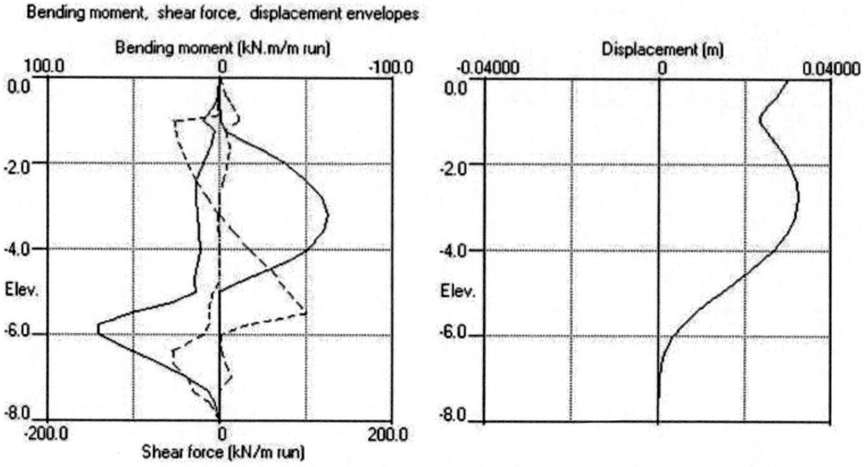

Fig. 7 Shear force, bending moment and deflection envelop (from WALLAP)

$$\text{Strut force} = 53\text{T}.$$
$$\text{Design force(P)} = 75\text{T}.$$
$$\text{Span} = 5\,\text{m}.$$

Adopted Section = 2 ISMC 400 Box.

Section Properties:

$$\text{Area (A)} = 125.86\,\text{cm}^2.$$
$$\text{Axial stress} = \text{P/A}.$$
$$\text{Axial stress} = 59.59\,\text{MPA}.$$

Check for Allowable Stress:

$$\text{Minimum radius of gyration(rmin)} = 8.09\ \text{cm}.$$
$$\text{Slenderness ratio} = \text{L/rmin}.$$
$$= 5000/(8.09 \times 10)$$
$$= 61.8$$

As per IS 800 (1984) Table 5.1

Permissible allowable stress in the section = 120 Mpa.

Hence adopted section is safe.

II. Micropile Design Data:

$$\text{Bending Moment} = 71\ \text{KN} - \text{m/m}.$$
$$\text{Shear force (P)} = 101\ \text{KN/M}.$$
$$\text{Adopted Section} = \text{Steel tube } 150/168.3/6.3\ \text{thk}.$$

The properties of the steel tube are shown in Table 3.

$$\text{Grade of steel} = \text{Yst } 310\ \text{Mpa(asper IS1 } 161 : 1998)$$
$$\text{Fy} = 310\text{Mpa}.$$

Section Properties:

$$\text{Area(A)} = 32.06\ \text{cm}^2.$$
$$\text{Moment of Inertia} = 1053.4205\ \text{cm}^4.$$
$$\text{Modulus of section} = 125.183\ \text{cm}^3.$$

Check for Allowable Bending Stress:

$$\begin{aligned}
\text{Spacing of piles} &= 300\ \text{mm} \\
\text{Moment in each steel tube} &= 21.32KN - \text{m}. \\
\text{Bending stress} &= M/Z. \\
&= 21.32 \times 10^6/125183 \\
&= 170\ \text{MPa} \\
\text{Allowable stress} &= 0.6 \times fy = 186\ \text{MPa}
\end{aligned}$$

Hence adopted section is safe.
The general arrangement of micropiles and the plan view is shown in Fig. 8.

7 Components of Micropile System

Figure 9 shows the components of the micropile section. Figure 10 shows the spacing and alignment of micropiles.

8 Grout Specification and Grout Quality Control

Grouting Material and Admixture

Table 3 Properties of Steel Tube

N.B	O.D	THK	WT.(G)	Area of cross-section (A)	Moment of inertia (cm^4) (MI)	Modulus of section (Z)	Radius of gyration (RG)	Grade of steel (GS)
150 mm	168.3 mm	6.3 mm	25.2 kg/m	32.0 cm^2	1053 cm^4	125 cm^2	5.73 cm	YST-310 MPa IS (1161:1998)

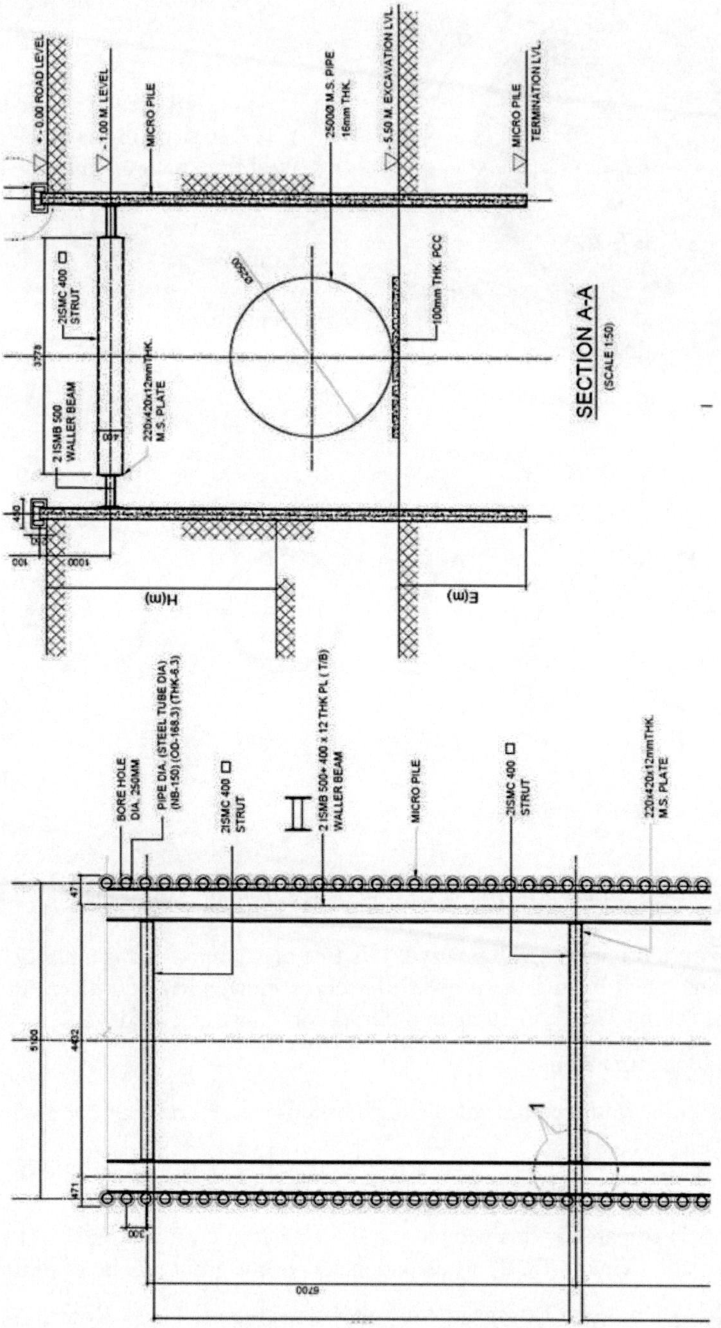

Fig. 8 General arrangement of micropiles section and plan view

Fig. 9 Components of a micropile system

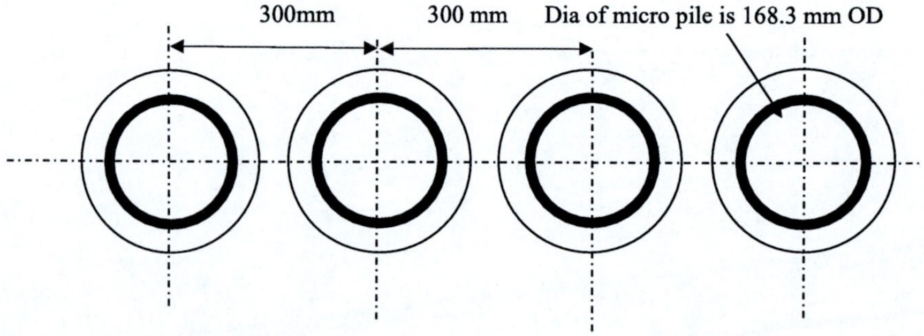

Fig. 10 Micropile spacing and alignment

The cement to be used for grouting is OPC obtained from approved manufacturer/ supplier.

The admixture if required will be used. It is free of Alumina and is a plasticizer and a retarding agent having the properties such as Flowability at a given water/ cement ratio, Setting Time and Strength of Grout for 7 days and 28 days.

Cement Grout and Admixtures

Plain cement grout with recommended admixture dosage will be used for grouting of micropiles.

Grout Mix

The grout mix is prepared with a w/c ratio of 0.4 with 20.0 L of water and 50 kg of Ordinary Portland Cement (OPC). Main parameters of the grout will be as follows:

Minimum strength of grout before stressing operations required is 30 N/mm^2

Water cement ratio = 0.40

20.0 liters of water

50 Kg of OPC

Non-shrink grout additives (Intraplast N-200 or cebex-100 or equivalent)

Laboratory and field tests were conducted to verify mixture, mixing efficiency, setting times and performance.

Water

Water available at site is made sure to confirm to IS 456:2000 and IS 3025.

Quality Control of GROUT (Ref BS EN 447)

Grout quality has been given importance since the anchorage depth grout is done immediately after installing the permanent casing. The grout has to flow the perforations of the permanent casing; hence care is taken for mix ratio and flow of the grout properly during all micropile grouting activities. Figure 11 shows the grout quality control tests conducted as per BS EN 447.

Grout samples of size 100 mm × 100 mm × 100 mm collected for every micropile batch and tested for compressive strength at 7 days and 28 days as shown in Fig. 12.

Mixing and pumping of grout done on special grout mixer with double tank are shown in Fig. 13, to make sure the grout is supplied continuously without interruption.

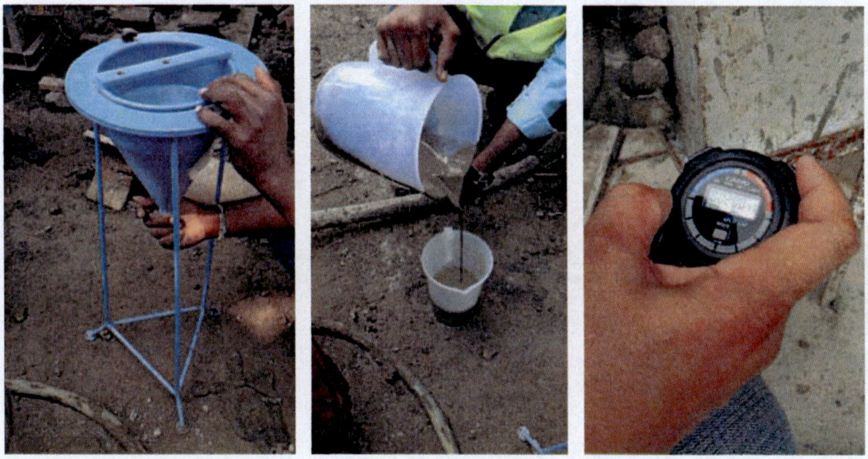

Fig. 11 Grout quality tests as per BS EN 447

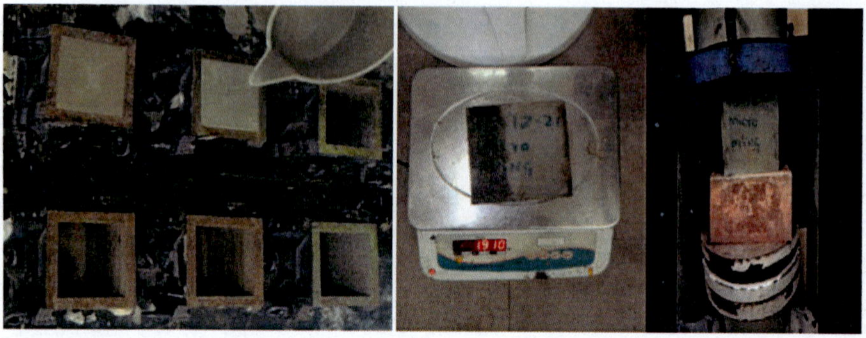

Fig. 12 Grout sampling for compressive strength test

Fig. 13 Grout mixing and pumping unit

9 Sequence of Construction

The micropile circular wall will act like a temporary formwork and permanent protection for pile line structure. The sequence of construction is as follows:

Step 1 Survey marking for micropile as per design spacing, i.e., 300 mm c/c.

Fig. 14 Fabrication of steel tubes

Step 2 Fabrication of steel tubes to the required length (Fig. 14) of 8 m and 7.5 m, as the 168.3 mm OD steel tubes in the industry are of standard length 6 m. The Welding of steel tube joints to make it to required length is done on proper bed for alignment and uniform joint process. Perforations are provided for the anchorage depth to make sure during the grouting work the grout flows to the annual area surrounded by the steel tube. The Steel tubes are provided with steel bar hooks as shown in Fig. 14 for handling with crane.

Step 3 Drilling of borehole (Fig. 15) is executed with tractor mounted heavy drilling rig 90HP with diameter capacities up to 1200 mm. In this project, the borehole has to be protected from collapsing from ground level till 2–3.5 m deep with a temporary steel casing. Figure 16 shows the general construction sequence of micropiles.

Step 2 Installation of temporary casing, installation of permanent casing (Fig. 17), partial grouting and concrete work.

Step 3 Installation of capping beam reinforcement (Fig. 18) and casting of concrete. The capping beam will be merged with back filled new road construction above it.

Step 4 Excavation up to a depth of 1.0 m, Installation of Steel Waler and Strut.

Step 5 Excavation continued till the required depth is reached (Fig. 19).

Step 7 PCC bed casted, 2.5 m dia pipeline installed, RCC structure to cap the pipeline is constructed and backfilled with sandbags as shown in Fig. 20.

10 Construction Issues and Challenges

The following issues and challenges were faced during the construction:

Fig. 15 Drilling of boreholes with rig

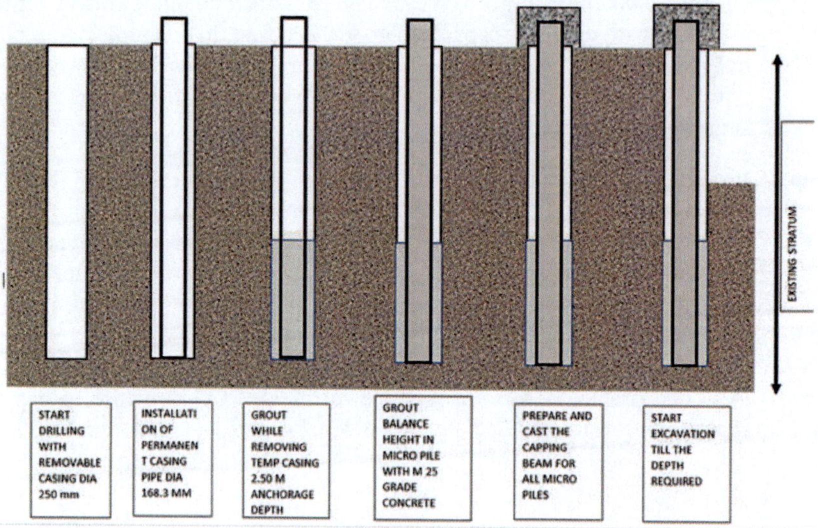

Fig. 16 Construction sequence of micropiles

Fig. 17 Installation of permanent casing

Fig. 18 Capping beam construction

Fig. 19 Excavation to the required depth with waler and strut installation

Fig. 20 PCC bed and pipeline installation

1. The entire length of permanent case pipe needs to be installed in one go. During installation, the height of crane booms was reaching the neighboring park trees, which needs to be protected. Hence, it was challenging to install the casing pipe without damaging the garden trees and power lines.
2. The process of micropile borehole drilling was a challenge as the drilling should not interfere the previously laid water/sewer line and underground electrical lines. However, in few rare cases, electric power lines were encountered, hence, work was stopped until bypassing such cables/water pipelines.
3. Equipment breakdowns was another major issue faced due to sudden changes of strata. Due to this, tooling and system of drilling process need to be looked carefully while executing.
4. The work scheduling was taken up in such way that work is done day and night basis. However, due to residential area, night work is restricted on majority.

11 Conclusion

The micropile walls provide solutions to the deep excavation problems where the conventional sheet pile wall execution is not possible due to difficulty in drilling the regular piles. Also, the micropile wall system with waler beam/strut mechanism helps to excavate for greater depths. With micropile wall system, deep excavations can be expedited without any compromise on the safety of construction/structure.

Project has a total length of 550 m wall on left and right side thus making total length of micropile wall to 1.10 km with number of micropiles totaling to 5000 approximately (29,000 RM). Progress of the project is halfway and expected to be completed by mid of 2023.

An additional micropile wall is designed for interconnection of SHAFT for driving and receiving PITS. Through the micropile wall, the pipeline will be erected safely.

Acknowledgements We are grateful to M/s Soma/ DM corporation management represented by Mr GV Rao, Sr. Vice President for giving us an opportunity to showcase our design and execution skills on this unique project. Special thanks to Mr MS Sarma, Technical Advisor for introducing our company to the management. Also, acknowledge with thanks engineering inputs from Shri Yatin.

References

IS 1161-1998: Steel Tubes for Structural Purposes
IS 456-2000: Plain and Reinforced Concrete—Code of Practice
IS 800-1984: Code of Practice for General Construction in Steel
Nagarajkumar BV, DFI 2020 Conference, Special Application of Micro-Piles for the construction of a bridge well foundation

Special Geotechnical Works for Metro Cairo (Egypt)

Mohamed Mostafa, Ashraf Wahby, Karsten Beckhaus, and Michael Baltruschat

Abstract Bauer Egypt was commissioned with the specialist geotechnical engineering works for 18 out of the total of 22 underground stations on Metro Line 3. Cumulated amounts of approximately 600,000 m^2 of diaphragm wall (D-wall) and 58,000 m^2 of silicate gel grout plugs were executed for open excavation pits. In addition to the daily challenges, particular efforts are worth highlighting: for recovering a tunnel boring machine (TBM) of the client, a diaphragm wall shaft 100 m deep using a natural clay layer acting as a bottom plug was constructed, to allow for horizontal drilling and subsequent soil freezing measures out of the shaft; also the deployment of low-headroom units to continue the D-wall works near and under existing bridges; and finally, special jet grouting works were executed at an intersection between an existing sewer and the TBM route.

Keywords Metro Cairo · Diaphragm wall · Silicate gel grout plug · Soil freezing · Jet grouting · Tunnel boring machine (TBM)

1 Introduction

Cairo, the capital of Egypt, is the biggest city of the Islamic hemisphere, with approximately 21 million inhabitants in Greater Cairo. Commuting has become a challenge for people already in the 80s of the last century and individual transport was by far not

M. Mostafa (✉) · A. Wahby
Cairo, Egypt
e-mail: m.mostafa@bauer.de

A. Wahby
e-mail: ashraf.wahby@bauer.de

K. Beckhaus · M. Baltruschat
Schrobenhausen, Germany
e-mail: karsten.beckhaus@bauer.de

M. Baltruschat
e-mail: michael.baltruschat@bauer.de

A. M. Krishna et al. (eds.), *Deep Foundations for Infrastructure Development in India*, Lecture Notes in Civil Engineering 373, https://doi.org/10.1007/978-3-031-37256-8_5

capable of covering the demands, and the capacity of the street infrastructure system is limited in this tightly populated city. The first tram opened in 1896 and this tram system formed an important part of public transportation. But sharing the streets, additionally with busses transporting approximately two billion passengers per year, called for the establishment of an underground infrastructure as being the efficient solution. Consequently, the Cairo Metro has become the most important public transport system. Two lines with a total length of 61.5 km are in operation and transporting approximately 950 million passengers a year by 2004. The National Authority for Tunnels (NAT), a Department of the Ministry of Transport, Arab Republic of Egypt, is the owner and in charge of all constructions. When Metro Line 1 was opened in 1987, it consisted of 4.7 km long tunnel section. In 1990s, the first tunnel under the river Nile was constructed for Metro Line 2. The works for Line 3 started in 2008. The total length is divided into four phases of which phases 2 and 3 with a length of 12 km are already in operation (Fig. 1). Bauer executed the most essential part of the geotechnical works for Metro Line 3, 18 out of 22 underground stations, and was awarded the geotechnical works for another six stations of the future Line 4. Feasibility studies are currently underway for two more Metro lines.

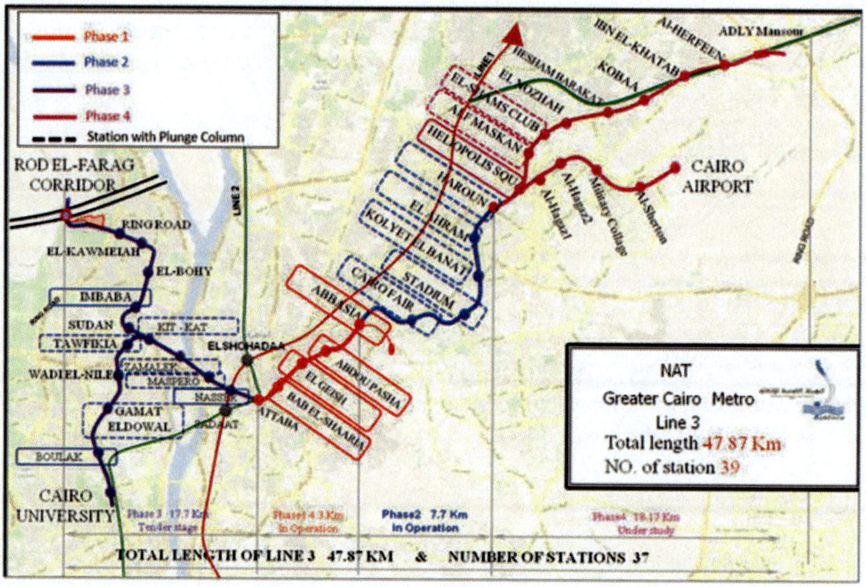

Fig. 1 Cairo Metro, Line 3—Stations awarded to Bauer are surrounded by a rectangle with the same color as the phase. Plunge columns were used in station where the boxes are dotted

2 Geotechnical Works for Metro Line 3

Bauer Egypt was commissioned with the specialist civil engineering works for 18 of the 22 underground stations on Line 3 (Fig. 1) and constructed a total of approximately 600,000 m^2 of diaphragm walls and 58,000 m^2 of silicate gel grout plugs. BAUER Egypt executed open excavation pits, for the start shaft of TBM, several Metro stations and ventilation shafts. Excavation and foundation levels for stations had been planned up to 32 m below groundwater level requiring retaining walls (diaphragm walls) with connection to deep-lying silicate gel grout plugs down to a maximum depth of 86 m, allowing for the excavation of the construction pit in dry conditions controlled by dewatering of limited seepage.

Mostly, the stations are located under existing roads. Depending on traffic, the concerning road had to be closed at least partly, if not completely, and traffic must temporarily be diverted. After the completion of D-Walls and silicate gel grout plug, the main contractor, a French-Egyptian Joint Venture, started excavation and construction using the top-down method with temporary struts, as shown in Fig. 2.

The tunnel boring machine (TBM), slurry-shield type by Herrenknecht, was of diameter 9.8 m and was driving from station to station with positioning in station. To safely connect TBM to the stations, or with ventilation shafts, sealing blocks of plastic concrete installed by the trench cutters to form a block prior to boring through with the TBM.

The excavation pits offered and executed by Bauer Egypt are based on a type of standard design, as shown in Fig. 4. The execution was based on the following general construction sequence:

- Construction of the vertical retaining walls by means of diaphragm wall technique.
- Installation of plunge columns.
- Construction of cut-off walls to divide a station box into smaller boxes.
- Construction of the sealing base as a silicate gel grout plug.

Fig. 2 Top-down construction with temporary struts

- Construction of concrete slabs while top-down excavating in stages. Works included installation of the temporary struts where needed by main contractor.
- Installation of the base raft slab by main contractor.

The surrounding diaphragm wall is designed for structural function in the upper section and for sealing function over its full depth. Retaining walls had been installed as reinforced concrete walls using several Bauer cutters. Soil conditions, as shown in Fig. 3 (The National Authority for Tunnels (NAT) 2004), at given sites had been investigated and found to being cultural fill, clay, silt, and sand. Since no natural impervious soil is existing in acceptable depth in typical Cairo soil conditions for Line 3, the construction pits usually also needed cut-off from groundwater at the bottom, by a horizontal grout plug using soft gel as construction material.

The grouting activities started after completion of the pit-enclosing wall, and the execution was done in two main steps: for preparation of the actual permeation grouting, vertical holes were drilled in a defined grid and grouting tubes were installed into the holes supported by a self-hardening bentonite-cement slurry. When this slurry, meant to seal off the upper boring from the injection level, gained sufficient strength, the actual grouting with soft silicate gel commenced, valve by valve, and layer by layer. The required silicate gel grout plugs for Cairo Metro Line 3 had thicknesses varying from 4.5 to 12.0 m, complying with the individual specifications

Identifier	Name	Description
L1	Fill	Silty SAND mixed with brick and rock fragments
L2A	Clay 1	Stiff to very stiff sandy, silty CLAY with sand layers
L2B	Clay 2	Firm to stiff silty CLAY
L3	Sandy silty CLAY	Mixed ground consisting of clay, silt & sand. Mainly cohesive
L4	Sand 1	Dense SAND
L5	Sand 2	Very dense SAND

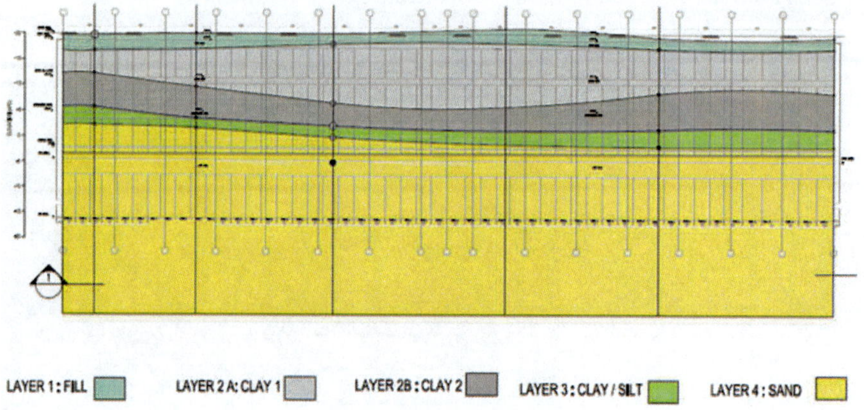

Fig. 3 **a, b** Typical geotechnical cross-section and soil description (The National Authority for Tunnels (NAT) 2004)

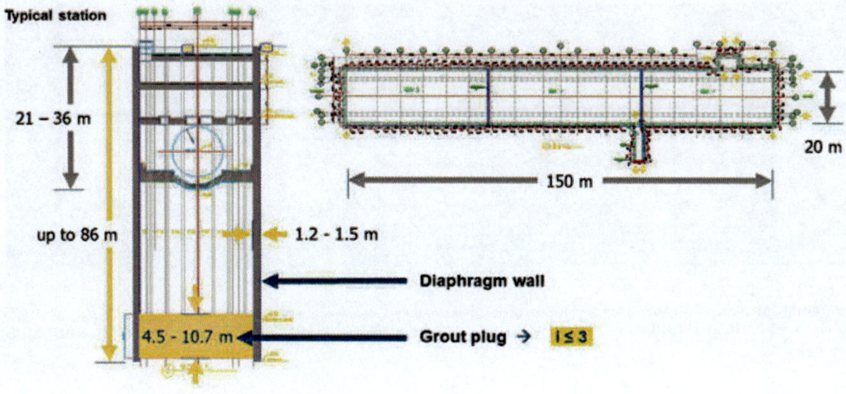

Fig. 4 General cross-section and top view

of site. To allow for grouting up to five layers, the same number of tubes with one valve each per hole was installed at the various layers' elevations accordingly. For mixing the silicate gel, sodium silicate and reactant are required so that the water becomes a gel. These components are mixed by continuous mixers in a so-called grouting container, with additional pumps up to 13 to serve the same number of injection valves at same time. Grout by Silicate gel is highly sustainable, as it requires less transportation produce silicate gel on site with less material consumption thus reducing carbon footprint. Before excavation, the water level is lowered inside by internal wells. By means of residual dewatering, the water level is kept at the required level below the bottom of the excavation pit.

3 Main Equipment

Diaphragm Wall (D-wall)

Up to five independently operating trench cutter systems were deployed. The trench cutters have cross-sectional area of 2.80 m × 1.20 m. These trench cutters were mounted on base carriers equipped with Hose Drum System (HDS) or Hose Tension System (HTS). In order to cut the homogeneous soils, all units are equipped with the same cutting wheels and tooth types. This approach, based on several projects in Cairo, ensured that the soil could be loosened in an optimized way. To ensure the high-quality standard at stations near/under existing bridges, BAUER deployed low-headroom units, as shown in Fig. 8b. During cutting operation, a closed bentonite slurry circulation ensures the continuous working process of the D-wall construction. Due to the removal of solid particles from the support fluid in the mud treatment plant, the excavated material can simply be moved off-site from a centralized area. After the separation, the slurry is available for reuse within the system. In order to operate

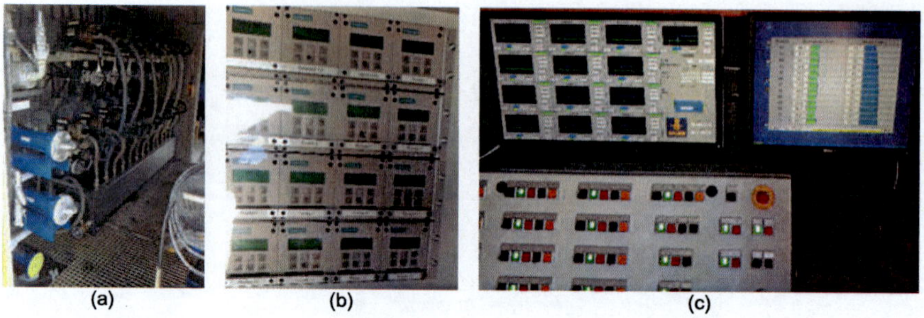

Fig. 5 a–c Soft gel container with installed pumps, control units and screen/computer to operate the works

Fig. 6 BAUER BG 28 during drilling up to a depth of 86 m for silicate gel grout plug

the cutter efficiently, 3,500 m^3 of bentonite slurry is stocked and circulated between the cutter, desanding plant, and storage tanks. Up to 1,000 m^3 of working slurry is processed every hour by the installed Bauer BE500 and BE550 desanding plants.

Permeation Grouting

Five BAUER BG 28 units with extended mast are shown in Fig. 6, five grout or soft gel containers each with grout pumps up to 13 as shown in Fig. 5a–c, and several mixing plants.

3.1 Jet Grouting

BAUER BG 28 drilling rig is shown in Fig. 7, Jet grouting pump and a complete mixing plant.

Fig. 7 BAUER BG 28 during jet grout works near existing building

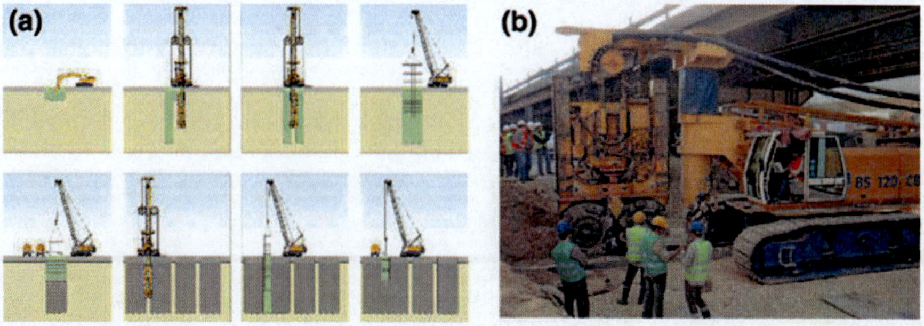

Fig. 8 **a** Construction sequences (schematic only) for diaphragm wall using cutter. **b** Low headroom CBC-cutter under an existing bridge

Diaphragm Wall

The underground retaining structure/wall constructed by means of the diaphragm wall technique formed of individual panels, each of panel length varying between 2.80 m and 6.80 m having nominal thickness of 1.20 m. In the pilgrim step system applied, only primary panels are constructed in the first step. In order to create a structural and watertight bond between the respective adjacent panels, primary panels are cut during the excavation of the secondary panels depending on the required depth.

Static calculation of the walls requires reinforcement in all panels, so prefabricated reinforcement cages are installed in the primary panels in an exact position, to avoid any subsequent contact between the cutter and the reinforcement during the trenching of the secondary panels, as shown in Fig. 8a.

4 Trench Stabilisation

At diaphragm wall construction sites in Egypt, the technically and economically most appropriate option is the application of a supporting bentonite slurry to stabilize the open trench. However, considering the conditions (e.g., the price and availability of bentonite, properties of the subsoil, storage and disposal, etc.), there may be reasons to consider alternatives. Polymers are used in diaphragm wall construction sites, e.g. in England (refer "Woodsmith Mine") or currently, for a Metro project in Manila in Philippines.

5 Permeation Grouting

Permeation grouting by means of silicate gel is used for the excavation pits at Cairo Metro project mainly due to the following reasons:

- In appropriate depth regarding safety against uplift the subsoils mostly consist of sand.
- The client as well as their technical consultants have specified this construction method in the tender documents.
- This construction method is very favorable in terms of sustainability, as cement consumption is very low.

The silicate gel consists of mainly water, sodium silicate, and reactant. The volumetric percentages must be defined for each construction site, depending on the technical requirements, availability of sodium silicate and reactant in the country.

A grout plug like in Cairo is in general executed by drilling and permeation grouting (for more details pls. refer to EN 12,715).

- 1st step: Drilling to the required depth using self-hardening slurry as supporting fluid.
- 2nd step: Installing of single PVC-lance with single valve, or multiple PVC lances with one single valve each. Depending on the required thickness of the grout-plug.
- 3rd step: Once the slurry has hardened, the silicate gel is injected into the soil by way of valve bodies.

The execution of silicate gel grout plug is based on successfully established principles:

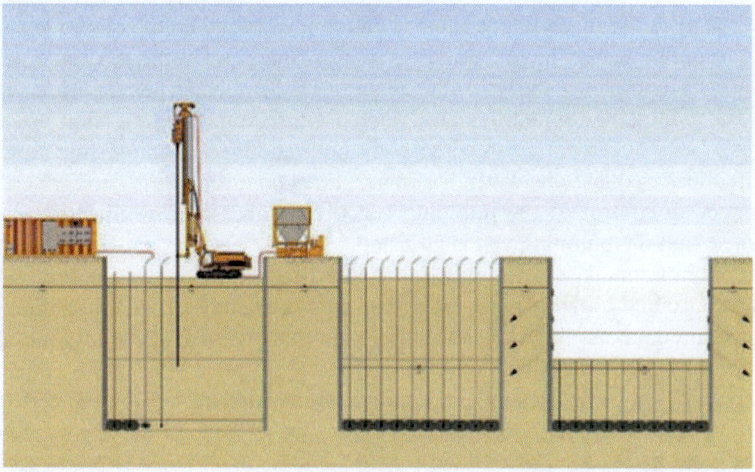

Fig. 9 Construction sequences (schematically only)

- Other activities within the construction site should be limited if they can interfere by their vibration, when drilling works are in progress, see Fig. 9.
- For each drilling point in the predefined triangular grid, as shown in Fig. 10a–c, grout pipes with one valve each per layer are to be installed at the center of the respective injection level.
- The individual injection levels are treated from top to bottom and within an injection level an injection front is pushed towards the surrounding (diaphragm) wall. The number and direction of these fronts essentially depend on the available area.
- The water tightness (permeability) of the excavation pit is validated by a pump test before starting excavation.

Specific parameters to be defined for each sealing base, considering, conditions like the feasible range of penetration by injection. The determination of the triangular grid (Fig. 10) as explained, expected drilling deviations are considered into the design. Based on this, the diameters of each gel body are determined for the project. The

(a) (b) (c)

Fig. 10 **a–c** Research and development in BAUER—showing spreading behavior and contact surfaces between successively injected silicate gel bodies in sand quantity of gel to be injected per valve is determined based on the effective porosity derived from the geotechnical report plus a safety margin. During the dynamic engineering that accompanies the work, the quantities to be injected per valve are adjusted if necessary. Increased quantities are injected via surrounding injection points if a single gel body could not be injected with the predefined quantity. Injection pressures are measured and checked for abnormalities but are not the decisive criterion for production control. If the pumping test is not successful, re-injections can be performed, and the pumping test is repeated.

Grouted horizontal cut-offs must be checked for safety against hydraulic uplift pressures. Grouted horizontal cut-offs for which the safety against hydraulic uplift is to be verified are assumed to be totally impervious with an imaginary membrane located at the proposed underside of the horizontal cut-off. In the analysis, the uplift pressure acting on the underside of the grouted horizontal cut-off is balanced against the total weight of all soil layers above the imaginary membrane (fully saturated by water or grout and partially saturated).

6 Jet Grouting

Before starting TBM works, a settlement prediction calculation was conducted for the bored tunnel section between Abbasia and Attaba stations. One imperative recommendation was that a Jet-Grouting soil treatment should be executed between the bottom level of a main Sewer and the future Tunnel that will cross the Sewer leaving a confinement between them of only 3.35 m.

The objective of the jet grouting between the existing Sewer and the CML3 tunnel is to allow the TBM to bore the tunnel in a safe way that there is no filling of the sewer with the slurry. In addition, the settlement and the differential settlements of the sewer tunnel should be limited to the minimum possible while the TBM is crossing below the Sewer. The jet grouting soil treatment zone is extended to 20 m along the sewer on each side of the crossing. As per the geometry and design of the treatment zone, a network of Jet Grouting points had to be installed with different inclinations in both directions and different depths. The design Jet Grout column was 2.0 m. In reality, the actual diameters reached up to 2.20–2.30 m.

The jet grouting process consists of disaggregation of the soil and its mixing with, and partial replacement by, a cementing agent; the disaggregation is achieved by means of a high-energy jet of a fluid with cementing agent. The main advantage of this process is that large solidified elements can be executed in the ground using a relatively small drill rod (drill rod diameter being approximately 15 cm), and the method is applicable within limited working space.

Construction sequence (Fig. 11) of the jet grouting columns,

- A string of jet grouting rods is drilled into the ground to the required depth by a rotary drilling rig. The lower end of the drill string is fitted with a nozzle holder and a laterally mounted jet grouting nozzle.

(a) (b)

Fig. 11 A and 11b. Jet grouting construction sequences and R&D in BAUER

- A jetting fluid (usually cement grout, depending on the type of jet grouting system) is pumped through the jet grouting nozzle at high pressure (400–600 bar). This produces a high-energy cutting jet that erodes the soil from its natural position and mixes it with the grout. The diameter of the column is determined by the density and type of soil as well as the jet grouting parameters, as shown in Fig. 10b.
- By rotating and simultaneously retracting the jet grouting drill string, the cutting jet describes a tightly spaced helix in the soil, resulting in a column-shaped space filled with grout and soil. The cement grout causes this mixture to set and solidify, resulting in the formation of load-bearing jet grouting columns.

Depending on site-specific circumstances, and the prevailing soil conditions (granular and/or cohesive), the system and sequence must be determined individually for each project to ensure the required quality.

Test Columns

Before commencing jet grouting operations, it is essential to construct test columns, if adequate suitability tests data are not available. The average diameter of each test column must be determined and compared with the diameter specified in the design. Based on the results of this comparison, the production parameters may have to be adjusted accordingly.

Quality Control

During the execution of the silicate-gel grout plugs and the jet grouting works, the following Quality Assurance measures are applied:

Permeation grouting (excerpt):

- Checking the drilling depth.
- Verticality measurement (selective), to check the deviations or tolerance limits.
- Control of self-hardening slurry: density.
- Control of soft gel: gelling time.
- Control of grouting parameters: pressure, volume.

Jet grouting (excerpt):

- Checking the drilling depth.
- Verticality measurement, to check the deviations or tolerance limits.
- Control of grout mix slurry: density.
- Control of grouting parameters: volume, pressure.

7 Soil Freezing

During the tunnel boring works for Cairo underground metro line 3, a collapse was caused by one of the segmental liners having failed during tunnel construction. Inflow of water and soil flooded the tunnel boring machine (TBM) resulting in a sinkhole at the ground surface. The TBM was buried at a depth of about 30 m under the street level.

After evaluation of different options considering jet grouting, ground freezing and mixed solutions using both techniques, finally the construction of a shaft directly in front of the TBM and horizontal ground freezing was chosen to rescue the TBM. Two rings of bore holes, with the holes tapering towards each other, were necessary for this. These holes were drilled from a previously constructed rescue shaft.

Bayer Egypt was immediately awarded contract for special tasks to rescue the Tunnel Boring Machine.

The scope of works is shown in Figs. 12 and 13.

- Stabilizing and strengthening the collapsed mass by using cement grouting.
- Construct a deep rescue shaft by installing a diaphragm wall shaft in a distance of approximately 2–3 m from the stuck TBM. Key dimensions are 18 m diameter, 1.2 m thick, and 95 m depth. Embedment into a low permeable clay layer for water cut-off, as shown in Fig. 15.
- Groundwater control by supplying and installing freezing works are shown in Fig. 17a–c. The soil around the buried machine along a distance of about 20 m is frozen to allow the main contractor to dismantle the machine and get it out in order to continue the tunnel boring works, as shown in Fig. 18.

The frozen soil body to be produced to act as sealing and structural member. As far as the structural function was concerned, the design provided for a gap with a width of 2.1 m (distance between the TBM and first stable tubbing ring) to be spanned. It was assumed that this stable arch had to be effective in this area for 3 months. In general, ground freezing technology was used for temporary groundwater control and excavation support in shaft construction and tunneling projects. The main application is, however, the creation of a water cut off (groundwater control) with a certain aspect of ground stabilization. Therefore, the methodology depends in general on soil material in conjunction with adequate volumes of natural water. The technology is environmentally friendly as it takes temporarily advantage of the physical properties of the soil and water without changing their quality and natural condition permanently. There will be practically no disturbance of the natural groundwater level and the ground conditions. The influence on surrounding structures is in general minor but

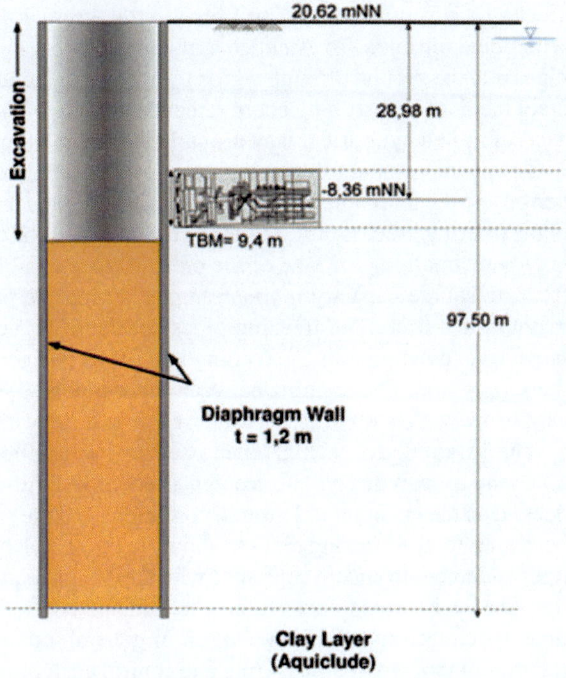

Fig. 12 The depth of the tunnel axis is at a depth of approximately 30 m and the groundwater level at approximately 2.5 m below ground surface

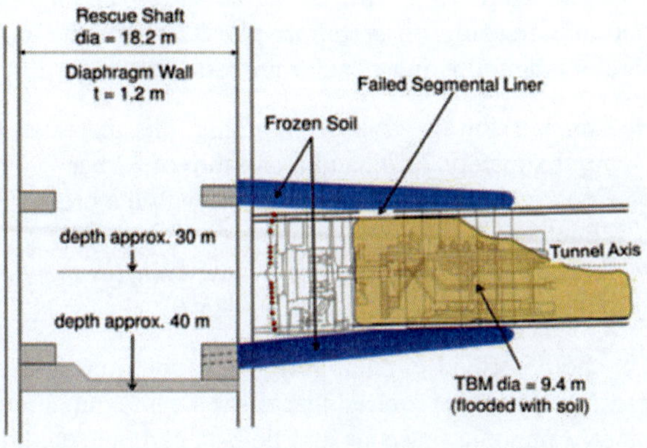

Fig. 13 Vertical cross-section of the rescue shaft, the stuck TBM and the planned frozen soil body with a length of approximately 25 m

basically depends on the soil as certain types of ground could cause severe heaving. The main disadvantage of the ground freezing is the high cost associated with the freezing process compared to other methods protecting against groundwater and creating soil stabilization. But very often ground freezing is the last resort when all other methods have failed or are technically impracticable. There are two principal artificial ground freezing procedures that are used to convert groundwater to ice, creating a strong watertight soil body in the ground: ground freezing with brine or with liquid nitrogen. To establish a frozen soil body, it is necessary to install freezing pipes into the ground through which the refrigerant medium can circulate. The refrigerant medium is absorbing energy/heat from the ground and the groundwater freezes as a result in a cylindrical way around the freezing pipes.

The actual ground freezing process is separated into two parts, the initial freezing period and the maintenance period. Initial freezing is the duration from the beginning of the freezing process until achieving the designed soil body. The maintenance period is only maintaining the size of the created frozen soil body. Therefore, the required freezing source capacity is much higher during the initial freezing period and the maintenance time. The freezing pipes consist of two concentric located pipes. The inner pipe (downpipe) is the feeding line of the refrigerant agent and connected with the supply line. The annulus between inner pipe and outer pipe is the return line in case of freezing with brine or exhaust line in case of freezing with liquid nitrogen.

The assembled freezing pipes need to be installed into the constructed boreholes and the annulus will be grouted afterwards. Drilling methodology and diameter depend on the encountered ground conditions and the specified freezing method. The boreholes for the freezing pipes as shown in Fig. 16a, b have to be installed within very tight tolerances to ensure appliance with the design requirements. Therefore, the position of each borehole needs to be checked and controlled to guarantee correct locations within the specified tolerances. In general, complex ground freezing projects require comprehensive measuring and control technology including throughout every stage of the freezing process.

Ground Freezing with Brine, as shown in Fig. 14. This method utilizes large portable freezing/refrigeration plants where the refrigerant solution is passing through the following circular process. The basic principle involves:

- Compression of gaseous ammonia; this increases the temperature of the gas to approximately 70 °C and a pressure of 12 bar.
- Cooling down of the gas to 35 °C with the pressure remaining at 12 bar, thereby liquefaction.
- Decompression of the liquid ammonia (evaporation), thereby lowering of the temperature to approximately −35 °C.

For the described circular refrigerant solution process, a separate circular water-cooling system is required that has to be integrated into the freezing or refrigeration plant. Ammonia (gaseous and liquid), carbonic acid, or fluorine can be used as the refrigerant solution. The third closed circuit contains the refrigerant agent that is pumped with a temperature ranging from −20 to −37 °C through insulated pipes to the freeze pipes in the ground to extract the head (energy) out of the ground to create the frozen soil body. This freezing process can take up to several weeks. The refrigerant agent is a watery saline solution mainly consisting of calcium chloride ($CaCl_2$), also called brine.

Figures 15, 16, 17 and 18 describe the pictorial description process TBM rescue (Fig. 19).

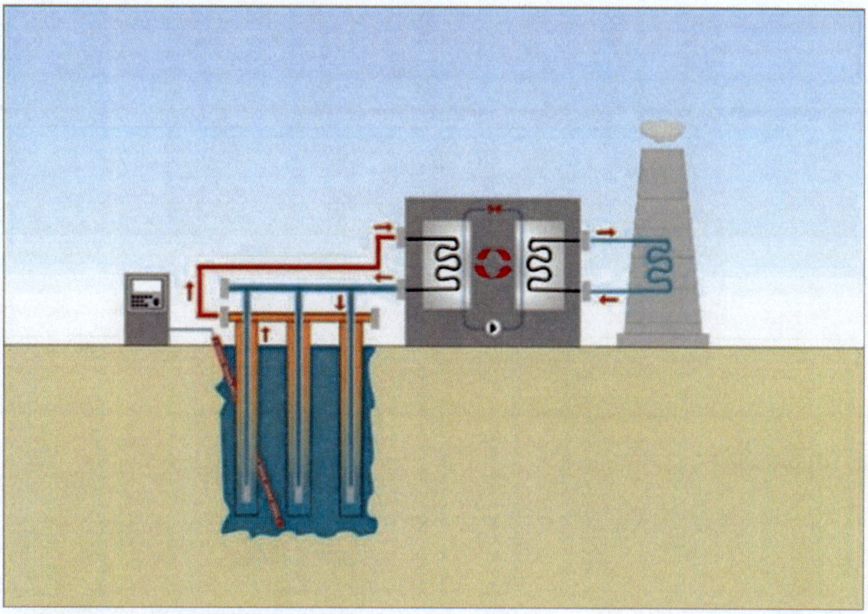

Fig. 14 Soil freezing with brine (schematically only)

Fig. 15 Installation of shaft by D-Wall, at a distance of approximately 2–3 m from the stuck TBM

8 Outlook or Concluding Remarks

After the successful implementation of the single-valve method for the execution of silicate gel grout plugs, and the completion of several excavation pits for Metro Line 3, Bauer was contracted for six more stations for new Line 4. Feasibility studies are currently underway for two more metro lines.

Fig. 16 Drilling rig in D-Wall shaft

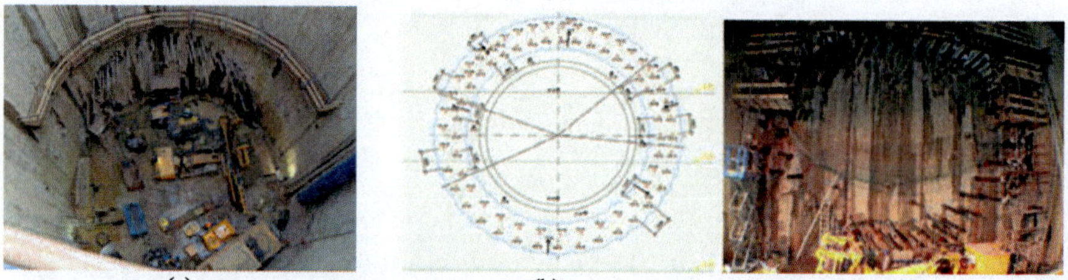

(a) (b) (c)

Fig. 17 a–c Drilling and installation of freezing pipes in diaphragm wall shaft

Fig. 18 Controlling of freezing pipes

Fig. 19 TBM shield rescued

Reference

The National Authority for Tunnels (NAT) (2004) Tender Documents—Cairo

Deep Excavation by Controlled Rock Blasting Considering Ground Condition and Nearby Structures in UG Metro Station—A Case Study

P. K. Ray, Devulapalli L. S. Rasagna, Sudip Kumar Koley, and Sainath M. Reddy

Abstract Bangalore Metro Rail Project Phase II is one of the technologically challenging projects, specially construction of underground metro rail system. The Bengaluru city is very congested and connection of metro rail system with existing transportation networks is equally challenging. In this Phase II of construction, the underground metro is being constructed by blasting hard rock formation and supporting the shear zone. Most of the subsurface rock formations comprises of Granitic Gneiss formations are of Precambrian age and shear zones that includes friable a thin weathered rock formation. With such subsurface geological conditions, the anticipated construction challenges include: (A) deep excavation of hard rock mass under old and sensitive structures and buildings, by controlled blasting; (B) variation in the rock grade from grade I to grade V; (C) weak rock mass at shear zone resulted in planar failure beneath the retaining wall; (D) water seepage along shear planes; (E) selection of a suitable method to conduct blasting near high-voltage electrical machinery or power cables. This paper describes the methods adopted to mitigate these technological challenges and to meet stringent environmental compliances. These include study and application of various geological, geotechnical parameters to develop a proper line drilling and controlled blasting techniques to minimize the ground vibrations and controlling the rock fragmentation and to optimize the equipment performance in terms of productivity and time schedules. The "Phase 2" software of "Rocscience" has been used and regression analysis was carried out

P. K. Ray · D. L. S. Rasagna · S. K. Koley (✉) · S. M. Reddy
Technical Services Department, ITD Cementation India Ltd., Kolkata, India
e-mail: koley.sudip@itdcem.co.in

P. K. Ray
e-mail: Prodyot.ray@itdcem.co.in

D. L. S. Rasagna
e-mail: devulapalli.rasagna@itdcem.co.in

S. M. Reddy
e-mail: sainath.mulinti@itdcem.co.in

A. M. Krishna et al. (eds.), *Deep Foundations for Infrastructure Development in India*, Lecture Notes in Civil Engineering 373, https://doi.org/10.1007/978-3-031-37256-8_6

by using MCD (Maximum Charge per Delay) and PPV (Peak Particle Velocity) to estimate with FOS (Factor of safety) for sensitive places such as nearby high-rise buildings and electrical transformers.

Keywords Controlled blasting · Regression analysis · Peak particle velocity · MCD (Maximum Charge per Delay) factor of safety · Shear zone · Line drilling

1 Introduction

Drilling and blasting is the most conventional, common and economic technique for rock excavation from the earliest days of invention of explosives to present days of modernisation. Although there have been significant developments in blasting technology, application of this technology for rock excavations is associated with a number of safety and stability-related problems to the surrounding rock mass as well as structures.

In general, the excavation specially on the hard rock by drilling and blasting method creates substantial noise and vibrations. Such phenomenon has an impact on the nearby localities and ongoing regular activities on the surrounding areas. The risk factor involved in blasting design needs to be assessed properly during design stage. The shortcoming of risk assessment on blasting design may lead to over-conservative design parameter assumptions, which will increase unnecessary project costs. On the other hand, if the vibration risk is under estimated, it can lead to unexpected damages to the buildings. This may cause public nuisance, delay in execution of the project and additional unknown costs. Therefore, it is desirable to consider planned risk management concept to make the project cost-effective. The legitimate risk management associated with blasting method needs to understand fundamentally in case of vibration propagation in the soil and rock and their interaction with structures. Continuous upgradation of blasting techniques in rock excavation resulting reduction of substantial project costs now a day which can be ensure through a proper system of vibration monitoring and data acquisition. Proper information on wave propagation in the ground and dynamic interaction of structures and foundations can be ascertained through this upgraded technology.

The blasting in the hard rock is likely to induce adverse impacts, i.e. flyrock, ground vibration and air overpressure. Due to induced blasting vibration, the shock waves will propagate via rock media, which may create damages to the adjacent structures. Proper design of blasting patterns, zone of influence and execution of the work with experienced workmanship could mitigate these effects.

Table 1 Controllable and non-controllable parameters influencing ground vibrations due to blasting (after Wiss and Linehan 1978)

Variable	Influence of ground motion		
	Significant	Moderately significant	Insignificant
Controllable parameters			
(i) Charge per delay	√		
(ii) Delay interval	√		
(iii) Detonator accuracy	√		
(iv) Burden and spacing		√	
(v) Stemming length			√
(vi) Stemming material			√
(vii) Charge length and diameter			√
(viii) Angle of blast hole			√
(ix) Direction of initiation		√	
(x) Charge depth			√
(xi) Bare and covered DF		√	
(xii) Charge weight per blast			√
Non-controllable parameters			
(xiii) Surface terrain	√		
(xiv) Type and distance of OB	√		
(xv) Wind and weather condition			√

2 Human Response to Ground Vibration

Human beings are more sensitive to ground vibration and noise than structures. People inside buildings will respond differently than people outside. People tend to complain about ground vibrations even below the accepted damage level because of many reasons. One of the most important factors is the presence of secondary sounds, such as rattling windows and doors. The threshold of perception for ground motion (without sound effects) is roughly 0.51 mm/s (Table 1).

3 Vibration Standards and Criteria to Prevent Structural Damage

Peak particle velocity has been globally used in practice for assessment of blast-induced damage to structures. Different countries adopt different standards depending on their type of industrial/residential buildings. Some investigators included frequency, along with the peak particle velocity for the damage criteria, also. A few suggested damage criteria based on peak particle velocity for different frequencies are given in Table 2. The importance of frequency component in the damage potential of a given ground vibration event is given in Fig. 1.

Table 2 DGMS (India) standards for structural damage criteria

Type of structure	Dominant excitation frequency, Hz		
	<8 Hz	8–25 Hz	>25 Hz
(A) Buildings/structures not belonging to the owner			
1. Domestic houses/structures (Kuchcha, brick and cement)	5	10	15
2. Industrial buildings	10	20	25
3. Objects of historical importance and sensitive structures	2	5	10
(B) Buildings belonging to owner with limited span of life			
Domestic houses/structures	10	15	25
Industrial buildings	15	25	50

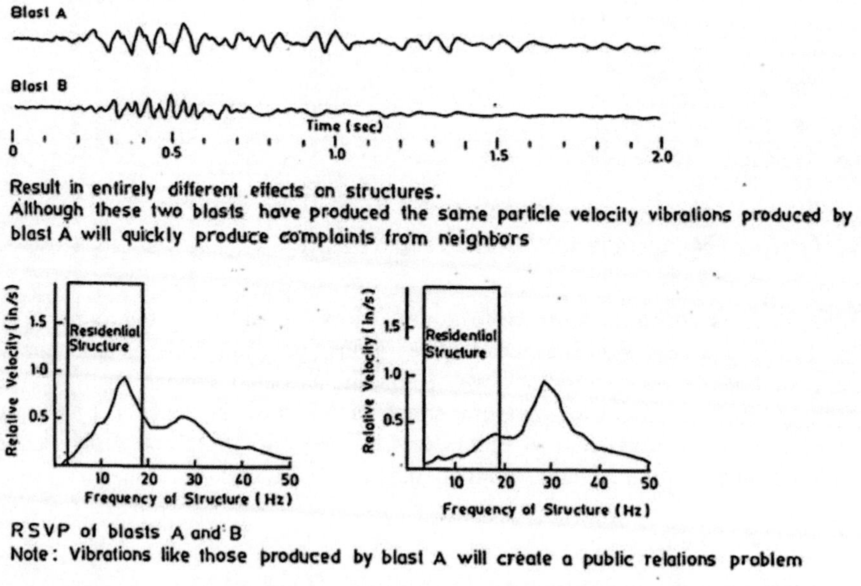

Fig. 1 Importance of frequency component as damage potential by comparing the blast data

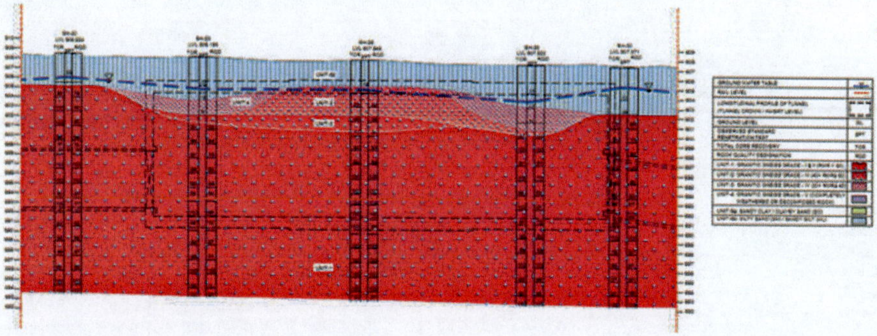

Fig. 2 Field geological data

3.1 Geology of the Area

Within the station area, geotechnical investigation was carried by drilling the boreholes at regular intervals up to the maximum depth of 25 m below ground level. The geology was found in the station area form 0.0 to 4.0 m is soil strata and followed by rock of different grade viz Grade-I to Grade-V. (Fig. 2)

4 Estimation of Peak Particle Velocity

The US Bureau of Mines has developed a mathematical model, which relates peak particle velocity with charge weight and distance (Eq. 1).

$$V_{max} = K\left(D/\sqrt{Q}\right)^{-b} \tag{1}$$

where

V_{max} = Peak Particle Velocity (mm/s).
$(D/\sqrt{Q})$ = scaled distance.
D = radial distance from blast to monitoring station (m).
Q = maximum charge per delay (kg).
K and b are site constants.

As the seismic transmission characteristics are unavailable and unpredictable in a new location, so in general practice, the required site constants (K, b etc.) are then determined by monitoring the ground vibration at different distances for known drilling and blasting parameters.

The data were collected from various metro stations in Bangalore with similar rock conditions with the present blasting sites for the sake of estimating the probable ground vibration and air overpressure levels that may be encountered at various distances during the actual blasting operations in the proposed station areas. With

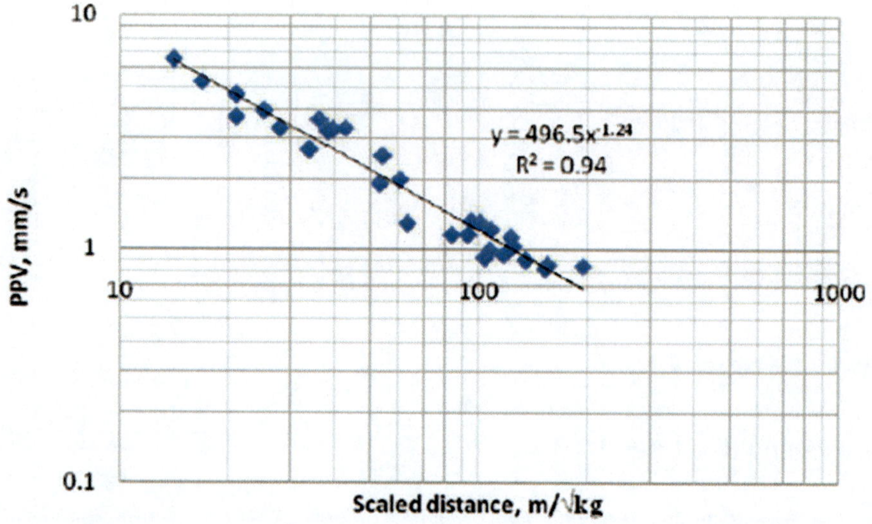

Fig. 3 Peak particle velocity versus scaled distance for an existing underground metro station, Bangalore

the data, Eq. (2) was developed, and Peak particle velocity versus distance for similar granitic rock mass at BMRCL Metro station site is shown in Fig. 3. The attenuation model shown in the figure indicates that the maximum peak particle velocity was observed as 3.9 mm/s for a distance of 22.5 m (with explosive charge of 0.5 kg). The dominant excitation frequency of vibrations was more than 25 Hz. In all the events, the vibration levels were less than 5 mm/s. Therefore, the same blast design parameters are good enough to start with station box excavations.

The derived site-specific predictor equation at 50% confidence level is given below:

$$V = 496.5\left(D/\sqrt{Q}\right)^{-1.14} \tag{2}$$

$$R^2 = 0.92$$

The study revealed that the frequency of ground vibrations was high (>25 Hz), and the suggested permissible vibration level based on DGMS norms is 5 mm/s. But keeping in view of the safety of some sensitive and critical structures in the vicinity, the safe vibration level is considered as 5 mm/s as there are chances of decrease of frequency at farther distances. Substituting the permissible level of ground vibration of 5 mm/s for different distances as per Eq. (2), the computed safe maximum charge per delay is given in Table 3.

If the proximity of critical structures (within 10 m), situations may arise where in blasting may not be possible due to vibration and air overpressure norms. Under these

Table 3 Computed safe maximum charge per delay

Distance, m	Computed safe maximum charge per delay, kg
10	0.06
15	0.14
20	0.25
25	0.39
30	0.56
35	0.76
40	0.99
45	1.25
50	1.55

circumstances, it is suggested to resort to alternative methods to rock excavation like mechanical breaking through rock splitters. Therefore, the blast designs suggested in this method statement are applicable only when the distance between the blast site and the nearest surface structures is 10 m and beyond.

However, this statement will be reviewed after conduction trial blasts and development of actual attenuation equation, as there is a possibility of getting different site-specific attenuation model. Although the threshold peak particle velocity (PPV) according to the DGMS circular is 10 mm/s for frequencies >25 Hz for objects of historical importance and sensitive structures, efforts are required to keep the PPV levels to about 5 mm/s (at 50% confidence level) at very sensitive locations like Prayer centres and heritage structures. Pre-splitting is one of the best options to reduce the vibration levels. A preliminary exercise was carried out to find out the critical peak particle velocity for different structures existing in the vicinity based on the structural and functional properties as well as the frequency of vibration. The preliminary threshold PPV values for all the structures are given in Tables 4 and 5.

Table 4 Preliminary threshold PPV values for sensitive structures

Name of structure	Permissible peak particle velocity (mm/s)
Schools and educational institutions	10
Hospitals	5
Heritage structures	2
Prayer centres (temples, mosque and churches)	5
General residential buildings depending of frequency	5–10
Civil structures like water tanks, towers	5

Table 5 Statistical data of conducted blasts

SL. No.	MCD (kg)	Distance (m)	PPV (mm/s)	Scaled Dist. (m/kg$^{1/2}$)	Log PPV
1	0.375	70	1.031	114.310	0.013
2	0.375	59	0.891	96.347	−0.050
3	0.250	68	0.772	136.000	−0.112
4	0.375	76	0.731	124.107	−0.136
5	0.250	90	0.528	180.000	−0.277
6	0.250	80	0.427	160.000	−0.370
7	0.375	48	1.381	78.384	0.140
8	0.250	60	0.832	120.000	−0.080
9	0.500	48	1.472	67.882	0.168
10	0.375	46	1.664	75.118	0.221
11	0.375	68	0.850	111.044	−0.071
12	0.375	46	1.140	75.118	0.057
13	0.375	54	1.057	88.182	0.024
14	0.375	55	1.233	89.815	0.091
15	0.250	70	0.725	140.000	−0.140
16	0.500	60	0.871	84.853	−0.060
17	0.500	48	1.660	67.882	0.220
18	0.375	55	0.936	89.815	−0.029
19	0.500	48	1.634	67.882	0.213
20	0.375	54	0.978	88.182	−0.010
21	0.375	55	1.243	89.815	0.094
22	0.375	57	0.964	93.081	−0.016
23	0.500	55	0.957	77.782	−0.019
24	0.375	54	0.856	88.182	−0.068
25	0.375	55	0.980	89.815	−0.009

5 Methodology

Line drilling is required to be carried out at the periphery of all sides of station box with large diameter deep drill drilling (Fig. 4).

The line drilling may be one of the techniques used for over break control. In line drilling, a single row of closely spaced, unloaded, small diameter holes are drilled along the excavation line. This provides a plane of weakness to which the primary blast can break and to some extent reflects the shock waves created by the blast, reducing the shattering and stressing in the finished wall. Line drilling is best suited to homogenous formations where bedding planes, joints, and seams are at a minimum. Line drilling has very limited application. The only place where it is applicable is in areas where even the light explosive loads associated with other controlled blasting

Fig. 4 **a** Line drilling holes; **b** Blasting hole and the explosive connection by using Nonel technology; **c** Use of rubber mats as final muffling over the drill blast hole

techniques may cause damage beyond the excavation limit or where line drilling is used between loaded holes to promote shearing and guide the presplit line.

The purpose of line drilling is to control the ground vibration and damage to the final wall or nearby structure due to production blasting. Precision in drilling is very important as any deviation from the periphery plane will have an adverse effect on the over-break. Then the blast design will be carried out with varying parameters line depth of hole, angle of inclination, charge per hole, delay pattern etc. and the vibration monitoring will be done using vibration/AOP monitoring equipment (Fig. 5).

A typical graph for the vibration (amplitude) (Fig. 6) in longitudinal, transverse and vertical direction is presented in a PPV versus time scale as per the seismograph.

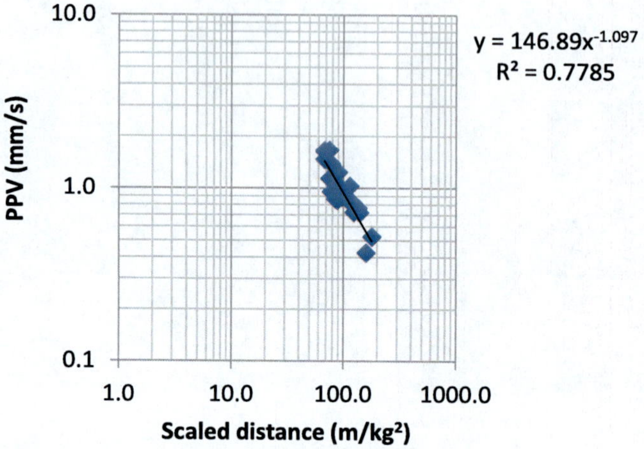

Fig. 5 PPV versus scaled distance for the under ground metro station

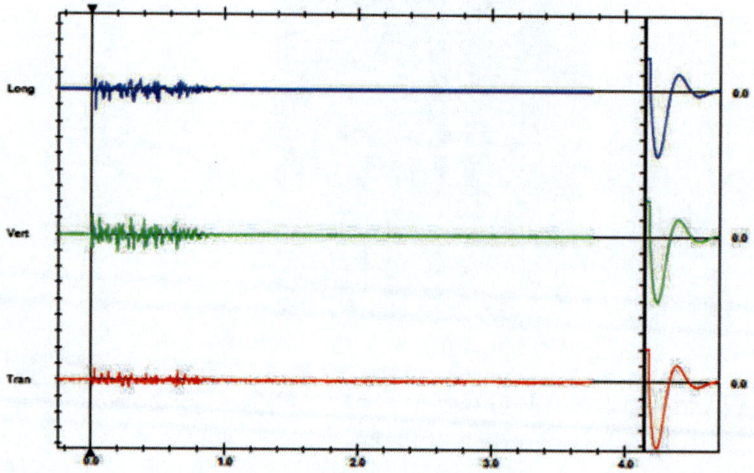

Fig. 6 Vibration (amplitude) graph from seismograph instrument

The outputs recorded in the seismograph near a structure are recorded and plotted in a velocity versus frequency graph considering the DGMS maximum restricted scale in Fig. 7.

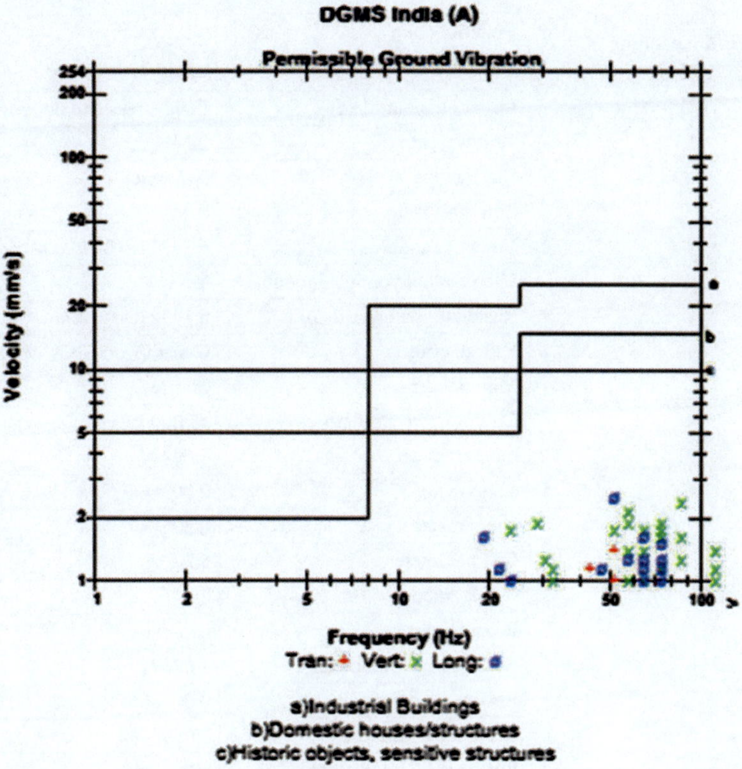

Fig. 7 DGMS limitation for the surrounding structures

6 Discussion on FEM Modelling

A 2D FEM model is prepared in Rocscience based on the geotechnical report just beside the blasting holes to cross-check the effect of blast-induced vibration on the surrounding slope of the rock excavation inside the station box. Rock properties for the present model are in Table 6.

Considering the interpret, it is noticed that except the face of the slope other areas are stable with a stability factor more than 1.25. It is under study for future development to generate a model to calculate the ground vibration in the form of PPV (mm/s) by providing the dynamic coefficients (like Raleigh coefficient) (Fig. 8).

Table 6 Material properties for the FEM modelling

Material color	☐
Initial element loading	Field stress and body force
Unit weight	22 kN/m^3
Elastic type	Isotropic
Poisson's ratio	0.4
Young's modulus	6.14327e+06 kPa
Use residual young's modulus	Yes
Residual young's modulus	6.14327e+06 kPa
Failure criteria	Generalized Hoek - Brown
Material type	Plastic
Compressive strength	45,000 kPa
mb parameter	4.69496
s parameter	0.003866
a parameter	0.505734
GSI parameter	50
mi parameter	28
D parameter	0
Residual mb parameter	1.67677
Residual s parameter	0.003866
Residual a parameter	0.0505734
Residual GSI parameter	50
Residual mi parameter	10
Residual D parameter	0
Dilation parameter	0
Tensile cutoff type	0
Materail behaviour	Drained
Porosity value	0.5
Static water mode	Ru
Ru value	0

7 Concluding Remarks

Underground Metro structures are required where there is a space constraint. Therefore, in those places, the blasting in hard rock (Granite) needs proper blast design, considering extra safety for flyrock outside the excavated area and the planning of proper field instrumentation monitoring. In this case, it is observed in the monitoring data that a maximum PPV of 1.3 mm/s is induced in a case of blasting. The maximum permissible value for a blast vibration as per DGMS is 5 mm/s. But for

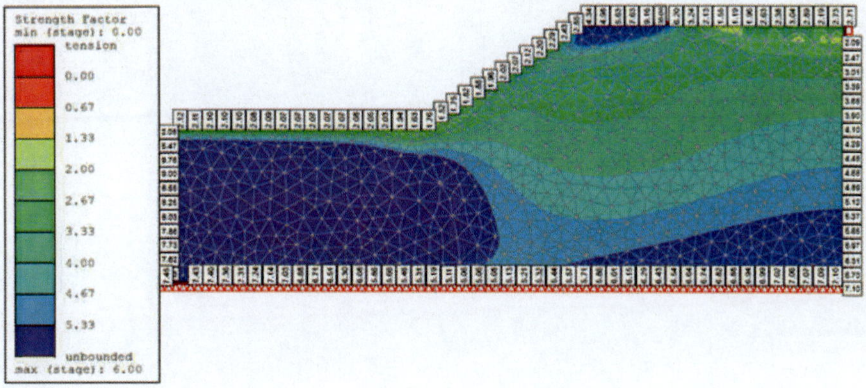

Fig. 8 Numerical modelling for the strength factor for surrounding slope at the blasting time

structural integrity of the structure, it is better to restrict the values within 2 mm/s. In the densely populated areas like Bangalore, the use of line drilling for a separation of bed rock profile with the surrounding foundation structures is very useful to minimize the vibration transfer in the form of PPV (mm/s). Moreover, future studies are still considered to develop a relation between the actual seismograph data from the field due to blasting with the numerical graphs after considering the line drilling for separation.

References

DGMS Circular No. DGMS (Tech)/(S & T) No. 7 of 1997 dated 29/08/1997
Dowding CH (1985) Blast vibration monitoring and control. Prentice Hall
Rocscience (2016) Software Manual. https://www.rocscience.com
Wiss JF, Linehan PW (1978) Control of vibration and blast noise

Safety Evaluation of Adjoining Structures in Urban Areas During Deep Excavation: A Case Study of Nagpur Metro

Shreya Vivek Hokam, S. W. Thakare, A. I. Dhatrak, and Vivek S. Hokam

Abstract Construction in congested urban areas has necessitated tall and massive structures. In urban areas, it is common to construct underground structure nearby existing buildings. Selection of an appropriate support structure system to be implemented in the excavation in an urban area is heavily dependent upon the geotechnical conditions and safety of the surroundings and neighbouring structures. Urban Rail, popularly referred to as Metro Rail, has seen substantial growth in India in the recent years. More cities are experiencing the need for metro rail to meet their day-to-day mobility requirements. The metro structures involved deep excavation for foundations, which induces the stresses and thereby alters the existing equilibrium and thus may affect the stability of neighbouring structures considering safety point of view. Hence, design and execution of deep excavations become one of the important challenging tasks of any geotechnical engineer in the modern era. Further, it is important to control the ground movement in order to prevent the failure of excavations. This involves specialized foundation technologies, usage of sophisticated equipment and new methods in underground construction, safety and monitoring systems. In this paper, the methodology used for protecting adjoining structures due to deep excavation for Nagpur Metro station is studied.

Keywords Deep excavation · Urbanization · Metro station · Safety

S. V. Hokam (✉) · S. W. Thakare · A. I. Dhatrak
Government College of Engineering, Amravati, Maharashtra, India
e-mail: hokamshreya@gmail.com

V. S. Hokam
Maharashtra Metro Rail Corporation Ltd., Nagpur Metro Rail Project, Nagpur, Maharashtra, India
e-mail: vivek.hokam@mahametro.org

A. M. Krishna et al. (eds.), *Deep Foundations for Infrastructure Development in India*, Lecture Notes in Civil Engineering 373, https://doi.org/10.1007/978-3-031-37256-8_7

1 Introduction

With the rapid urbanization in India, effect of construction of buildings/structures with deep excavations on their adjacent buildings has been a pressing issue, mainly related to the soil displacements. Due to the presence of urban structures and public facilities adjacent to the excavation sites, the use of a support system with good and safe performance is essential. Therefore, various support systems can be proposed for stabilizing the excavations and their surroundings. Selecting an appropriate support system in urban areas that can increase the stability of excavations has been a long-time challenge for design engineers. When the new desired excavation depth goes deeper than the foundation of the surrounding buildings, the existing structures are subject to stress, vibrations and displacements.

Deep excavations are frequently used to create infrastructure or underground spaces (such parking lots, stores, or cellars) in many cities in heavily populated areas of the world. Due to the quantity and quality of space that stays above ground, underground development supports the quality of life in cities. The constraints under which these projects must be created have become more difficult in recent years because to the growing demands on space for multiple purposes, including transportation, housing, electricity lines, sewage, etc. Despite the fact that these facilities are designed and built with a lot of care, their construction is not without issues. On the other hand, there will be issues with damage, delays and cost overruns with many underground construction tasks. Various steps are attempted to reduce building damage and annoyance to nearby inhabitants.

Phenomenal growth of construction activities in developed and developing countries, like India, has resulted in enormous pressure on civil engineers especially in metro and cosmopolitan cities, to construct deep underground structures for high-raised buildings, towers, tunnels etc. besides the care that are to be given to underground utilities and adjacent structures. Construction of underground structures are a challenging task for any civil engineer, more specifically when unfavourable multi-complex soil condition exists. In such constraints, the traditional method of underground construction, like braced excavations is becoming ineffective in terms of design and execution.

Deep excavations with supporting systems play a crucial role to create additional usable space in specified areas. Further, deep excavations are frequently constructed in close proximity to the existing buildings and underground utilities. The deeper the depth of the excavations, the greater is the time for construction for varying sub-soil conditions which further more pose challenges to the construction engineer. The excavation induces the stresses and thereby alters the existing equilibrium and thus the underground becomes unstable. Many factors affect the stability of excavation such as soil strength, groundwater, building surcharges, construction methods and also construction techniques.

In various congested locations, there has been a growth in the need for underground space, which can be used for underground parking, railway tunnels/metro stations, road tunnels, building renovation, etc. Underground construction requires significant deep excavations. Most of the time, excavation sites are quite close to already-existing buildings or facilities and issues with deep excavations and their consequences on nearby structures are getting worse. To lessen the effect of deep excavation on nearby structures, various approaches or procedures are used.

The paper presents the system of methodology adopted by Nagpur Metro authorities for safety of adjoining building and structures while execution of metro structures with deep excavation.

2 Literature Review

2.1 General

Extensive review of literatures published in the past periods on excavation supports is carried out in order to understand the trends and the key developments in deep excavation systems. This literature review's primary goal is to evaluate the scientific and technological advancements that have taken place in this field. The literature related to support system for deep excavation are discussed in detail as follows.

Lings et al. (1991) presented and compared the results of a deep excavation made in Gault Clay using top-down techniques to design hypotheses. Measurements were made of pore pressures, heaven and earth pressures. They discovered lateral stresses were significantly reduced by diaphragm wall design, and deflections, prop forces and bending moments were also significantly reduced compared to predictions.

Garvin and Boward (1992) described the five-level underground parking building in Pittsburgh that was built using cut-and-cover techniques. For the 7–8 m excavation, a diaphragm wall with tiebacks was employed as support. Throughout construction, movements and groundwater levels were observed. The slurry wall technology worked effectively and allowed for dewatering inside the excavation with the least possible impact on groundwater levels outside the excavation. There was no trouble with the 80-year-old sandstone building next to excavation.

Giffored and Wheeler (1992) detailed how a five-level underground parking garage in Baltimore's excavation was supported by a structural slurry wall. Construction procedures were made simpler by the deployment of a slurry wall. The slurry wall had a fair amount of water resistance, and motions were reasonable. The pre-trenching operation was largely responsible for the settling.

Schoenwalf et al. (1992) explained how a structural slurry wall performed when it was utilized in the top-down building of a seven-level (24.4 m deep excavation) garage for the Post Office Square project in downtown Boston. The authors contrasted finite element analysis' predictions of measured movements. Less than 8 mm of movement was recorded during building the diaphragm wall, and there was a slight rebound of the earth during the concreting of the panels. Movements of the structural garage floor slabs caused by shrinkage and creep were observed.

Winter et al. (1992) reported how a 10 m-deep excavation next to the Epiphany Church in Washington, D.C., supported by a structural slurry wall with tiebacks, performed. The wall instalment close to the church had movements that were 3 mm or less. A very small amount of displacement was noted throughout the excavation.

Ou and Chiou (1993) studied the data from 10 case histories of deep excavations in Tapei were examined to determine the peculiarities of ground surface settlement during excavation. They looked at the maximum wall deflection as well as the connection between ground settlement and wall deflection. Near the bottom of an excavation, there was frequently the greatest lateral wall deflection.

Wong and Patron (1993) demonstrated the effectiveness of deep excavations for Singapore's Central Expressway (CTE) Phase II project. For each of the varied support systems employed and soil conditions encountered, they looked at lateral wall deflections, soil settlements and prop loads. They discovered that excavations with a combined thickness of soft soil layers overlying stiff soils typically had lower maximum wall movements, and they came to the conclusion that adding the first layer of struts near the top of braced walls is effective in minimizing movements for walls embedded in stiff soil.

Bhatkar et al. (2016), described the nature of surface settlement and wall deflection during extensive excavation in soft soil. On the Hooghly River bank in an eastern metropolis of India, excavation work was done. This excavation was done as part of building an underground metro station. To forecast the behaviour of the soil during excavation, numerical analysis was done. The best excavation technique was thus decided upon.

Rybak et al. (2018) outlined the common damages that occur during the building of supporting walls (sheet pile drive, piling and development of diaphragm walls). The case study's construction site was in southern Poland.

Ayasrah et al. (2021) established a number of numerical simulations to look into how the adjacent pile cap foundations of the Garage El-Attaba building would be affected by the Greater Cairo Metro-Line 3-Phase-1 tunnel construction. Following a detailed review of the simulation results, it was discovered that tunnelling caused more axial stresses, bending moments, axial settlement and lateral deflection.

3 Deep Excavations Hazards

It's crucial to lessen how construction projects affect already-built projects, and vice versa. Due to the adjacent presence of houses and other engineering structures, deep excavations in urban settings necessitate specific considerations. The neighbourhood's existing structures may be considerably impacted by deep excavations for construction. For both the environment's safety and the safety of the workers inside excavations, effective shoring system design and construction are essential. Prior to beginning the design process, a deep excavation analysis is frequently necessary. The theoretical analysis of deep excavations incorporates models of the earth's elasto-plastic behaviour, the behaviour of the interface between the land and retaining structures, as well as the excavation process.

3.1 Influence on Adjacent Buildings

Deep excavation risk in construction is a problem that is frequently disregarded. Deep excavations in populated regions are fraught with danger from a variety of sources. The evaluation of risk sources for disposal may necessitate additional steps and costs for the design and construction of the excavation enclosure, which the investor is compelled to bear. The more the trench depth increases, the more challenging it will be to complete the work while also taking into account hazards to their stability from themselves or from nearby building projects.

Large potential ground deformations in and around the excavation, which could harm nearby structures and utilities, are a key worry with deep excavation operations. When lateral and vertical ground motions impinge on a structure's foundation, they cause angular distortion and lateral extension strain, which become the damage criteria. The installation of a support system and the lateral displacement of the excavation wall are two causes of ground movement.

Patterns of ground movement are settlement at the ground surface and lateral displacements of the ground surface (Fig. 1). It is important to understand how the ground movements due to excavations influence nearby structures. The response of buildings to excavation-related ground movements is dependent on the source and pattern of the ground movements, the type and condition of the structure, and the mitigation measures employed to protect the building.

The primary factors that influence the deformation of the system and the retained ground include type of subsoil, lowering of groundwater level, depth and width of excavation, stability of the bottom of excavation, stiffness of the support system, construction technique, surcharge load, ground movement and quality of investigation and design.

Fig. 1 Different patterns of ground movement

3.2 *Events of Structural Failures Due to Deep Excavation of Adjoining Structures*

Many historical events of structure failure due to deep excavation have been recorded in many countries. Structure failure events pose a significant threat not only to human life but also to the environment and in general to economic development. Thus, it is essential to investigate properly these failures and to find out technical solutions to reduce their risk of occurrence. Recently, project safety draws increasing attention from the public authorities. This is because damages resulting from structure failure can lead to terrible disasters with tremendous loss of life and properties, especially in densely populated areas. Two potential structures failure events are reviewed in this section.

4 Protective Support Systems for Deep Excavation

Various slope protection or shoring methods such as Soil nails, Soil anchors, Rock anchors, Contiguous Bored Piles (CBP), Secant piles or Sandwiched piles, Sheet pile wall, Diaphragm wall, CBP/Sheet pile/Diaphragm wall with anchors or struts or soil nails etc. are available and are being adopted in the construction industry. However, economical design of supporting system without compromising on the safety is of paramount importance. Selection of a suitable method of support system for an excavation depends on several factors such as the depth of excavation, soil profile, sensitivity of the adjacent structure, water table and duration of protection in the case of temporary shoring etc.

Among the several methods of support system, the popular methods that are invariably preferred and adopted by construction contractors, in the recent times, are diaphragm wall, contiguous pile, soil nail and secant pile, etc.

5 Scope of Study

In this paper, the effect of deep excavation of commercial building near Kasturchnad Park Metro Station constructed by Maharashtra Metro Rail Corporation Ltd. has been studied on adjoining MSEDCL building. This commercial building is having two basements and 23 upper floors. The location of station and commercial building is shown in Fig. 2.

The construction of metro station along with commercial building was started in September 2017. The commercial building is surrounded by road, flyover and one office building of MSEDCL (Fig. 4). Kasturchand Park Metro station has been commissioned since August 2021. The excavation level was about 9.5 m below natural ground level. In order to avoid safety hazards, Maharashtra Metro (Maha-

Fig. 2 Location of station and commercial building

Metro) had taken precautionary measures for the protection of the nearby structures by providing secant piling prior to start the construction activities as shown in Fig. 3. The secant piles having 1200 mm diameter and spacing of 1000 mm centre to centre were provided up to the depth of 13 m along the periphery of commercial building. Afterwards construction activities of commercial building with raft foundation were started.

Secant pile walls are formed by constructing intersecting reinforced concrete piles. The secant piles are reinforced with steel rebar and are constructed by either drilling under mud or auguring. Primary piles are installed first with secondary (male) piles constructed in between primary (female) piles once the latter gain sufficient strength with pile overlap. The piles interlock or overlap to tolerances designed to achieve an effective interlock between the two.

The main advantages of Secant Pile Walls are increased construction alignment flexibility, increased wall stiffness compared to sheet piles and can be installed in difficult ground (cobbles/boulders) and also less noisy construction as shown in Fig. 3.

Fig. 3 Secant piling

Fig. 4 Location of commercial building and MSEDCL building

As on date, construction of two basements, ground floor and first floor has already been completed as shown in Fig. 4.

6 Numerical Modelling by MIDAS GTS NX 3D

MIDAS GTS NX is a Geotechnical and Tunnel Analysis System that is founded on the expert analysis and exceptional graphic technologies. MIDAS GTS NX enables the users to intuitively generate complex geotechnical models. Such modelling capabilities are armed with very strong analysis features, powered by TNO-Diana solver providing the fastest and most comprehensive in analysis solutions. In densely built area with high land value, the trend is towards building higher and deeper, often adjacent to existing structures. Such projects involve the adoption of tight ground movement criteria which are nowadays commonly checked using 3D finite element analysis. The general work sequence of GTS NX includes geometry modelling, mesh generation, analysis condition, analysis and post-processing and result evaluation.

6.1 Problem Definition

The analysis in the present study consists of modelling of deep excavation in soil bed considering different soil characteristics with varying depth and its properties. Figure 5 shows typical plan and section of the excavation along with the adjoining building. It is aimed to find out the settlement of the adjoining building due to deep excavation in soft soil by the analysis in MIDAS GTS NX. Following parameters of excavation and adjoining structure are considered for the analysis:

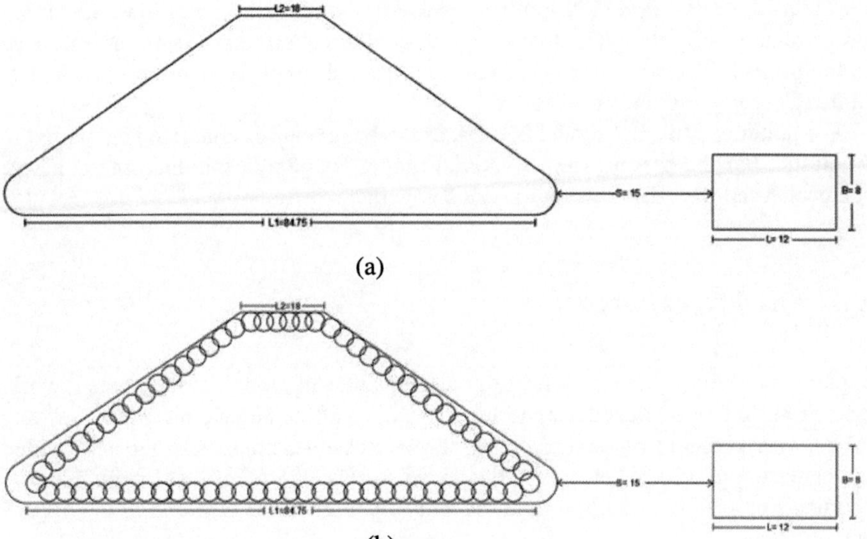

(b)

Fig. 5 **a** Typical plan, **b** Typical plan with secant pile

i. L1 = length of bottom excavation = 84.75 m
ii. L2 = length of upper excavation = 18 m
iii. L = length of the adjoining building = 12 m
iv. B = width of the adjoining building = 8 m
v. S = spacing between excavation and adjoining building = 15 m
vi. d = depth of excavation = 10 m
vii. diameter of secant pile = 1.2 m
viii. Length (Ls) = Depth of excavation + 2 m.

6.2 Model Development in MIDAS GTS NX 3D

In this step, the geometries of soil, excavation and adjoining building were developed according to the required dimensions as shown in Fig. 6 developed in MIDAS GTS NX 3D. For soil model, 3D-line function was used to create its outline and then 'Extrude' function was used to achieve required depth of soil model. Soil model of size 100 m × 100 m × 30 m, adjoining building of 12 m × 8 m × 10 m was selected for the present study.

6.3 Assigning Properties

Material properties for the models are imported manually. For soil, the model selected for the analysis is 'Mohr Coulomb' model. MIDAS GTS assumes that the stress–strain relation of 'Mohr Coulomb' model is Elasto-Perfectly-Plastic and is used in the project. The soil considered for this research work is medium clay, hence, Mohr–Coulomb model is best suited.

For building, 'Elastic' model is used, as the behaviour of concrete is assumed to be elastic. For the present study, material properties of soil, building and concrete segments used in MIDAS 3D are given below in Table 1.

6.4 Mesh Generation

Separate meshing is provided for raft, building and soil model. In the present work, auto mesh function is used for meshing. Number of elements chosen for meshing varies with variation of dimension. As far as possible, finer meshing is provided by increasing number of elements to achieve more accurate results. Figure 6 shows geometry of adjoining building and Secant Pile after Mesh Generation in MIDAS GTS NX 3D.

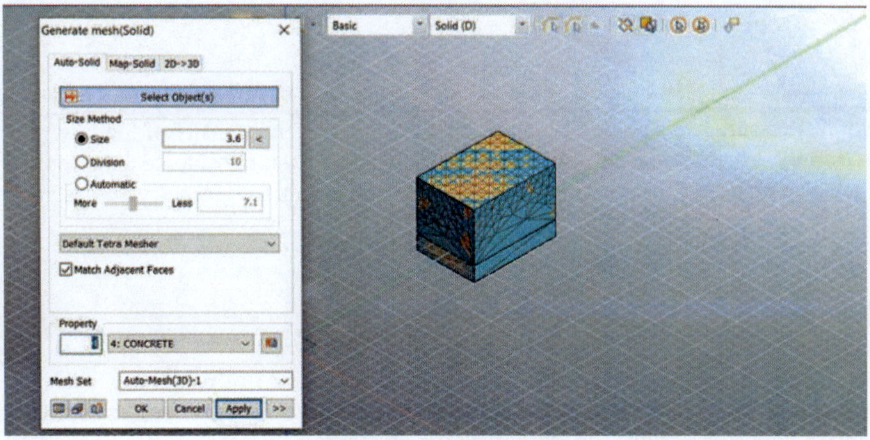

(a)

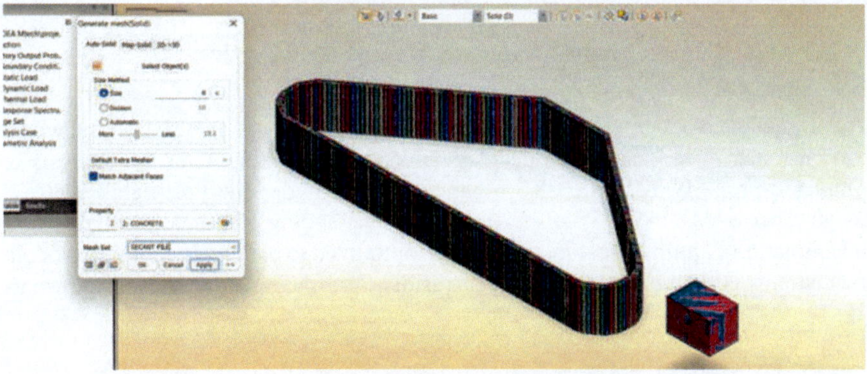

(b)

Fig. 6 Geometry of **a** Adjoining building **b** Secant pile after mesh generation

Table 1 Properties assigned to soil layer for analysis

Description of soil	Unit weight	Elastic modulus of soil	Poisson's ratio	Angle of internal friction	Cohesion	Saturated unit weight
Symbols	γ	Es	υ	Ø	c	γs
Unit	kN/m^3	kN/m^2	–	degree	kN/m^2	kN/m^3
Soil bed	18	50,000	0.3	12	31	19
Concrete	24	2.7×10^7	0.15	–	0	–

6.5 Application of Load

In this step, vertical load intensity is applied on the adjoining building. In the present work, 'Pressure Load' type of load is used to apply given intensity of load. In this, uniformly distribution load (pressure type) of 50 kN/m^2 was considered to be applied on the top surface of the building. Also, boundary condition for geometry was defined.

6.6 Defining Non-linear Analysis

The steps for solving the geometry have been minimized in non-linear analysis. It analyses the geometry in just one step by activating the mesh set at once, the self-weight taken in this set has linear variation along the depth.

7 Analysis, Post-processing and Result Evaluation

Model generated in previous stage is analysed in this step. In the present work, the non-linear static analysis is decided as described earlier. After the analysis is completed properly, MIDAS GTS organizes and provides the post-data result for design process. MIDAS GTS offers numerous outstanding graphical result displays and animation-capturing features, which are optimized to display large-scale models and complex construction stage analysis results. A probe result window is shown in Fig. 7.

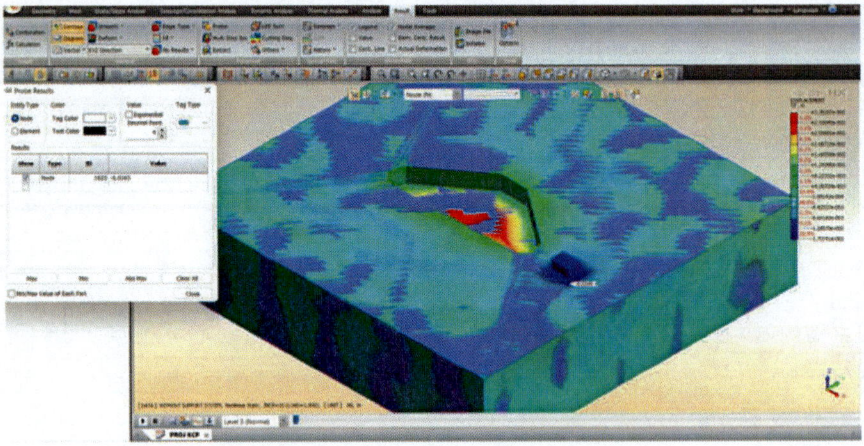

Fig. 7 Probe result window

8 Analysis, Post-processing and Result Evaluation

Model generated in previous stage is analysed in this step. In the present work, the non-linear static analysis is decided as described earlier. After the analysis is completed properly, MIDAS GTS organizes and provides the post-data result for design process. MIDAS GTS offers numerous outstanding graphical result displays and animation-capturing features, which are optimized to display large-scale models and complex construction stage analysis results. A probe result window is shown in Fig. 7.

9 Results and Interpretation

In the present study, analysis is carried out for every individual case, by applying vertical load in increment. As per the probe results of analysis, the settlement at the edge point of the building below soil bed by carrying out non-linear static analysis calculated as 165 mm without support system and 28 mm with support system, i.e., secant piling.

It is interpreted that with support system, i.e., by using secant piling, the settlement of the building is less as compared to the excavation without any support system.

It is observed that by providing secant piling along the periphery of commercial building, the settlement values drastically lowered down with respect to the results obtained without the provision of secant piling. Also, as on date, the existing structure of MSEDCL building is in excellent condition without any signs of settlement, cracks, etc.

10 Conclusions

Selection of an appropriate support structure system to be implemented in the excavation in an urban area is heavily dependent upon the geotechnical conditions of the excavation site and safety of the excavation surroundings and neighbouring structures.

Maharashtra Metro has taken proactive measures for providing support system in the form of secant piling due to which no settlement or any distress has been observed in the adjoining MSEDCL building as well as surrounding pavement.

References

Ayasrah M, Qiu H, Zhang X (2021) Influence of Cairo Metro tunnel excavation on pile deep foundation of the adjacent underground structures: numerical study. MDPI, Symmetry J 13:426

Bhatkar T, Barman D, Mandal A, Usmani A (2016) Prediction of behavior of a deep excavation in soft soil: a case study. Int J Geotech Eng 10–19

Garvin R, Boward J (1992) Using slurry walls to protect a historic building: a case study. Slurry Walls: Design, Construction and Quality control ASTM Special Topic Publication 1129:117–127. 37

Giffored D, Wheeler J (1992) Concrete slurry wall for temporary and permanent foundation wall at gallery at harborplace—Baltimore Maryland. Slurry Walls: Design, Construction and Quality Control ASTM Special Topic Publication 1129:151–163

Lings M, Nash D, Ng C, Boyee M (1991) Oberved behavior of deep excavation in gault clay—A preliminary appraisal. In: Proceedings of the 10th European conference on soil mechanics and foundation enginnering, vol 10(2), pp 467–470

Ou CY, Chiou DC (1993) Three-dimensional finite element analysis of deep excavation. In: Proceedings of eleventh southeast Asia geotechnical conference, Institution of Engineers Malaysia, Kuala Lumpur, Malaysia, pp 769–774

Rybak J, Ivannikov A, Kulikova E, Żyrek T (2018) Deep excavation in urban areas—defects of surrounding buildings at various stages of construction. In: MATEC web of conferences on building defects, p 146

Schoenwalf D, Whiteman R, Abbott E, Becker J (1992) Post office garage project—A case study of instrumented slurry wall performance. Slurry Walls: Design, Construction and Quality Control ASTM Special Topic Publication 1129:361–376

Winter E, Skep N, Tallard G (1992) Slurry wall performance adjacent to historic church. Slurry Walls: Design, Construction and Quality Control ASTM Special Topic Publication 1129:164–171

Wong LW, Patron BC (1993) Settlements induced by deep excavation in Taipei. In: Proceedings of 11th southeast Asia geotechnical conference, Institution of Engineers Malaysia, Kuala Lumpur, Malaysia, pp 787–791

Innovative Techniques and Testing for Geotechnical Investigations Monitoring, Performance, and QA-QC of Deep Foundations

Forensic Evaluation of Bored Piles for a Factory Project at Jhajjar

Ravi Sundaram, Sorabh Gupta, and Sanjay Gupta

Abstract A geotechnical forensic evaluation of piles for an industrial facility in Haryana was performed to assess the quality of the piles and to evaluate piles installed using two different types of rigs. Low-strain pile integrity tests were performed on all piles to assess the pile length and possible presence of structural defects. Coring was done through questionable piles for the clinching evidence of concrete quality and pile length. Probe holes adjoining piles helped confirm the bulging of piles at shallow depth. The paper highlights the importance of implementing a well-planned quality assurance program to confirm that all piles installed are of the desired quality.

Keywords Pile quality · Quality assurance of piles · Low-strain pile integrity tests · Pile-concrete coring

1 Introduction

At a factory site in Jhajjar, Haryana, 246 bored cast-in-situ piles (240 working piles and 6 test piles) were cast for the main building/factory area using two different types of rigs, hydraulic rotary rig and truck mounted rig (TMRs). After casting about 150 piles using hydraulic rotary rig, truck mounted rigs were used to speed up the piling activity. After installation of the piles, the client had doubts on quality of piling work executed at site with regard to the structural integrity of the installed piles, particularly those installed using the TMRs. The pile layout and the rigs used are illustrated in Fig. 1.

R. Sundaram (✉) · S. Gupta · S. Gupta
Cengrs Geotechnica Pvt. Ltd., Noida, Uttar Pradesh, India
e-mail: ravi@cengrs.com

S. Gupta
e-mail: sorabh@cengrs.com

S. Gupta
e-mail: sanjay@cengrs.com

A. M. Krishna et al. (eds.), *Deep Foundations for Infrastructure Development in India*, Lecture Notes in Civil Engineering 373, https://doi.org/10.1007/978-3-031-37256-8_8

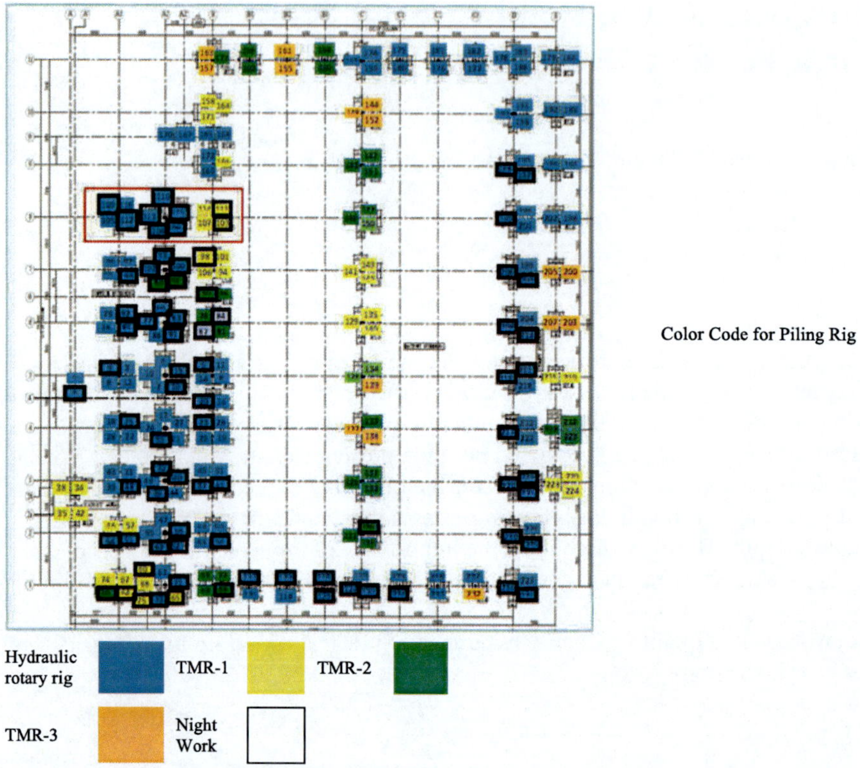

Fig. 1 Pile rig usage layout

Figure 2 illustrates a few piles exposed for testing and evaluation prior to casting of pile caps.

Fig. 2 Exposed piles

2 Tests Performed

The contractor had performed a few initial static load tests on test piles. To evaluate the condition of pile quality, the authors carried out a detailed program of confirmatory geotechnical investigation and pile testing as given below:

1. A confirmatory geotechnical investigation was performed that included three boreholes. The results were compared with the previous geotechnical investigation conducted at site.
2. Low-strain pile integrity tests (PIT) were performed on all the piles in accordance with ASTM D 5886-16 (2016) so as to assess the pile quality, its likely length and to identify the presence of any defects.
3. Concrete coring was done through selected piles to assess the concrete quality and the actual pile length.
4. Probe holes were drilled adjoining selected piles to assess the soil adjoining the pile and to determine the likelihood of bulging of pile.

3 Site Stratigraphy

Based on the borehole data, the surficial soil consists of clayey silt up to about 2–3 m depth, underlain by silty sand to about 5–6 m. Below this, sandy silt of low plasticity is encountered to 10.5 m depth, followed by silty sand to 14.45 m depth. SPT values range from 4 to 13 to 2.0 m depth and from 10 to 26 to 5.0 m depth. Below this, the SPT values range from 20 to 38 to 10.5 m depth and from 32 to 71 to 14.45 m depth. A pictorial summary of the borehole profiles is illustrated in Fig. 3.

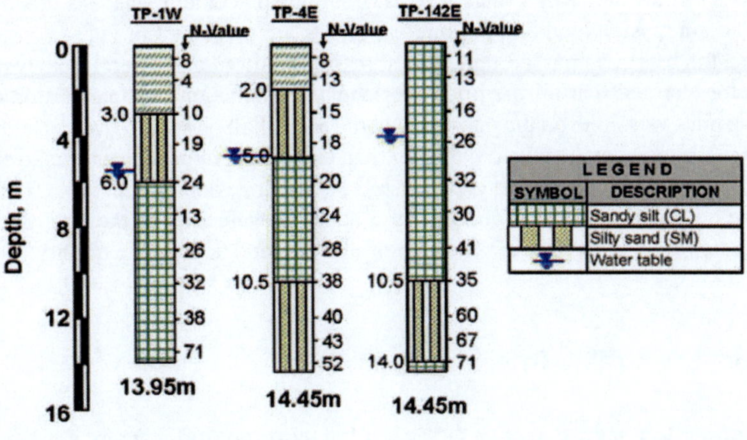

Fig. 3 Typical boreholes profile

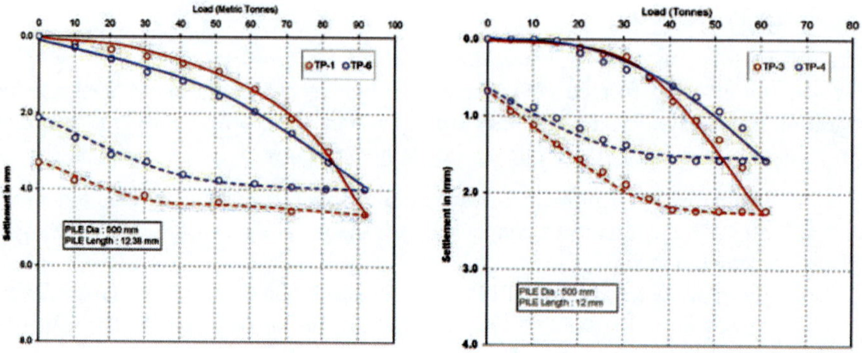

(a) Compression Load Test on Test Piles

(b) Pullout Tests on Initial Piles

Fig. 4 Load-settlement curves—Typical tests on initial piles

Although there were some differences in the soil classification in comparison to the data of previous investigation performed by the client, the trend of SPT values is similar.

4 Pile Load Tests

Initial static pile load tests were conducted on 500 mm diameter 12.4 m long test piles installed just outside the work area prior to starting the installation of the working piles. The piles were designed for a compression load of 35 tons and pullout capacity of 23 tons. All piles were safe under the design load. Under the maximum applied load of 2.5 times the design load, the maximum displacement was less than 5 mm. Typical compression load and pullout test results are presented in Fig. 4.

Since the test piles passed under compression, pullout and lateral load tests, the contractor claimed that all the piles are deemed to be safe. The installation of the working piles was done on the premise that the piles shall be safe. However, although the test piles were having capacity higher than the design value, excavation to expose the working piles indicated several defects at shallow depth (see Fig. 2). Therefore, the owner requested the authors for a detailed evaluation of the quality of pile construction to confirm that all piles are of uniform and acceptable quality.

5 Low-Strain Pile Integrity Tests

Low-strain Pile Integrity Testing (PIT) is a low-cost tool to locate major defects in piles and estimate pile lengths of bored and driven piles installed at site. It can serve to enhance the confidence level of the engineer on the quality of piles installed (Gupta et al. 2008). The test is carried out by the Pulse-Echo (or Sonic Echo) Method, in accordance with the guidelines given in IS: 14893-2021 and ASTM D 5882-16.

PIT is a non-destructive integrity test method for foundation piles and effectively evaluates pile shaft integrity testing by low-strain surface-impact methods (Rausche et al. 1988). It can detect potentially dangerous defects such as major cracks, necking, soil inclusions or voids. It can also be used to assess the pile length (Gupta et al. 2015).

To assess the pile quality, PIT was done on all the piles installed at site. To perform the test, the pile top surface is prepared in order to form a clean, sound and hard concrete for testing. Figure 5 presents photographs of the test being done on two typical piles.

The accelerometer is fixed to the top of the pile using a special wax. Signal enhancement by repeated hammer blows on the pile head helped to minimize the effects of noise or other disturbances (see PIT Collector User's Manual 2005). Some typical PIT plots are presented in Fig. 6. Typical results are summarized in Table 1.

Some of the PIT results were inconclusive. Inconclusive PIT data (inconsistent results, absence of clear toe response, etc.) suggest that the pile structural integrity is doubtful. The absence of clear, repeatable PIT data would generally indicate poor pile top preparation, weak or discontinuous pile concrete, presence of shallow defects, highly non-uniform pile cross-sections and other structural or construction-related problems.

It may be noted that while TP-1 passed the static compression load test, pile had large variations in cross-section and clear bulging was evident below 5 m depth as per PIT results. TP-4 was safe under pullout loading; however, PIT suggested that

Fig. 5 Pile integrity tests in progress

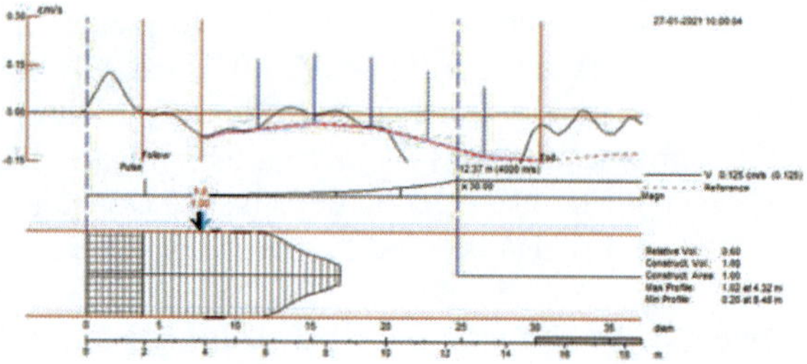

Pile 126: Major defect at about 6-8 m depth, either due to reduction in pile cross-section or concrete discontinuity at this depth

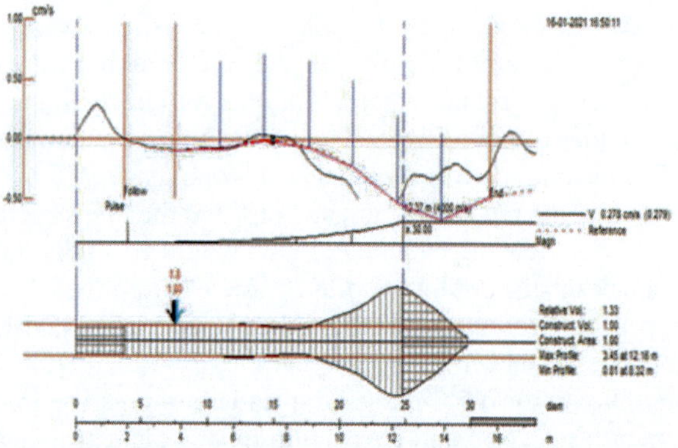

Pile 142: Major increase in pile impedance below about 10 m depth below test level. Pile length seems OK

Fig. 6 Typical pile integrity test results together with profile analysis

pile length is less than planned. Overall, the PIT indicated that about 17 working piles were likely to have major defects.

6 Pile Coring

A few selected test piles were evaluated further by coring through the piles. Experience has shown that coring through the concrete (Gupta et al. 2017) can effectively evaluate the concrete quality and depth and also assess the condition of the pile toe. Figure 7 illustrates pile coring in progress.

Typical concrete cores collected from TP-4 are illustrated graphically in Fig. 8. Figure 9 summarizes the recovery and RQD values together with the interpreted pile profile from PIT results for two typical piles. Table 2 presents the interpretation based on the concrete coring through the piles.

As per the results of crushing strength tests on concrete core sample obtained from pile coring, the density of concrete for most samples is lower than the standard

Table 1 Typical results of low-strain pile integrity tests

Pile number	Type of rig used	PIT data interpretation/Test conclusions
TP-1	Hydraulic rotary rig	Large variation in pile cross-section with depth. Increase in pile impedance below about 5 m depth
TP-4	Hydraulic rotary rig	Early toe response. Pile may be short (9–10 m). Further investigation is required
26	Hydraulic rotary	Major defect (reduction in pile impedance) at about 6–8 m depth, which may be either due to reduction in pile cross-section or concrete discontinuity at this depth
72	Truck mounted rig	Reduction in pile impedance between 6 and 9 m depths, followed by major increase in impedance
94	Truck mounted rig	Major defect (reduction in pile impedance) at about 6–8 m depth, which may be either due to reduction in pile cross-section or concrete discontinuity at this depth
142	Truck mounted rig	Increase in pile impedance below about 10 m depth below the test level
182	Hydraulic rotary rig	PIT could not be performed due to improper pile top preparation or presence of weak concrete at the test level

Fig. 7 Pile coring in progress

value of 2.44 g/cc. Also, the results of the crushing strength of the samples are on the lower side indicating that the quality of concrete is poor.

Fig. 8 Pile coring results: TP-4. *Note* Soil encountered below 10.5 m depth

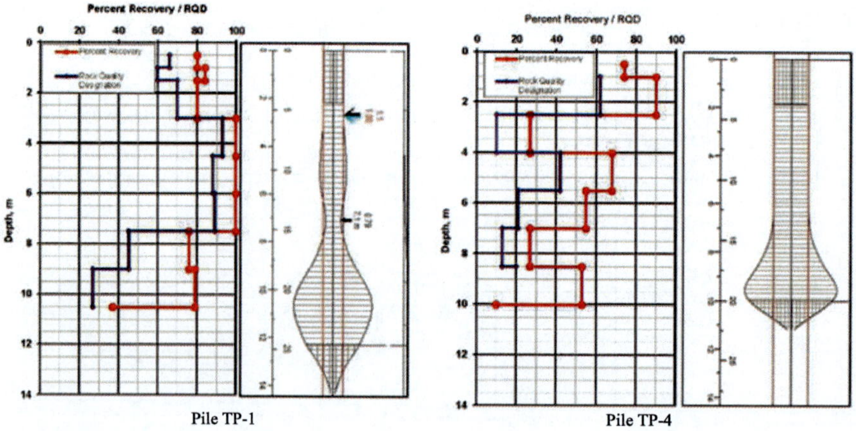

Fig. 9 Concrete core recoveries and RQD values correlated with PIT profile analysis

7 Probe Holes Around Piles

Around selected piles in which bulging was suspected, probe holes of 150 mm diameter were drilled by auger at four diametrically opposite locations. The bores were drilled about 20 cm away from the edge of the pile. The purpose was to evaluate possible bulging of the piles (increase in pile cross-section) at deeper depths, as indicated by the PIT results. The probe holes were extended up to about 14–15 m depth below ground level, which is beyond the anticipated length of the pile.

If concrete is encountered in the probe holes, this will indicate bulging of the pile. If concrete is not encountered, either the PIT impedance increase is due to other reasons (variations in concrete density or soil resistance, for example), or the bulging may be less than 200 mm (and hence not captured by the probe bores). Typical probe hole results for two piles are presented in Fig. 10. Table 3 presents the interpretation based on the probe-hole data.

The probe holes clearly indicated bulging at Pile 182. At TP-1, coring indicated weak concrete of somewhat lower density that corroborates well with the probe hole

Table 2 Summary of pile coring results—Observations and interpretations

Pile number	Design pile length, m	Pile length from PIT, m	Actual pile length determined from coring, m	Observations and interpretations
TP-1	12.37	12.3	9.75	The actual length of the pile is less than the anticipated length of 12.37 m. Necking observed at 7.1 m depth
TP-4	12.37	9.0–10.0*	9.0	Soil was met below 9 m depth. The actual length of the pile is less than the anticipated length. Pile integrity results suggested pile length of 9–10 m
TP-6	12.37	12.37	9.0–10.5	The actual length of the pile is less than the anticipated length of 12.37 m. Also, steel reinforcement was encountered at about 6.0–9.0 m depth

*PIT results show reduction in pile length

data. On the basis of PIT results, the authors are of the opinion that at Pile 142 also, concrete quality may be poor.

8 Condition of Some Exposed Piles

Photographs of some exposed piles shown in Fig. 11 confirm that the pile quality at shallow depth leaves much to be desired. Bulging, poor concrete quality, exposed reinforcement, lack of concrete cover, etc. were observed in many cases.

9 Conclusions

The study conclusively proved that several piles at the project site had major defects. The combination of pulse echo low strain integrity test, pile coring and probe holes around the piles are effective methods to identify pile lengths as well as possible presence of defects/discontinuities. The combination of these tests is an effective forensic approach to assess the quality of the piles installed.

The authors emphasize here that the results of static load tests are not sufficient evidence that the piles installed at the project site are of acceptable quality. A well-planned quality assurance program is required to ensure that every pile is evaluated and assessed.

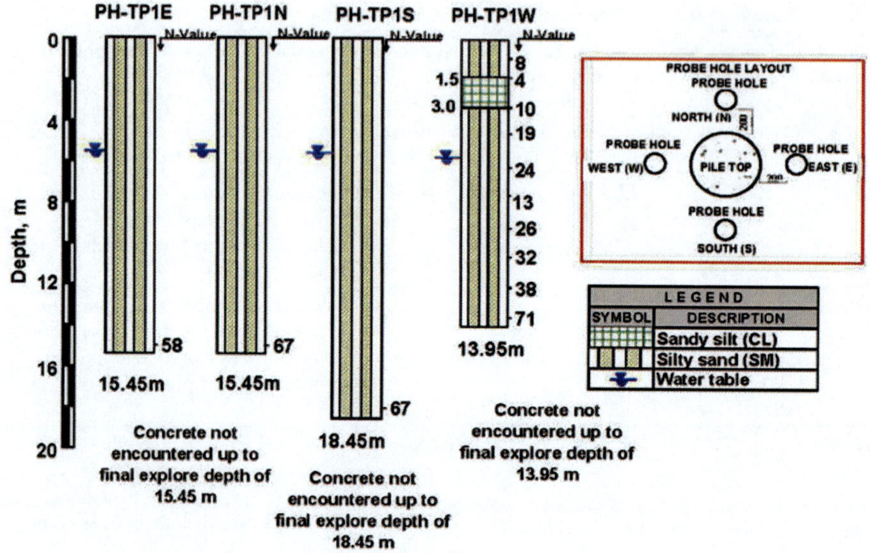

Pile TP-1

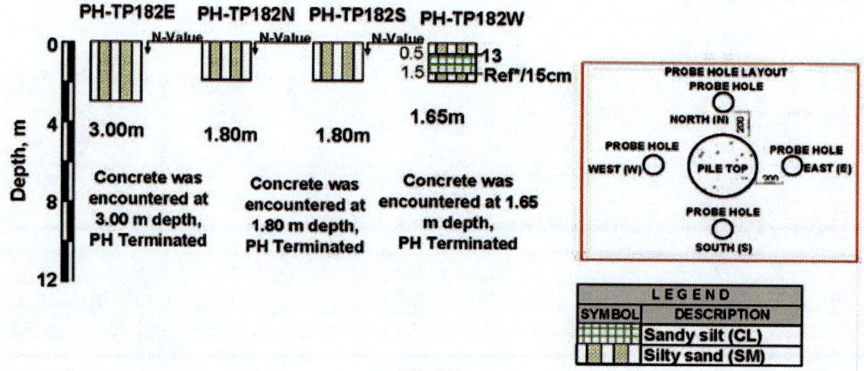

Pile 182

Fig. 10 Probe-hole profiles

Low-strain pile integrity tests are an effective and economical way to identify piles of questionable integrity. Such piles may be investigated further by high-strain dynamic load tests or static load tests. Probe holes adjoining the piles can help identify bulging at shallow depths. Coring through the pile-concrete may be resorted to if substantial reduction in pile length or major defect/integrity is suspected.

The scope of this study was limited to forensic evaluation for assessing pile quality. The client was advised to take appropriate measures including installing additional piles in the pile groups with defective piles.

Table 3 Summary of results of probe holes around selected piles

Pile number	Probe hole results				Overall conclusions
	North	South	East	West	
TP-1	No concrete was encountered	No concrete was encountered	No concrete was encountered	No concrete was encountered	No bulging of pile indicated by probe hole; concrete density may be less
142	No concrete was encountered	No concrete was encountered	No concrete was encountered	No concrete was encountered	No bulging of pile indicated by probe hole
182	Concrete was encountered at 1.80 m depth	Concrete was encountered at 1.80 m depth	Concrete was encountered at 2.0–3.0 m depth	Concrete was encountered at 1.65 m depth	Concrete was encountered at 1.8 m depth suggesting an increase in pile diameter (bulging) at about 1.5–2.0 m depth

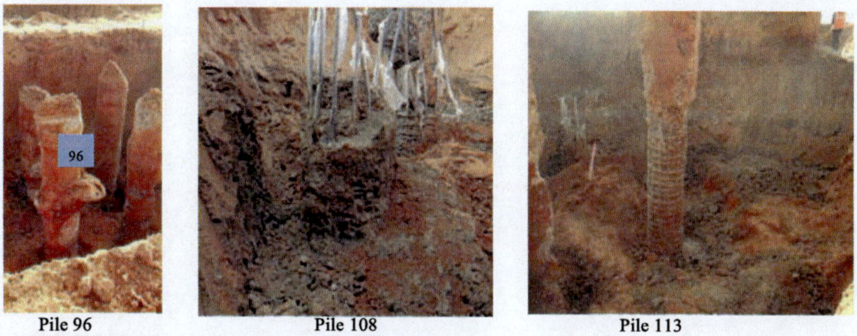

Pile 96
Bulging at shallow depth in Pile 96

Pile 108
Poor concrete quality in Pile 108

Pile 113
Exposed reinforcement in Pile 113

Fig. 11 Condition of some of the exposed piles

References

ASTM D 5882-16 (2016) Standard Test Method for Low Strain Integrity Testing of Piles. American Society for Testing and Materials. Standard Test Methods for Low Strain Integrity Testing of Piles, ASTM D 5882-16

IS:14893-2021. Indian Standard on Non-Destructive Integrity Testing of Piles (NDT)- Guidelines. Bureau of Indian Standards, New Delhi

Gupta S, Sundaram R, Gupta S (2008) Pile integrity testing for monitoring pile construction. In: Proceedings, geosymposium 2008, Indian geotechnical society Delhi Chapter, New Delhi, pp 173–178

Gupta S, Sundaram R, Gupta S (2015) NDT techniques for determining depth of foundations. In: Proceedings, Indian geotechnical conference IGC-2015, Pune

Gupta S, Sundaram R, Gupta S (2017) Identifying defects in deep bored piles—Case studies from India. In: Proceedings, 17th international conference on soil mechanics and geotechnical engineering, (17th ICSMGE), Seoul, Korea, pp 2637–2640

PIT Collector User's Manual, October (2005)

Rausche F, Likins GE, Hussein MH (1988) Pile integrity by low and high strain impacts. In: Third international conference on the application of stress-wave theory to piles, Ottawa, Canada, pp 44–55

"Digital Specialist Foundations"—Using the Digital Twin During Project Execution

Marcus Daubner

Abstract This paper explains how BAUER Spezialtiefbau GmbH uses digital methods on projects—starting from work preparation to completion of works. More precisely, the advantages of BIM-based planning during work preparation and the automated and structured data management—both machine production data and the manually recorded quality-relevant reports—during execution phase are displayed. In this context, the benefit of the digital twin to optimize processes, to reduce defect costs, for a more effective reporting system, and the further usage of as-built information in the BIM or GIS process are described.

Keywords Digitalization · BIM · Digital twin

BAUER Spezialtiefbau GmbH is pursuing the approach to systematically record all relevant data from planning, production, and quality control, as well as their integral interconnectedness and automated analysis. A central issue is to subsequently use the planning data during the construction process and the automated re-feeding of the production data in the as-built information. As the data are now used across the individual processes, the BIM context is subsequently extended. Therefore, the digital construction processes within the Bauer Group are combined under the heading "Digital Construction" (Fig. 1).

In the specialist foundation sector, digitalization of the documentation process offers without doubt a vast and hardly used potential for optimization. The effort it takes to compile production proofs for the specialist foundation elements embedded in soil is a substantive reason to introduce the "digital construction site". To manage these information and documents, extensive documentation and verification have to be carried out, especially when it comes to construction methods with a high number of individual elements. So far, these data are manually recorded and/or filled in reports. It seems natural to consistently digitalize this process, to record the data in an automated way and to generate aggregated KPI (key performance indicators).

M. Daubner (✉)
BAUER Spezialtiefbau GmbH, Schrobenhausen, Germany
e-mail: Marcus.Daubner@bauer.de

A. M. Krishna et al. (eds.), *Deep Foundations for Infrastructure Development in India*, Lecture Notes in Civil Engineering 373, https://doi.org/10.1007/978-3-031-37256-8_9

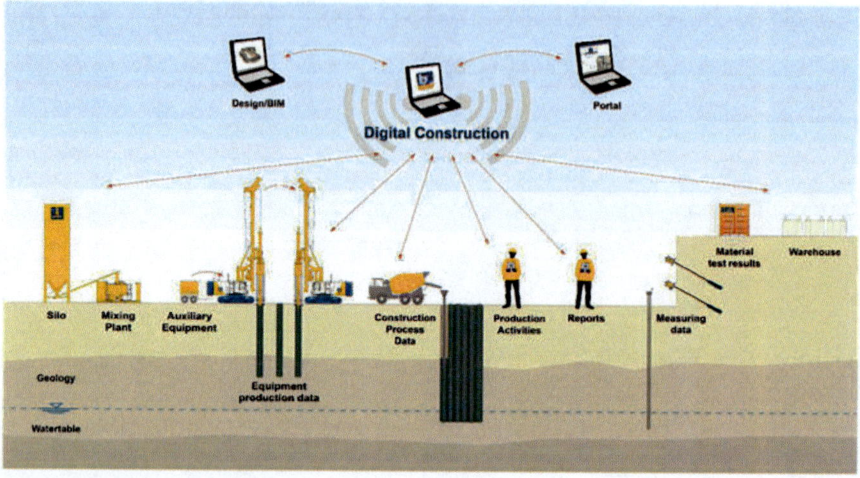

Fig. 1 Chart "Digital Construction"

The software components at present available on the market are mainly used to carry out the planning process or the calculation process. Bauer tries to extent this scope and to also use the available data during work preparation and project execution.

1 Using BIM Planning During Work Preparation

One use case is the planning of the construction sequence and the rig positioning. By connecting the detailed schedule with the 3D model, the phase drawings are compiled in an automated way. In Fig. 2, the weekly schedule is displayed. This visualization is quite helpful during work preparation to identify anticipated geometric collisions and to avoid them.

During the execution phase, this phase drawing is updated weekly, and drawings are automatically created. Critical situations that result from postponements can be identified based on the current phase plans and used for coordination between the construction methods involved, and also to explain the situation to the client.

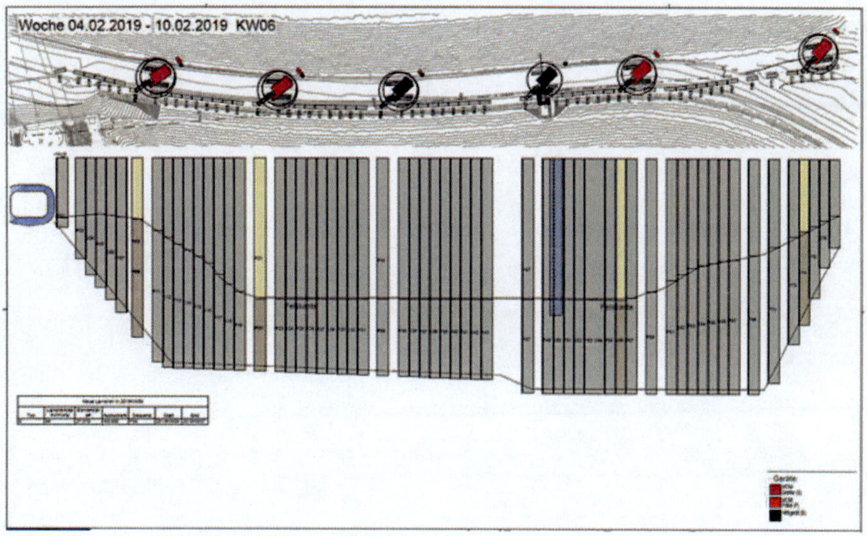

Fig. 2 Automated weekly phase planning derived from 4D BIM model displaying top view of rig position and side view of performance forecast

2 Using a Production Data and Process Management System for Specialist Foundations to Generate a Digital Twin

Advancing to recording and using production and quality data during execution phase still fails—especially for specialist foundation jobs—as the software needed to display the production process is currently not available on the market. This is insofar astonishing as machine-driven production dominates the specialist foundation sector and is thus predestined for digitally recording the production parameters as generated in the production process.

The planning data are in general further used on-site via BIM viewers. The backflow of information from the construction process is carried out by manually filing the production data directly into the planning models. Here as well, the software components are not available, which transfer the production information in the models for the individual construction methods in an automated way.

Considering the potential offered by a digital site, BAUER Spezialtiefbau GmbH developed an integral data management system to process production and quality data whose central tool is the software b-project. The essential task of b-project is to digitally record production data in a structured way. To save and manage data in Excel spreadsheets is an established and very flexible way to record and analyse parts of the construction process, and it is nowadays still used on many sites. A more advanced added value gained from the documented data is, however, not possible, which is due to—amongst other reasons—the limited, project-specific definition of the Key

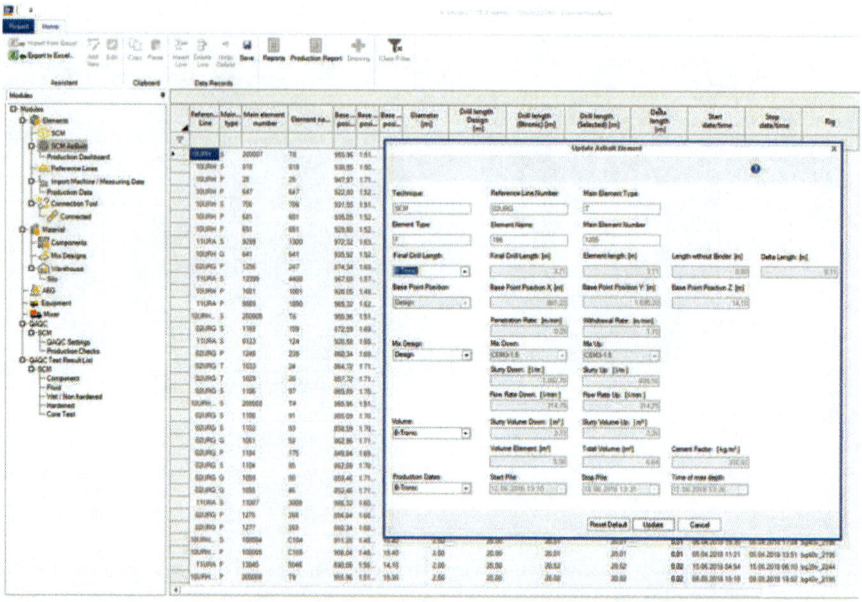

Fig. 3 User interface of b-project—detailed information on constructed element

Performance Indicator (KPI) and also because the data are not structured via clear keys. Individually gathered information is therefore only usable with great effort for cross-evaluation and generally end up in a data tomb after project completion due to the system applied.

The management software used by BAUER displays the following aspects of data application (Figs. 3 and 4):

- Standardized transfer of the planning data to the construction site, to make it available for various use cases of the site.
- Reading the digital machine production data in the management system.
- Mobile Apps providing filled-in reports for the site manager and digitally transmitting the manually recorded production and quality data in the database.
- Subsequent, automated consolidation of the construction data.
- Integrated, web-based production and quality evaluations for QA/QC to ensure a high-quality product.
- Drawings and visualizations as needed on site.
- Generation of reports, production reports or lists for the quality assurance process, for process optimization or for controlling.
- Interface to return the digital production and quality data from the management system in the planning systems.
- Further usage of recorded data in BIM- and GIS-based documentation process.

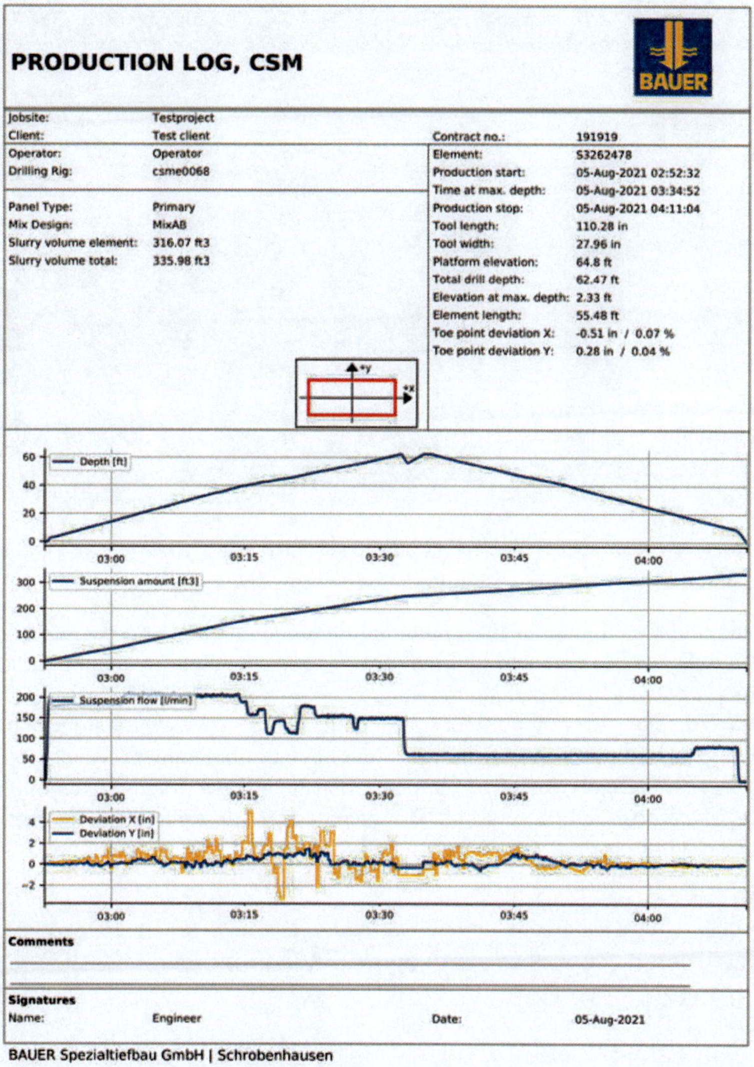
PRODUCTION LOG, CSM

BAUER

Jobsite:	Testproject		
Client:	Test client	Contract no.:	191919
Operator:	Operator	Element:	S3262478
Drilling Rig:	csme0068	Production start:	05-Aug-2021 02:52:32
		Time at max. depth:	05-Aug-2021 03:34:52
Panel Type:	Primary	Production stop:	05-Aug-2021 04:11:04
Mix Design:	MixAB	Tool length:	110.28 in
Slurry volume element:	316.07 ft3	Tool width:	27.96 in
Slurry volume total:	335.98 ft3	Platform elevation:	64.8 ft
		Total drill depth:	62.47 ft
		Elevation at max. depth:	2.33 ft
		Element length:	55.48 ft
		Toe point deviation X:	-0.51 in / 0.07 %
		Toe point deviation Y:	0.28 in / 0.04 %

Comments

Signatures			
Name:	Engineer	Date:	05-Aug-2021

BAUER Spezialtiefbau GmbH | Schrobenhausen

Fig. 4 Production report example generated by software "b-project"

Interconnecting the construction data, for example from the concreting process with the quality data from the material tests, is the basis for the automated compilation of various cross-evaluations. This automation significantly reduces the documentation and monitoring effort and helps to display the optimization process in a standardized, automated, and thus more efficient way.

The software devised for the demands of the site enables various visualization possibilities and facilitates the identification of optimization potential via freely

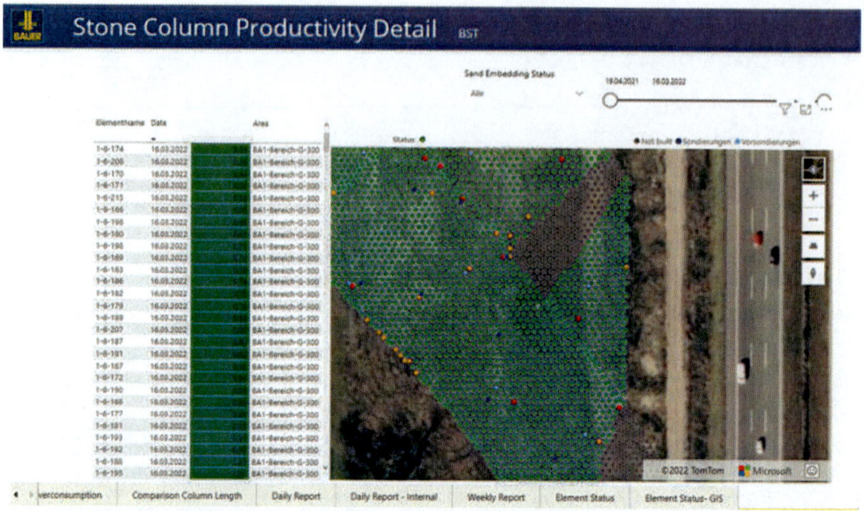

Fig. 5 Example visualizations of the optimization potential to reduce waste (design vs. as built)

selectable multiple filters. At this stage, the digital twin shows its potential. These visualizations can be presented by highlighting them in colour in a chart but also in a graphical analysis.

Figure 5 displays the potential for optimization in 2D top view for stone columns. Here, the grey circles represent elements that have not been installed yet, coloured circles describe already installed elements that also have potential for optimization, e.g., overconsumption of grout. The green circles describe elements that were constructed within technical and economic boundaries.

3 Web-Based Access to Aggregated Measuring and Production Data

A further part of the integral approach is the web-based, location-independent presentation of the evaluations and KPI's. This functionality allows for the timely access to aggregated performance and quality data, also outside the site office, both for internal project stakeholders and for the client or its representative. The standardized presentation of data enables a quick overview of and insight into the current state of performance of projects. It is the basis for further and more in-depth evaluations and process optimizations.

The production data can be applied to report the site progress (see Fig. 6) or to get a prediction of the material stock. In this case, the material will be sufficient for 37 days (see Fig. 7). This forecast from the digital twin helps the site staff to avoid idle time due to missing material.

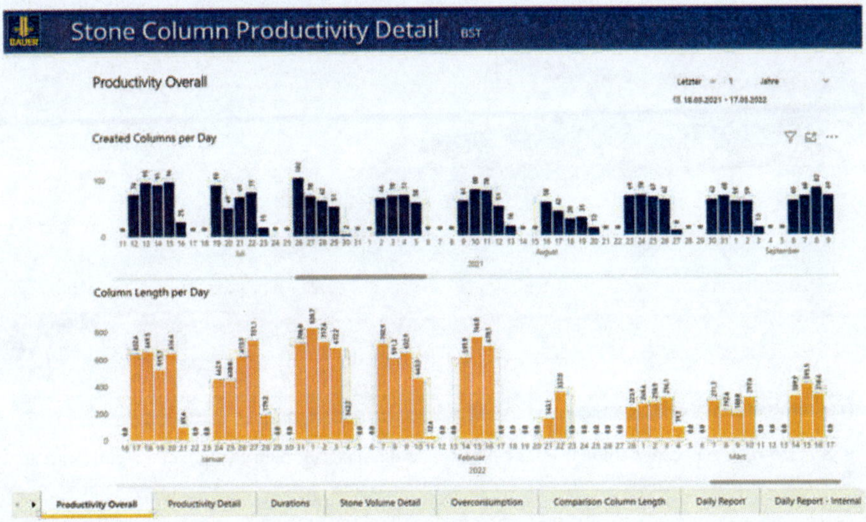

Fig. 6 Performance evaluation of a construction method in web-based digital construction portal

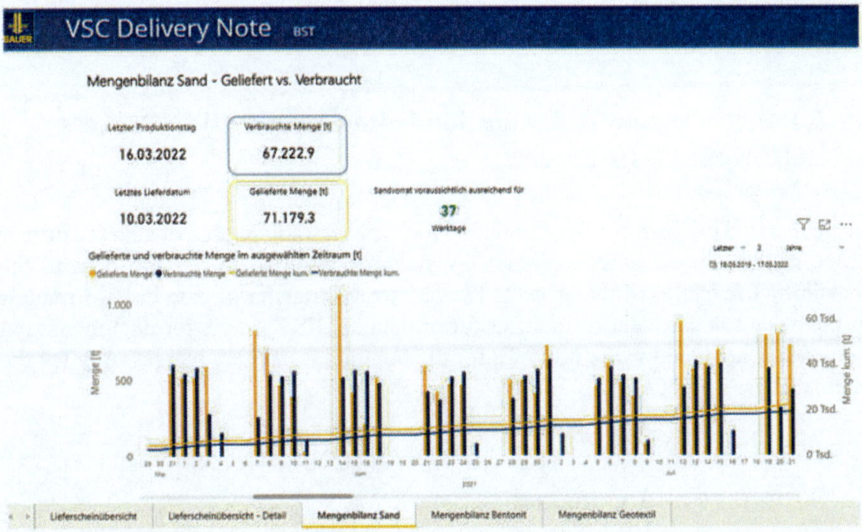

Fig. 7 Prediction of sand in the warehouse (stock vs. production)

By closely cooperating with site staff, more than 150 visualizations have been compiled to identify the quality condition, the performance, but also the optimization potential. Being able to use these evaluations on other projects is the advantage of these standardized, construction-method-related visuals. When starting the works on site, the site staff can immediately use a wide variety of detailed evaluations. The

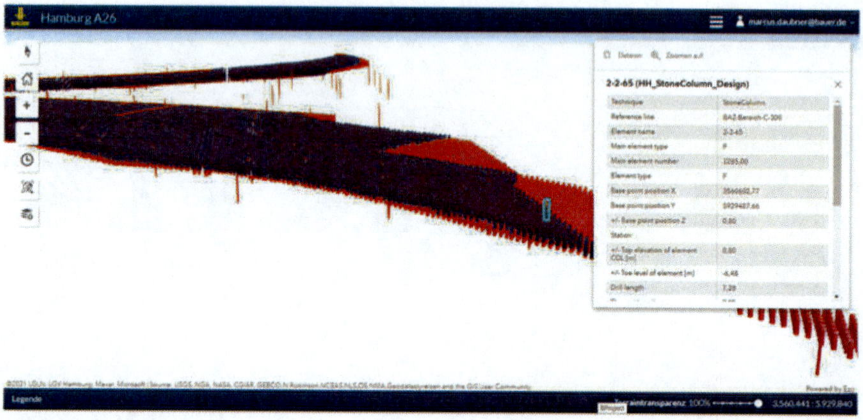

Fig. 8 Geo-referenced documentation of planned and actual geometry (planned = yellow, actual = blue) in the GIS database

decision-making process is significantly accelerated due to this timely visualization. This premature avoidance of error costs is an important aspect when it comes to execute the works both economically and with high quality.

4 Further Usage of Information from Construction Process in BIM and GIS Process

The generated data are not only available for the execution phase but also for further usage in subsequent processes such as documentation or to use the data in the remaining life cycles of the project. The automated transfer of data in BIM models of the design department or the transfer of data in GIS systems for documentation purposes is an added value not only for the contractor but also increasingly for the clients (Figs. 8 and 9).

5 Summary and Conclusion

The structured and centralized information management of production data with a digital twin allows for a standardized analysis of the processes and the product quality. These analyses may be used for increasing the effectiveness when compiling the production documentation, to optimise performance and for coming to a decision in due time during ongoing projects. The structured analysis of data is the basis for a detailed assessment of construction parameters in the long run and thus contributes

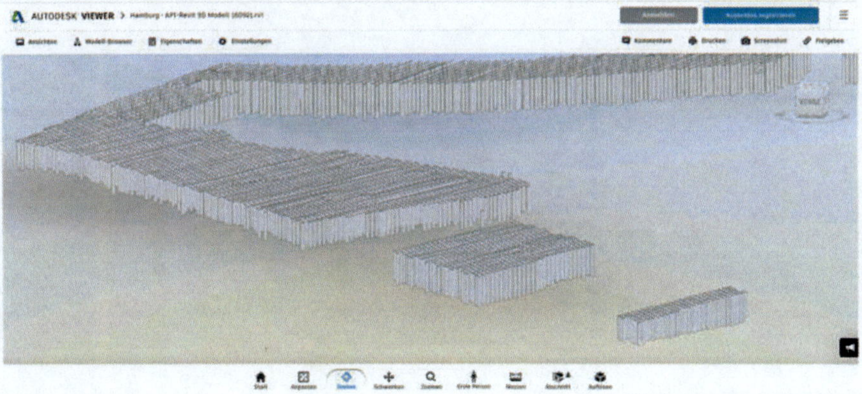

Fig. 9 Import of as-built information into a BIM Software

to reducing the project risk when it comes to assessing performance, product quality and process sequences.

As a consequence, production information is identified as being part of the value-added chain in a modern construction company and thus has to be reasonably used as a further step towards professionalization.

Geotechnical Monitoring of Deep Shoring Adjacent to Train Infrastructure

Jakub Szary

Abstract Excavation and shoring adjacent to the above-grade Gilmore train station and elevated guideway was undertaken in 2019. A perimeter secant pile cut-off wall was constructed to a depth 37 m (121 ft) below ground surface consisting of 1.0 m (3.3 ft) diameter drilled piles spaced 750 mm (2.5 ft) on-center, tied back with soil anchors, and reinforced with W610 steel sections every fourth pile. A bracing/jacking system was also constructed to support three free-standing guideway caissons within the excavation. An automated instrumentation program was installed to monitor movements of shoring walls and train infrastructure in real time and consisted of a robotic total station, inclinometers, tiltmeters, strain gauges, extensometers, and anchor load cells. Baseline monitoring was able to establish diurnal and seasonal movements within the structure, strongly correlated with temperature. The system actively monitored various shoring elements and train infrastructure in real time and alerted the project team when values exceeded predetermined threshold values. Remote reconfiguration of the automated system allowed for more frequent monitoring during jacking of Pier 29 in March 2022. Monitoring data were in agreeance with on-site manual surveying results and jacking dial gauges and proved to be a reliable monitoring system for construction applications.

Keywords Geotechnical instrumentation · Total station · Tiltmeter · Inclinometer · Strain gauge · Extensometer · Load cell · RST instruments · World sensing · Trimble

1 Project Background

The Gilmore Place development site is located in Burnaby, British Columbia, Canada and consists of a rectangular lot, 28,328 m^2 (7 acre) in area, with an above-grade train station and caisson-supported guideway system bisecting the site diagonally from the southwest to the northeast. The project consists of two phases and will incorporate

J. Szary (✉)
Municon West Coast Monitoring Ltd., Vancouver, BC, Canada
e-mail: jakub@municon.net

A. M. Krishna et al. (eds.), *Deep Foundations for Infrastructure Development in India*, Lecture Notes in Civil Engineering 373, https://doi.org/10.1007/978-3-031-37256-8_10

Fig. 1 Construction progress May 2021 (left) and December 2021 (right)

the existing train infrastructure into the final design of the structure. Once complete, the project will be home to the tallest tower in Western Canada standing at 64 stories or 216 m (709 ft). Construction began in 2019 and Phase 1 shoring and excavation work was completed in December 2021 (see Fig. 1).

A perimeter secant pile cut-off wall was designed and constructed to a depth of up to 37 m (121 ft) below ground surface and consisted of 1.0 m (3.3 ft) diameter drilled piles spaced 750 mm (2.5 ft) on-center, tied back with soil anchors. Each pile was filled with concrete with a specified strength of 10 MPa (1450 psi) at 28 days, with every fourth pile reinforced with a W610 steel section. Due to the excavation extending below the bottom of the existing guideway supporting caissons, additional secant piles and a bracing/jacking system were also designed to support three free-standing guideway piers (Piers 29, 30, and 31) and to enable jacking of the piers should excessive movements occur during construction (see Fig. 2).

Soil conditions at the site consisted of mineral fills, overlying bog/swamp deposits of compressible peat, organic silt, and firm low plastic silt up to 7 m (23 ft) below ground surface, overlying very dense glaciated till-like soils including a bedding of glaciofluvial sand and gravel containing a confined aquifer with artesian pressures of up to 4.0 m (13 ft), overlying tertiary bedrock at depth.

2 Instrumentation and Monitoring Plan

A robust, automated instrumentation program was designed and implemented in coordination with the design team. The system was designed to monitor various shoring elements and train infrastructure in real time and to provide immediate alerts (via text message and email) to the design team and stakeholders should values exceed predetermined thresholds. All instruments were connected to an Automatic Data Acquisition System (ADAS) that enabled data to be collected remotely. Instrument dataloggers communicated via radio signal to a central base station located adjacent to the site where data were transmitted via cellular connection to off-site servers. Data

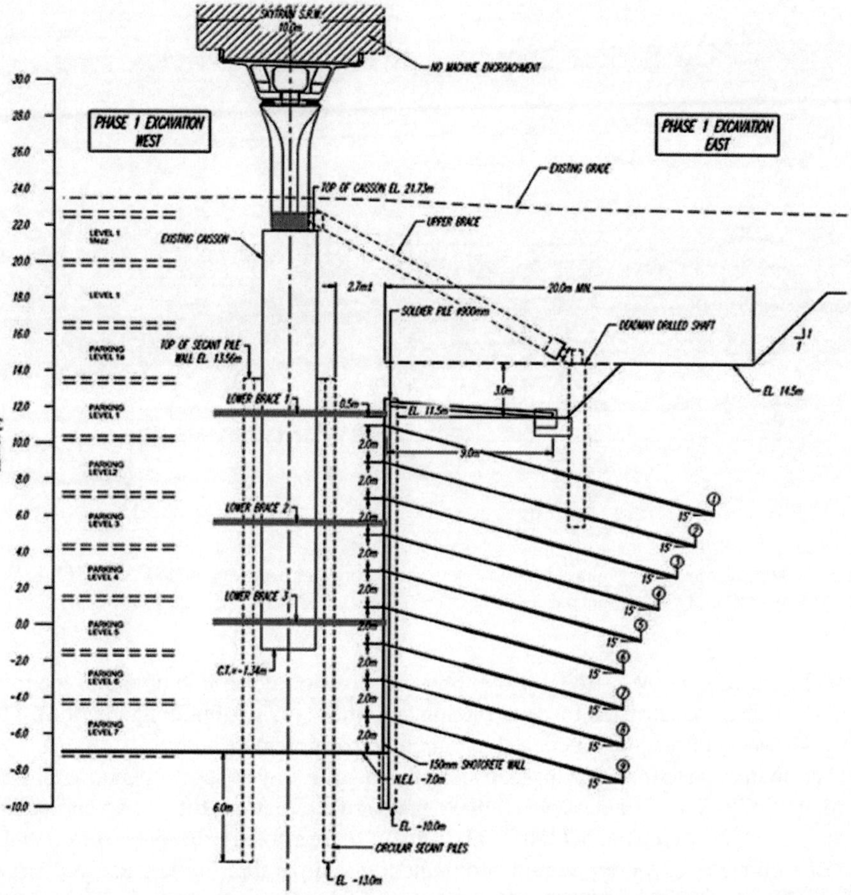

Fig. 2 Typical free-standing pier shoring section

were recorded in approximately 1-h intervals and presented on a privately accessible website made available to the project team. Several instruments were installed prior the start of construction activities to obtain baseline data (see Fig. 3).

An automated motorized total station (AMTS) was included in the monitoring program to provide 3D displacements at locations of monitoring prisms throughout the site. The system operates autonomously and provides frequent, high-precision measurements several times per day. Survey prisms were affixed to various structures and shoring elements throughout the site including the guideway, guideway supporting piers, train station, and shoring walls. Data were collected in a local coordinate system and processed to provide relative displacements perpendicular and parallel to the orientation of the guideway. Several off-site reference prisms were required in locations considered stable and outside the zone of influence of construction activities. Reference prisms are used to calibrate each epoch and allowed for

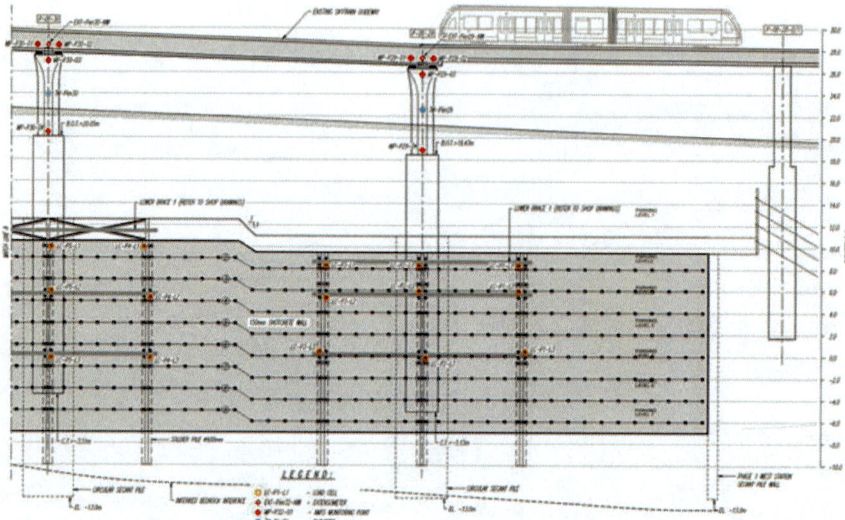

Fig. 3 Typical monitoring plan elevation view indicating the locations of load cells, tiltmeters, AMTS targets, and extensometers

movement of the AMTS to a different on-site position when line of site to the total station became obstructed by construction activities (such as building erection). The AMTS was mounted on a pedestal and included a protective cover.

Tiltmeters were installed on each station and guideway supporting pier within the project limits. The tiltmeters were able to measure inclination (tilt) in two perpendicular axes with a precision of 0.001°. The sensors were affixed to the piers such that the measurement axes were oriented perpendicular and parallel to the guideway structure. Vibrating wire extensometers supplemented the tiltmeters to measure structural expansion joints between guideway sections at pier locations.

In-place inclinometers (IPIs) were installed in 17 pile locations adjacent to the train station and guideway piers. IPIs are designed to measure lateral soil movements in two perpendicular axes when remote and continuous monitoring is required. Multiple IPIs are assembled with pivots allowing measurement of inclination over discreet, configurable intervals. The chain of IPIs is inserted into a vertical 70 mm (2.8 in.) diameter grooved casing and is suspended by a collar at ground surface. Wheel assemblies centralize the pivot points and data from each sensor are collected through a connectorized signal cable that chains together the string of sensors. IPIs were spaced in 1.5 m (5 ft) intervals within the casings.

Vibrating wire strain gauges and load cells were utilized to monitor the bracing elements supporting the three free-standing piers. Strain gauges were welded onto each pipe brace and were configured to allow measurement of strain in both tension and compression. Strain data were used to calculate load changes within the braces and proved vital in ensuring their capacity was not exceeded when pier jacking

Fig. 4 Typical load cell (left) and strain gauge (right) installation

was attempted. Load cells were installed during anchor tensioning and allowed the project team to monitor anchor loads in areas where the bracing structure connected to the shoring wall. Load cell anchors were locked off at 50% of design loads so design engineers were able to monitor incremental load increases as excavation work advanced (see Fig. 4).

3 Baseline Monitoring

The AMTS, tiltmeters, and extensometers were installed in the spring of 2019 and allowed for a baseline monitoring period of approximately 1 year prior to the start of major excavation work. The data indicated consistent diurnal and seasonal fluctuations in all three data sets. Extensometer data indicated increases in the structural expansion joint distance during colder periods of the year, suggesting contraction of the guideway section pair. During the warmer months of the year, the expansion joint distance was noted to decrease, suggesting expansion of the guideway sections. Displacements of ±11 mm (0.43 in.) were observed over the course of the baseline monitoring period and were strongly correlated with temperature changes (see Fig. 5).

Similar behavior was observed with tiltmeters and the AMTS system. Tiltmeter data fluctuated between 0.1277 and −0.1788 degrees and showed significantly greater deviation during the colder winter months of the year (see Fig. 6). AMTS perpendicular and parallel movements varied between ±5 mm (0.2 in.). Two AMTS monitoring points were located on either side of guideway expansion joints and their movements were consistent with extensometer data.

The baseline monitoring data were critical in understanding the behavior of the train infrastructure prior to commencement of excavation work. The observed movements were observed in multiple (independent) sensors and were strongly correlated with temperature. Project monitoring thresholds were initially established very

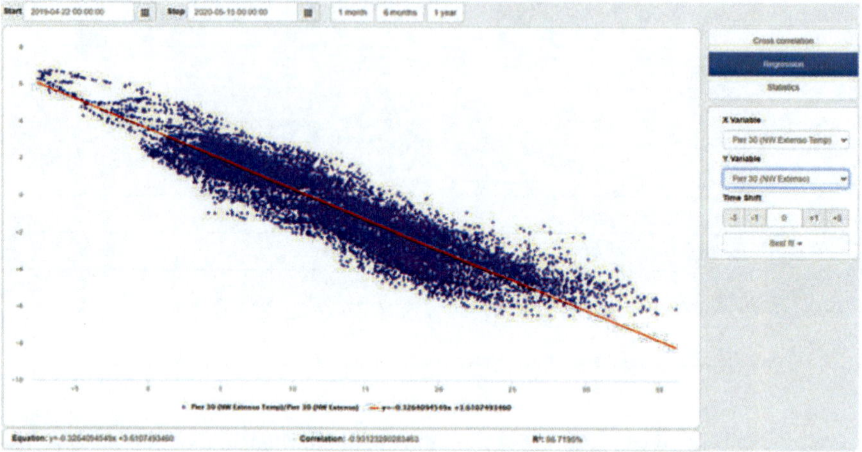

Fig. 5 Temperature correlation for Pier 30 northwest extensometer ($R^2 = 0.87$)

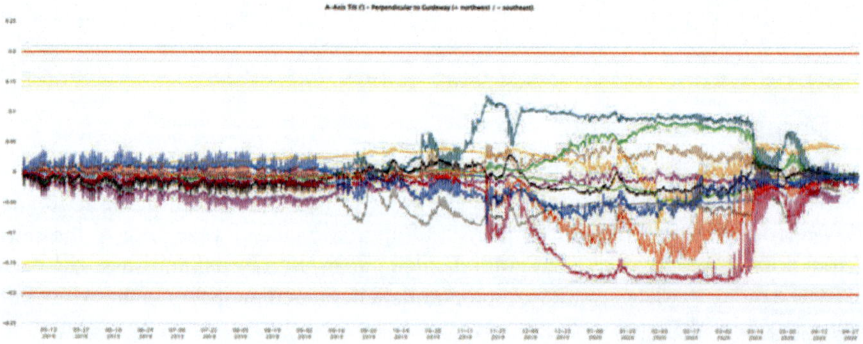

Fig. 6 Baseline A-axis tiltmeter data between April 22, 2019 and April 30, 2020. Large thermal effects are noticeable during the baseline period, especially during the winter

conservatively by the design team and stakeholders such that the observed temperature fluctuations would have caused threshold exceedances. After review of the baseline data, thresholds were increased to allow for diurnal and seasonal temperature variations in the data and an additional set of thresholds were established where temperature-corrected data were used. Data viewed as a 24-h rolling average provided a means to reduce diurnal effects when interpreting data.

4 Construction Monitoring

Remote monitoring continuously recorded data during excavation and construction work. Coordination was required with the construction team to install inclinometers, load cells, strain gauges, and additional AMTS monitoring prisms as excavation and shoring work advanced. Inclinometer installations required piles to be completed with casings installed in W-beam webbings by the piling contractor. Load cell installation required coordination with the geotechnical representative and was completed during anchor testing. Load cells were also used to confirm anchor testing results in the field. Strain gauge installation required welding and was coordinated during construction of steel members throughout the site.

Diurnal and seasonal effects continued to be observed as additional instrumentation was installed. These effects were particularly apparent in the larger, steel bracing members of the three free-standing piers. Loading varied up to 400 kN (90 kip) daily due to thermal contraction and expansion of the pipe braces. Thermal fluctuations were observable in load cell data. Anchor loads varied by approximately 10 kN (2 kip) daily. Inclinometer sensors experienced more stable temperature conditions due to being located below grade and better insulated from diurnal and seasonal temperature fluctuations. However, inclinometers located within the free-standing piers exhibited daily movements of 3 mm (0.12 in.) due to greater temperature exposure (see Figs. 7, 8, and 9).

The primary challenge relating to the remote monitoring system is line of site considerations with the AMTS system. To function correctly, the AMTS requires unobstructed view of monitoring targets and reference prisms. Excavators, drills, and even heavy rainfall occasionally obstructed the view of targets throughout the project. Reference prisms also became obstructed, resulting in global abnormalities or *shifts* in the monitoring data due to inaccurate epoch calculations. Data would return to previous values during subsequent measurements when the reference prism(s) were

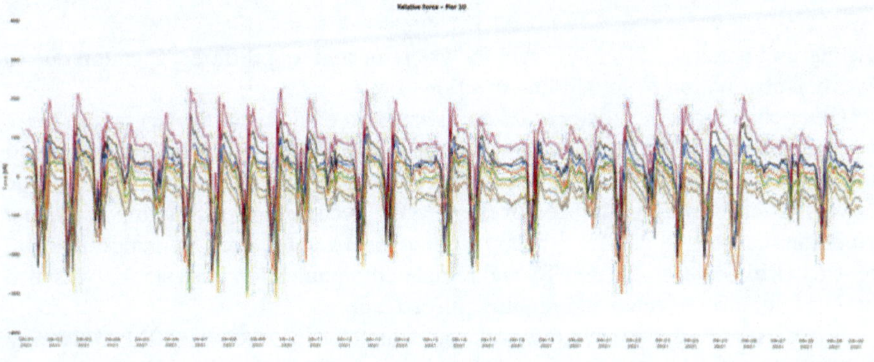

Fig. 7 Pier 30 strain gauge data (September 2021). Loads varied between ±200 kN (45 kip) daily due to thermal effects

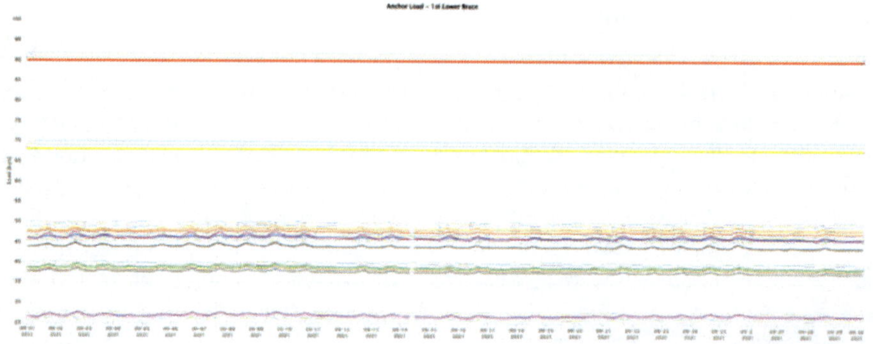

Fig. 8 Lower bracing load cell data (September 2021). Daily thermal effects are observable as *ripples* in the data

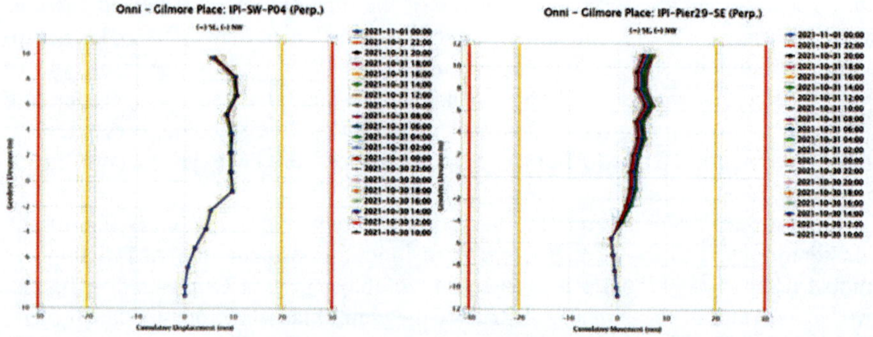

Fig. 9 Perpendicular inclinometer movements over 24 h for location IPI-SW-P04 (left) and IPI-Pier29-SE (right). No noticeable thermal fluctuations observed in IPI-SW-P04 (located below grade). Diurnal movements of 3 mm (0.12 in.) were observed in IPI-Pier29-SE located within the free-standing Pier 29 (greater exposure to temperature fluctuations)

visible. In September 2021, the AMTS was relocated to a different location due to eventual obstruction from building erection.

Other challenges included contractor damage to equipment. Survey targets were occasionally sprayed with shotcrete, communication cables were severed by excavation equipment, and inclinometer sensors were damaged due to anchor drilling. Due to very strict leveling tolerances of the AMTS, occasional construction activities/vibrations caused the AMTS to become out of level and required manual re-leveling on-site. Grid power to the AMTS and remote communication base station was also periodically disconnected and required investigation.

Phase 1 shoring and excavation work was completed in December 2021. Guideway movements were limited to 15 mm (0.6 in.) and within the Level 1 threshold of

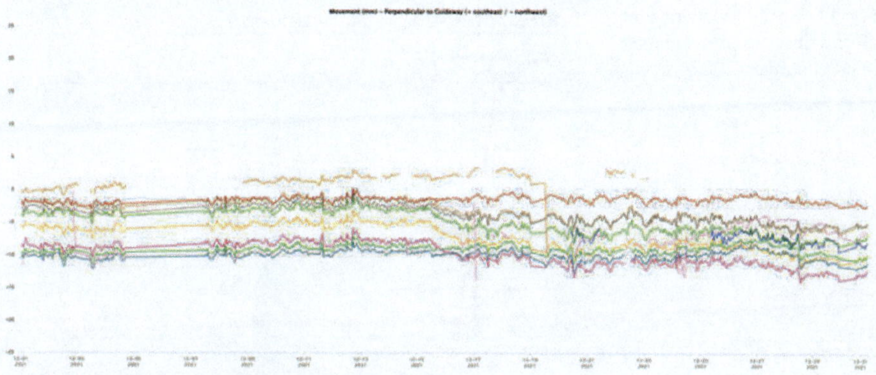

Fig. 10 Perpendicular AMTS movements at completion of Phase 1 shoring and excavation work (December 2021). Pier and guideway movements were limited to 15 mm (0.6 in.) and within the Level 1 threshold of ±25 mm (1 in.)

±25 mm (1 in.). No noticeable trends were observed in the load cell data, indicating good anchor performance. Inclinometer data also indicated minimal movements within the shoring wall adjacent to the bracing structure and were within the Level 1 threshold (see Fig. 10).

5 Brace Removal

Building construction adjacent to the three free-standing piers continued into January 2022 and followed a bracing removal sequence developed by the design team. A reinforced concrete wall was constructed around the secant piles of each free-standing pier and was locked-in with parkade slabs as construction advanced, transferring loading from the bracing elements to the building structure as bracing elements were removed.

As construction advanced, monitoring data indicated additional movements were being experienced by the piers and guideway structure. On February 23, 2022, guideway movements at Pier 29 had increased to 28 mm (1.1 in.) towards the northwest. Inclinometer data at P1 slab elevation indicated movements of 27 mm (1.1 in.) had occurred in the opposite direction towards the southeast (see Fig. 11).

It was not well understood what mechanisms were causing the observed movements. The consultant team hypothesized that slab pours coincidentally occurred at inopportune times such that the pier's position was locked-in during contraction of the bracing elements caused by diurnal thermal movements. Other factors that may have had an impact were thought to be parkade slab shrinkage during curing, and train/wind loading at guideway elevation.

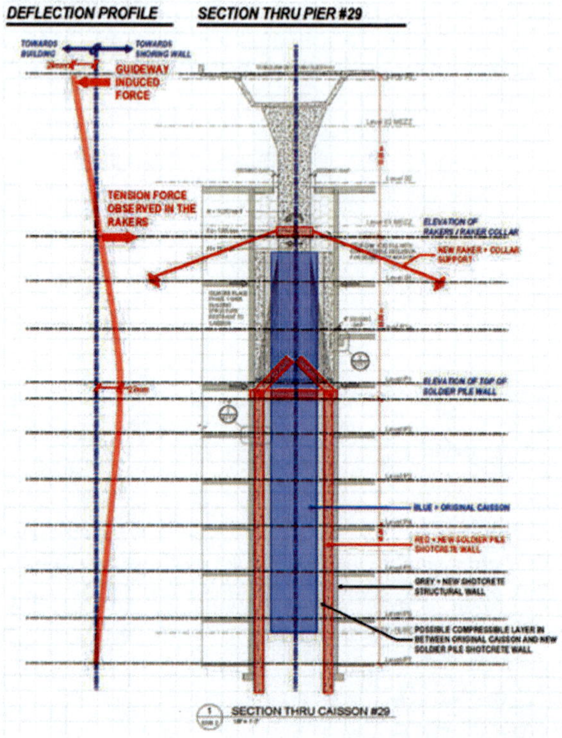

Fig. 11 Section view of Pier 29 and observed movements in late February 2022

In March 2022, movements of Pier 29 continued to incrementally increase as parkade floors were poured and bracing elements were subsequently removed. Movements of the guideway were approaching 35 mm (1.4 in.) from baseline. It was decided to proceed with jacking of Pier 29 in an attempt to adjust the pier's position closer to its original location. At this stage in construction, all lower braces had been removed and only the upper rakers remained (see Fig. 12).

The monitoring system was remotely reconfigured to allow for more frequent AMTS and strain gauge data collection during jacking. The AMTS measurement frequency is limited by the number of survey prisms it is programmed to measure. Monitoring of non-essential survey prisms was temporarily disabled, resulting in a 5-min monitoring frequency. Strain gauge dataloggers were also reconfigured to allow a monitoring frequency of 30 s. This required the installation of a second communication base station to relieve the original system of communication bandwidth. The jacking procedure consists of lift-off, increasing loads in 200 kN (45 kip) increments, and holding each application of load for 15 min. Jacking loads were not to exceed 1780 kN (400 kip).

The remote monitoring system successfully monitored the movements of the pier and stress levels in the bracing elements during jacking. Movements of the pier were confirmed in the field via manual surveying and were in agreeance with the

Fig. 12 Pier 29 jacking set-up using 4 × Simplex R556 jacks per brace

remote monitoring system. Raker loads were also confirmed in the field by jack gauges and were consistent with the online data. The pier was successfully jacked by approximately 5 mm (0.20 in.). Loading in the east and west rakers was increased to approximately 1500 kN (337 kip) and 950 kN (214 kip), respectively. No further jacking was conducted to allow for diurnal changes in bracing load due to thermal fluctuations (see Fig. 13).

Movements of ±40 mm (1.6 in.) and 30 mm (1.2 in.) were observed during lower brace removal of Piers 30 and 31, respectively. To reduce the effects of concrete shrinkage impacting Piers 30 and 31, delay strips were used to allow the majority of the parkade slabs to cure prior to locking in the piers. Removal of the upper rakers was also delayed allowing for additional parkade slabs to be poured around each pier.

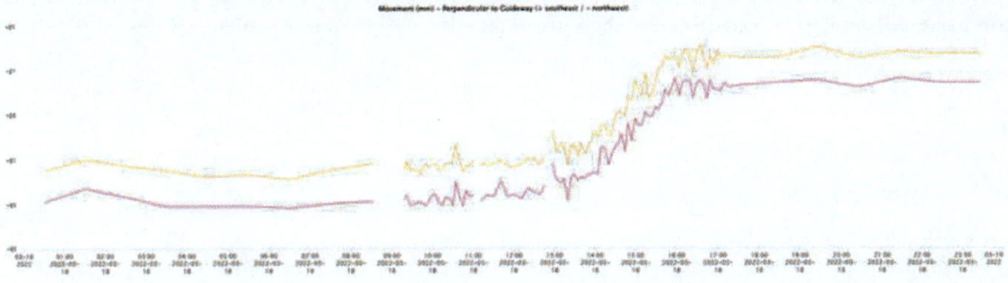

Fig. 13 Pier 29 perpendicular movements during jacking. The AMTS system was remotely re-configured to allow measurement of the pier every 5 min

On June 18, 2022, the final brace was removed from Pier 30 with minimal additional movements observed, suggesting concrete shrinkage was a factor in the additional movements observed after brace removal.

6 Conclusions

The remote instrumentation program employed during this project proved to be a reliable system for monitoring shoring and train infrastructure elements. The system was able to accurately monitor movements, loads, and strain, providing the project team with critical data regarding the performance of the shoring system and train infrastructure in real time. Diurnal and seasonal movements were able to be quantified during the baseline monitoring period and were strongly correlated with temperature changes. Thermal effects continued to be observed during construction. Daily load changes of 400 kN (90 kip) were observed in the steel bracing elements supporting three free-standing piers within the excavation. Anchor loads varied by approximately 10 kN (2 kip) daily.

The monitoring system was able to alert the project team via text message and email when values exceeded predetermined thresholds. The system was able to be reconfigured during pier jacking operations to provide critical data at more frequent intervals. AMTS measurement frequency was able to be reduced to every 5 min by disabling monitoring of non-critical survey targets. Strain gauge dataloggers were reconfigured and allowed for a monitoring frequency of 30 s. AMTS and strain gauge data accuracy was confirmed in the field by manual surveying and jack dial gauges. Jacking operations were able to adjust Pier 29's position by 5 mm (0.20 in.) and loading in the bracing elements was increased to 1500 kN (337 kip). The system continues to monitor shoring and train infrastructure elements and is expected to remain in place for several years during excavation and shoring works for Phase 2 of the development.

Acknowledgements I would like to extend my special thanks to the following individuals for their contribution to the project and assistance with this presentation: *Matt Kokan* (*M.A.Sc., P.Eng.*), *Alex Gossen* (*B.Sc., P.Eng.*), *Farshid Bateni* (*Ph.D., P.Eng.*), *Kazunori Fujita* (*B.Sc., P.Eng.*), *and Geoff Poh* (*P.Eng.*). I would also like to acknowledge GeoPacific Consultants Ltd., Onni, Glotman Simpson, Allspan, and TransLink for their effort and dedication to this project.

A Review Low Strain Integrity Testing Practices in India and Complexities Involved in Evaluation of Test Results

Ravikiran Vaidya and Vishal Dhamecha

Abstract Low Strain Pile Integrity Testing has been used in India since 1990s and several have tried to demonstrate the utility and benefits of this method. However, the use of this technology was limited till 1998 as not enough data maybe were not available to justify its use and application on major infrastructure and real estate projects. The first author has done significant work in this method resulting in its use in practically all sectors of the industry that use concrete bored pile foundations as a supporting element. The paper explains various signal enhancers or modifiers involved in processing of low strain test data and its implication on the final output. It also describes how the method can lead to abuse or can provide poor interpretation if incorrectly used. The current effort is to demonstrate that data collection and evaluation become complex with changes in soil stratum and pile profile. The paper highlights the present testing practices and provides a guideline for the classification of pile integrity into few simple categories in Indian context rather than writing complex interpretations.

Keywords Low strain integrity testing · PIT · Pile foundation · Deep foundation

1 Introduction

Cast in situ place concrete bored piles also known as drilled concrete shafts are produced by drilling holes of various diameters in the ground and filling with concrete. Compared to driven piles, they may have advantages such as lower vibration and noise. However, since direct inspection of the final product is impossible, some doubt always persist regarding the structural integrity of bored concrete piles. Integrity tests, excavated foundations, failed static tests, extracted cores have revealed

R. Vaidya (✉) · V. Dhamecha
Geo Dynamics, Vadodara, Gujarat, India
e-mail: ravikiran@geodynamics.net

V. Dhamecha
e-mail: vishal.dhamecha@geodynamics.net

A. M. Krishna et al. (eds.), *Deep Foundations for Infrastructure Development in India*, Lecture Notes in Civil Engineering 373, https://doi.org/10.1007/978-3-031-37256-8_11

defects in bored piles on numerous occasions. For this reason, concerned engineers and/or owners require that as part of a standard quality assurance program, low-strain pile integrity testing be conducted on all or a certain percentage of the piles at a project site. In some cases, more emphasis is laid on testing of piles with unusual casting records (Liang and Rausche 2011).

Various non-destructive test methods are available to evaluate the structural integrity of the piles (ASTM D5882-2016). This includes the Pulse Echo Method, the Frequency response method, Gamma Approach, etc. The most commonly used method today in India is the Pulse Echo Method as it is simple and reliable if properly conducted and interpreted.

The Low Strain Integrity Test (LSIT) method using Pulse Echo technique that is standardized by various versions of ASTM D5882 has seen increased use since early 1990s. The method has found more acceptance in India after 1998 when the first author successfully demonstrated its application on several projects. The recently available IS: 14893-2021 (2021) also provides significant information on the method, its application, interpretation and limitations.

The Pulse Echo technique that is described in the paper primarily involves impact of a lightweight hammer ranging from 0.4 to 5 kg to generate a stress wave that travels from point of impact to the pile bottom and returns back to the top. The travel of this waveform is recorded by an accelerometer. Changes to this wave are observed when there are changes in cross-section, soil resistances, cracks or noise due to rebars. Since there are several parameters that affect the final output, the results are generally interpreted by an Engineer who has good knowledge of wave mechanics coupled with some knowledge of soil and concrete. If for example, there is a decrease in impedance at a certain distance below the pile top, then this is displayed by the stress wave as a velocity change in the same direction as the direction of the impact pulse. The impedance is defined by EA/C where E is the elastic modulus, A is the pile cross-section area and C is the wave speed inside concrete. Thus, even if an impedance reduction is observed, an ambiguity still remains whether such a reduction is a weaker zone of concrete defined by a reduction in E or a reduction in cross-section defined by A. For simplistic purposes, both constitute a defect as it is a requirement that the pile is homogenous and uniform along the entire cross-section. A bulge is normally not a defect although it is not preferred for initial pile load tests. For pile diameters of 1 m and above, it is likely that the defect may be in one quadrant and thus may not get detected with limited data collection if such data are collected from other quadrant. The paper describes the basics of testing equipment, data collection and the features inside the software that may have a bearing on the final output. The paper also provides a classification that can be used for pile acceptance/rejection.

2 Testing Equipment

There are various manufacturers of equipments used worldwide for Low Strain Pile Integrity Testing. Three devices are needed for performing of low strain integrity test, Accelerometer, Hammer (with or without force Sensor) and Processor (PIT Equipment) as shown in Fig. 1. Various sizes and weights of hammers varying between 0.4 and 5.0 kg are used based on the depth and the diameter of the pile. Smaller hammers have less energy and higher frequencies whereas larger hammers apply greater energies and lower frequency to the pile top (Liang and Rausche 2023). Note that a smaller hammer may detect defects near the top and a heavier hammer may miss defects near the top but will more clearly identify defects near the bottom of the pile. However, due to uncertainty in the actual length and wave speed, irrespective of the hammer weight, the low strain method in general will not detect defects in the bottom 5–10% of the pile especially for long length predominantly friction piles. The method may, however, in some cases identify weaker pile bottom condition for rock socketed piles even though the length evaluation remains uncertain.

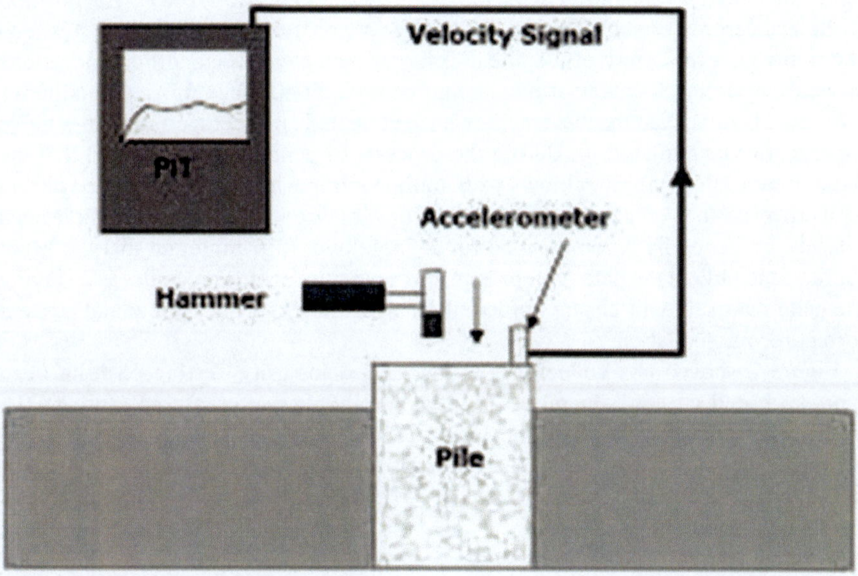

Fig. 1 Schematic diagram of low strain integrity testing

3 Data Acquisition and Evaluation

One of the foremost requirements for good evaluation is proper data collection. The first basic requirement is that the pile top should have sound concrete free from any weak material, laitance, soil, mud, etc. There should not be any micro cracks on the pile top that may affect the results. The pile top should be easily accessible, reasonably flat and the reinforcement either bent outside or in such a position as to allow the tester to fix the accelerometer and provide impacts without any hindrance. Grinding in several cases creates micro cracks and hence a reasonably uniform surface without grinding is acceptable. A compromise in any of these requirements may result in improper data collection and subsequently poor interpretation.

The testing can be done once concrete achieves 75% strength (ASTM D5882). Early testing may result in lower wave speed and create doubts on concrete quality. It is important that the tester is aware of the pile type, diameter, length, soil stratum, etc. before proceeding for testing. This will ensure that there are no unknowns during data collection or interpretation. The testing can be started by attaching an accelerometer or motion sensor with the Petro Wax or any bonding material on the top of pile and collect data with several blows to the top of the pile (Nayak 2020–2021). The position of the accelerometer should be such that it is away from the rebars and it is also preferably near the centre of the pile. For larger diameter piles, additional locations for every 500 mm diameter of pile should be attempted. Since this is a qualitative process, it is important that several blows are collected and averaged so as to remove uncertainties or human error during the process. In general, it is expected that the tester is aware that improper blows with nonlinear impacts should be rejected during such process. An average of 3–6 blows that constitutes one data set and multiple such datasets are generally recommended for a single pile. It is important that the tester continues to collect the data until proper interpretative records are collected. Having adequate data sets will ensure sufficient data for the expert to review and make a correct recommendation.

Figure 2 shows data collection of Low Strain Integrity Testing on reinforced concrete bored cast in situ pile. The hammer should preferably be close to the accelerometer sensor although the ASTM D5882 allows a distance upto 300 mm between the sensor and the hammer. Thus, proper position of the hammer, accelerometer, number of data sets, and surface preparation all play a very important role in the data collection process.

For data processing and interpretation, several signal enhancers or modifiers are provided in the software. The paper describes the data processing using PIT-W software although other software and allied equipment providers also have similar features as these are generally a standard requirement of ASTMD5882. The modifiers most commonly used are wave speed (WS), Magnification (MA), Magnification Delay (MD), Low Pass Filter (LO), High Pass Filter (HI), Pivot (PV) and Wavelet (WL). Each parameter has different use and different purpose; however, their basic purpose is to generate a final output that best represents the quality of pile. Figure 3 shows the example of good pile that has a first peak indicating start of the pile and

Fig. 2 LSIT data collection in progress

second peak at 10.65 m indicating termination of the pile (i.e. change in medium or material property). A clear toe response is evident at 10.65 m, and the computed wave speed is 3600 m/sec for this length of the pile which is an input inside the software. Apparently, the pile has no major anomalies and can be classified as good pile. Generally, in the Indian scenario, a wave speed of 3500 and 4200 m/sec can be expected although this will also depend on density, percentage reinforcement, quality control at site, grade of concrete etc. This variation in wave speed can be in much closer range based on site-specific quality control.

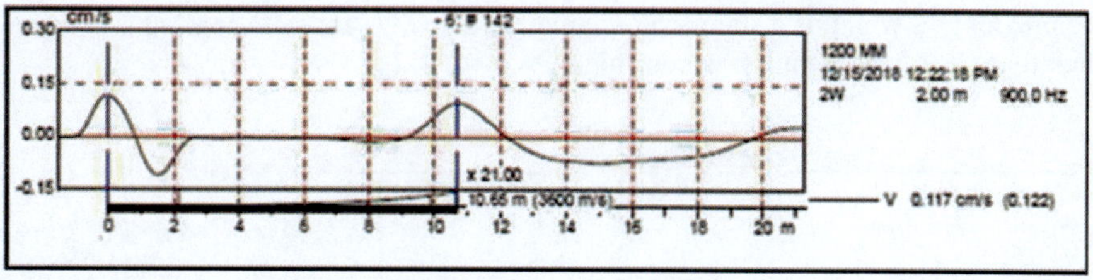

Fig. 3 Typical LSIT record after interpretation

4 Low Strain Testing Software Features and Interpretation

4.1 *Use of Magnification (MA) and Magnification Delay (MD)*

The impact of the hammer should always be such that the wave has sufficient energy to reach the bottom of the pile and reflect back to the top. However, absorption of energy due to soil resistance, pile non-uniformities and/or pile material damping reduce clear reflections as the wave propagates along the pile length. This reduction is more rapid towards the end of the pile where the soil is more stiff. As a compensation for this signal reduction, a magnification factor (MA), exponentially increasing with time is applied to the record which helps identify and measure reflections from the lower pile portion and near the pile bottom. To prevent record distortions and misuse, the magnification shall start from approximately 20% of the pile length and exponentially increase with time. The starting time of MA is known as magnification delay (MD). The MD should generally always be a default value of 20% and under no circumstances MD should exceed 50% of pile length. An exception may be allowed only when the pile has a large free-standing length more than 50% of its total length The MA should be chosen such that its value is at least equal to the length of the pile or a multiple of the pile length. Magnification greater than 100 is not recommended. Also, for piles of different lengths at the same site, the MA should be chosen such that they produce the same multipliers at the same depths. In certain cases, for more understanding of an individual pile, the MA can be a value such that the magnitude of reflections from the pile bottom is approximately equal to the input pulse (Ravikiran and Garland 2013).

Figure 4 shows integrity record for a pile that shows a faint toe response when MD is 20% of the pile length but the MA is only 10 which is less than 50% of the pile length of 23 m. Although the pile does not show any defect, a toe response identified by ± cycle along *x*-axis is not clearly visible. However, refer to Fig. 5, the same pile is now amplified with MA equal to 46, which is twice the pile length. The record now shows a clear response from the pile bottom. In this case too, the MD is 20% of the pile length, which is a standard practice. Hence, MA and MD applied should be such that clear interpretation is possible.

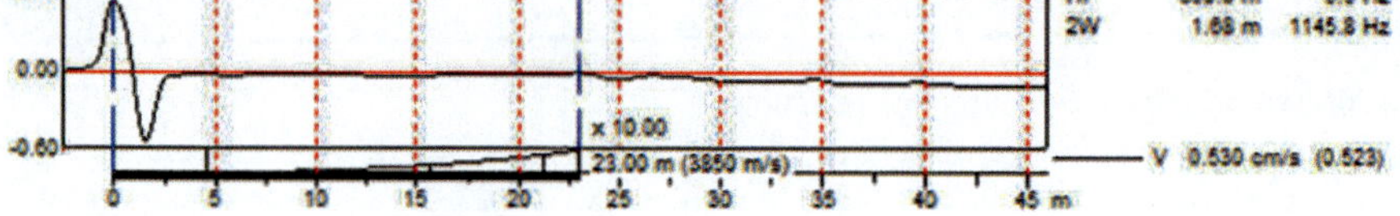

Fig. 4 LSIT data show MA < L

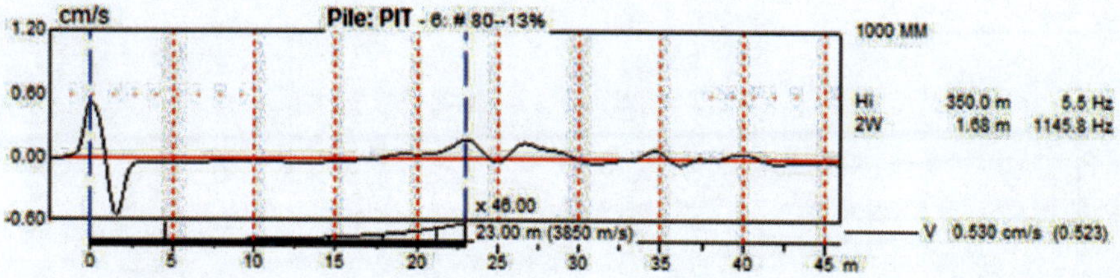

Fig. 5 LSIT data show MA = 2 * Length of Pile

Just as a proper MA only can clearly identify toe response as explained above, an improper use or abuse of magnification delay (MD) may result in making a defect within the pile. Figure 6 shows a data that looks uniform pile with a clear toe response. However, a look at the scale below the graph shows that the MA is applied only from 15 m, implying MD is about 90% instead of a default value of 20%. Refer to Fig. 7 which now corrects this improper use of MD. The data with the proper MD now show a clear defect at 7.5 m. Thus, using correct signal, modifier is an important requirement before proper interpretation.

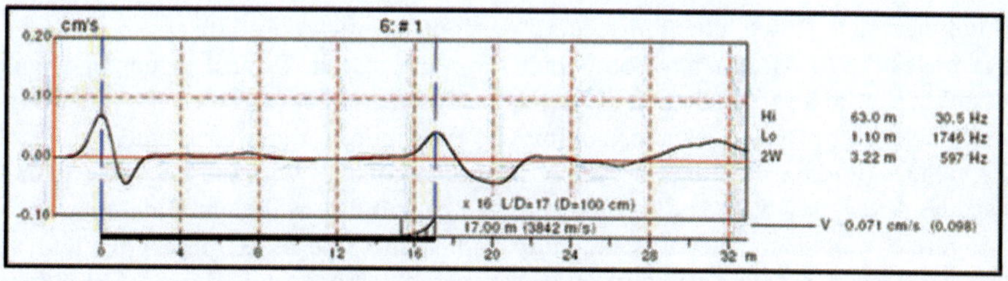

Fig. 6 LSIT data showing MD = 90% of L and MA < L

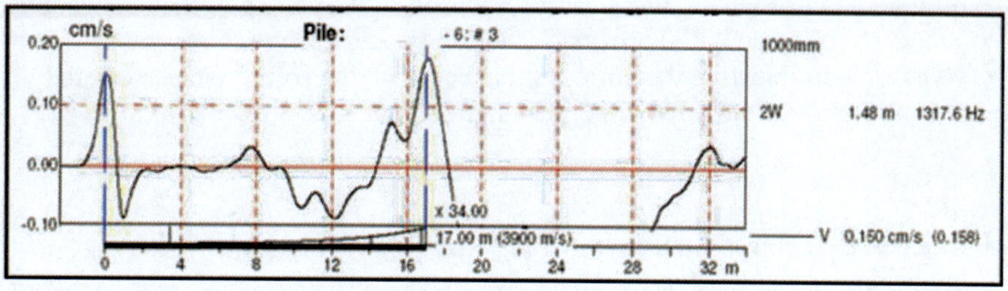

Fig. 7 LSIT data show MA = 2 * L and MD = 20% of L

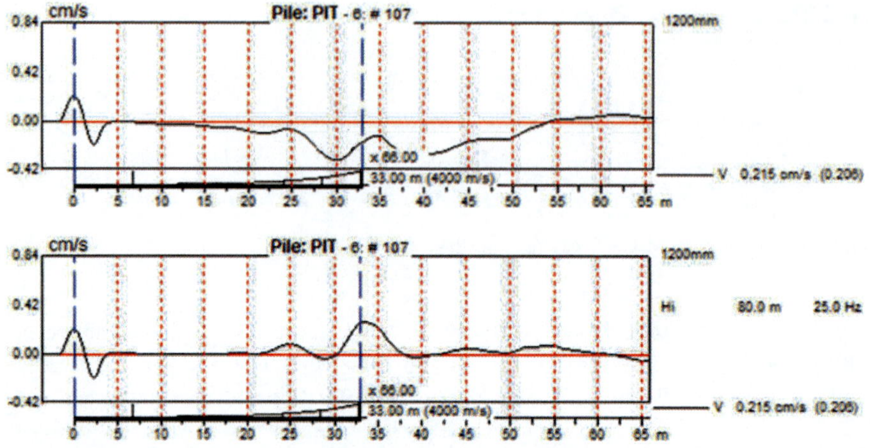

Fig. 8 LSIT graphs without and with HI pass filter (HI = 80)

4.2 Use of High Pass Filter (HI)

HI helps remove slowly changing record portions usually caused by soil resistance or removes the low-frequency component from the record. Typical values of High Pass filter should be zero where it is not required to be used or can be 15–20 times the input pulse width for longer piles and maybe even more for shorter length of piles. The value should be such that the velocity curve co-insides with the *x*-axis as much possible. A very low value between 0 and 30.5 m will distort the graphical record.

Figure 8 represents such a case study with zero value of HI pass filter and a corrected response with a value of HI = 80, which is more than 20 times the pulse width. Once a proper value is input, the pile shows clear defect which otherwise was not evident with an incorrect or "zero" value. The HI can be any value of zero or 30.5 or more and no specific basis can be assigned to it as along as the basic curve is corrected to co-inside with the *x*-axis. Thus, the experience of the tester will play a very vital role in selecting the correct value based on the soil resistance offered to the travel of the waveform along the pile length.

4.3 Testing of Micropiles or Piles with Reinforcement Closely Spaced

Micropiles are commonly installed in Mumbai for load bearing as well as shoring wall applications. These bored micropiles are typically 273 mm in diameter with an outer casing and having 4–8 number of 25 mm rebars or equivalent and filled with concrete. Permanent casing is provided in most cases up to starting of weathered

Fig. 9 Picture of closely spaced micropiles

rock or stiff soil. The lengths of these piles are typically from 8 to 15 m. This implies an L/d ratio from 30 to 55 with closely spaced rebars.

Figure 9 shows a picture of a project site with closely spaced bored micropiles and rebars projecting for significant height above cutoff level. It is a challenge to evaluate integrity for such small diameter piles where the diameter is small, percentage reinforcement is high, piles are closely spaced and rebar extending for longer depths above concrete. The difficulty is to find space to place the accelerometer and also a reasonably flat surface within the rebars for hitting the pile with the hammer. Another problem is that the wave reflecting from the pile concrete gets super-imposed with reflections from rebars creating a ringing effect or a distorted wave. The software does provide filters commonly known as Wavelet or LO PASS to remove high-frequency component or ringing effect of the rebar to make it appear smooth. LO PASS is generally not recommended in current practice. Also, WL can be generally a value not more than 4 m in most cases l. Excessive use of WL may only distort data and not provide conclusive findings. Figure 10 shows data on such a microbored pile. As evident, the waveform does not provide any conclusive findings on the pile integrity inspite of using WL as 3. The first author has evaluated several hundred data sets and most of them provided similar results. Thus, testing of small diameter microbored piles will in general not provide conclusive findings on pile integrity and instead high strain dynamic load tests on 5% of the piles should be considered to confirm integrity and pile capacity.

4.4 Incorrect Data Collection/Interpretation or Abuse of Data

Another case study shows a graph in Fig. 11, which is basically a sinusoidal curve with no apparent input peak and a clear output peak but was reported as having no integrity problems. The data show that amplification was only 4.4 for a length of

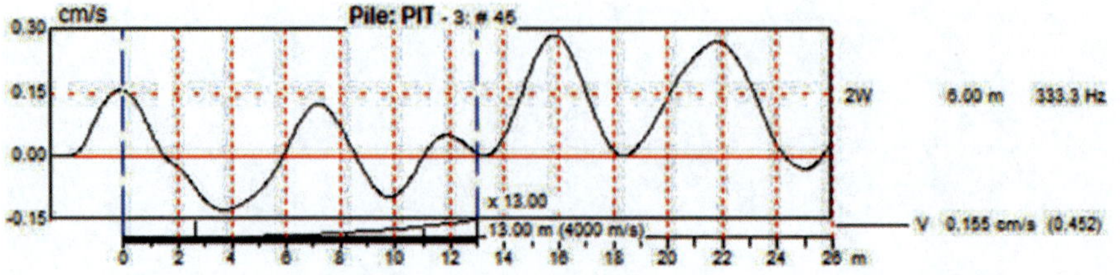

Fig. 10 LSIT on micropiles

9.7 m. The filter applied is not clear and apparently 42 blows seem to be averaged. The data were referred to the authors and since there was no clear graph, a re-test was recommended. The re-test done by the authors now clearly shows a defect at 7 m with proper data collection and processing as shown in Fig. 12.

To further verify the findings of low-strain testing, a high strain dynamic testing was also conducted on the same pile. Figure 13 presents the results of such a test. As evident, the pile shows a significantly high a settlement of 26.5 mm and a major impedance reduction at pile bottom to the extent of 73%. This shows the risks associated by employing testers who are not competent to conduct or interpret test data due to poor knowledge. The case study also highlights that mere acceptance of the pile in the test report has no relevance unless it is backed by proper data collection and processing.

Hence, it is important for the end user to not only merely believe the conclusions written in the table but also understand and read the graph. A catastrophe may occur or structural safety may be compromised, if several defective piles in a pile group or in a project are certified as acceptable by inexperienced or pliant testers.

Pile	Details	Reflectogram	Remarks
P/79	11/06/2019 C:3950 Amp:4.4 F:4 9.7 Avg:42	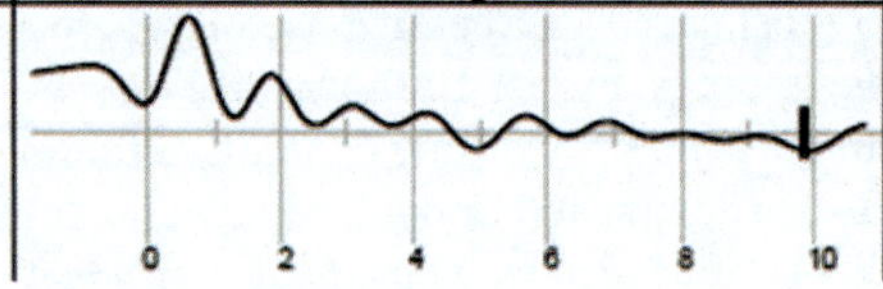	No Integrity Problem

Fig. 11 Incorrect LSIT data and interpretation

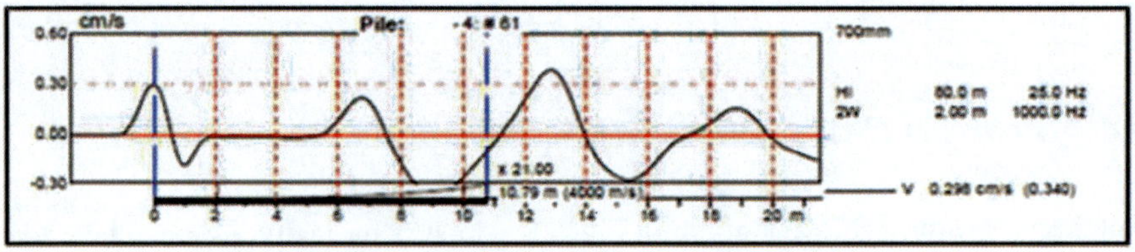

Fig. 12 Proper data collection with a re-test

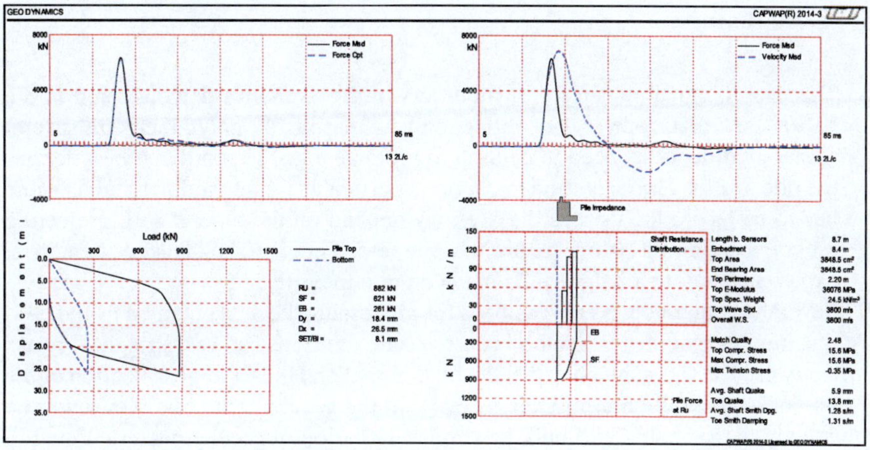

Fig. 13 HSDPT confirms defect as seen in LSIT

The first author has reviewed several testing reports provided by various testers in India and wide discrepancies in comments were observed. This also leads to further confusions in language, interpretations, end purpose, etc. and in several cases does not serve the intended purpose of the test report. That the low strain integrity test is a basic tool to evaluate pile integrity is well known and trying to make a herculean effort or to read too many things from the graph only leads to more confusion. Thus, in the opinion of the author, the conclusions can be provided in the following categories after mentioning the limitations of the test.

Table 1 Low strain integrity testing categories as per toe response

Category A	Piles those are acceptable with a clear toe response (response from pile bottom)
Category B	Piles that show clear Indication of serious defect (Impedance reduction) and toe response are not apparent. Such piles may be subject to further investigation or may be underrated, rejected based on site-specific experience, design requirements etc.
Category C	Clear indication of small defect but toe response is apparent. If more than 15% of the piles at a project are in this category, further investigation in the form of static/high strain dynamic load testing is recommended to identify the reason and then take suitable action based on the findings of such results
Category D	Inconclusive piles. Very high soil resistance or bulges may absorb most of the hammer energy impact and thus a clear toe response may not be evident. Such piles are termed as inconclusive. A decision on such piles can be taken on the basis of construction records, overall findings at the project site, criticality of the structure above it, load on the pile etc. A load test may be considered in certain cases before acceptance. If more than 10% of the piles at a project site are inconclusive, then it may warrant a detailed investigation on the reasons for such a conclusion before acceptance of such piles

5 Conclusion

1. The use of signal enhancers or modifiers in the examples demonstrated in the paper show that, even if data collection was reasonably good, there are several signal modifiers that need to be used to produce a LSIT record that best defines the pile health. The discretion to choose a particular signal modifier and its value has to be left to the user as the use may depend on the type of soil, projecting rebars, location of testing within the pile, etc. Thus, the LSIT needs expertise, experience before a judgment is made on pile integrity.
2. Several signal modifiers are required for processing the data obtained from LSIT, this implies that the test method is more state of art instead of a straight forward analysis. There is subjectivity in the final output and thus a standardization of the output is generally not possible as even on the same pile no two blows will be exactly the same but will only be similar. Location of accelerometer on the top of the pile and variability in hammer impact, weight of hammer is also one of the reasons for the signals not being exactly the same even when repeated with the same tester. However, the results will still be very similar if the test is properly done and significant variability between different testers is not expected. This implies that final interpretation is still expected to remain the same or similar even considering variability in data collection and processing.
3. Abuse of the method as explained in Para 4.4 is easily possible by testers who are either inexperienced or who are forced to produce reports to merely satisfy project requirements rather than a genuine feedback on the quality of the pile. If majority of the piles at a project are bad, then mis-interpretation or poor interpretation can lead to a potential catastrophe. However, a proper test is expected to reveal significant information on pile integrity which otherwise is unknown.
4. The method is not applicable to bored concrete micropiles as noise from closely spaced rebars interferes with the reflections from concrete. Hence, if there is a genuine requirement to assess micropiles, limited piles be tested first and only if proper output with a toe response is available, then testing be continued. High Strain Dynamic testing upto 5% of the piles should be considered instead of low strain integrity testing for micropiles.
5. It is strongly recommended that the Contractor, Engineer and the Owners all select an experienced tester with adequate knowledge. It is essential that all the stakeholders or the reviewer of the report study the graphs and the findings in detail before the piles are finally accepted or rejected on the basis of the low-strain findings.

References

ASTM D5882-2017 (2017) Standard test method for low strain impact integrity testing of deep foundations

IS:14893-2021 (2021) Low strain non destructive integrity testing of piles

Liang L, Rausche F (2011) Quality assessment procedure and classification of cast in place pile shaft using low strain dynamic test

Nayak NV (2020–2021) Foundation design manual, 8th ed

Rausche F, Likins G, Hussain M (1994) Formalized procedure quality assessment of cast in place shafts using sonic pulse echo methods

Ravikiran V, Garland L (Nov 2013) Guidelines for successful high strain dynamic load tests & low strain integrity tests for bored piles. Deep Foundation Technologies for Infrastructure Development in India, Mumbai, India

Challenges to Working and the Impact of Existing Codes with Respect to Technical Innovations in Deep Foundations

State-of-Art Working Platform for Heavy Equipments Used for Deep Foundation Construction

R. Arunsiva, S. Sivaraman, B. Venugopal, and M. Kumaran

Abstract Being a rapidly developing country, India is undergoing a massive infrastructure phase change remarkably. The construction industry is already set to boom, but the time and space constraints necessitate the employment of heavy plants and machinery to take the foundations deeper than ever. Operating such machinery on soft or poor ground conditions without a good working platform leads to uncertainties with massive loss of men and machinery. Plenty of such cases can be observed on construction sites across the globe. Unfortunately, forming a temporary working platform is given least to no importance in the construction industry, particularly ground engineering. This glitch can only be addressed through standard guidelines or codal recommendations on designing and implementing temporary working platforms issued by regulatory bodies. The present paper highlights the importance of providing the working platform, existing design method for working platforms such as TWf 2019:02 of Temporary Works forum, Building Research Establishment BR470 and a brief discussion on the design and construction of the geosynthetic reinforced working platform with a case study is also included. A pitch has been made on the importance of Indian codes/guidelines for the design and construction of temporary working platforms. These guidelines will ensure the safety of the men and heavy machinery and, in addition, increase the productivity of foundation construction irrespective of weather conditions and also become part of the regular construction scope.

Keywords Construction site · Soft ground · Piling rigs · Temporary working platform · Geosynthetics

---

R. Arunsiva (✉) · S. Sivaraman · B. Venugopal · M. Kumaran
L&T GeoStructure, Chennai, India
e-mail: arunsiva.r@lntecc.com

S. Sivaraman
e-mail: sivaraman.pandiyan@lntecc.com

B. Venugopal
e-mail: bvg@lntecc.com

M. Kumaran
e-mail: mkn@lntecc.com

A. M. Krishna et al. (eds.), *Deep Foundations for Infrastructure Development in India*, Lecture Notes in Civil Engineering 373, https://doi.org/10.1007/978-3-031-37256-8_12

1 Introduction

India is already on the envisioned track to surpassing a five trillion dollar economy. It is accomplished through the expected growth of the construction sector at a Compound Annual Growth Rate (CAGR) of 10% as per the Global Construction Market report (2022). Also, the Indian construction sector is expected to be the third-largest construction and infrastructure market by 2025 at the global level, making it one of the country's largest employment engendering sectors, in turn engaging more significant human and mechanical resources ever than before. This requires an increased responsibility from the construction industry towards higher productivity without compromising on the safety and healthy workplace environment. The deep foundation industry generally operates with the movement and operation of rigs, cranes, heavy-loaded trucks, plants, and machinery in the bare land/temporarily formed platform/road throughout all weather conditions in a year. Globally, the deep foundation construction sites encounter issues of the toppling of rigs and cranes or near misses of such. These incidents trigger severe injuries or even fatalities to the operators, and ground crews, heavy damage to the equipment and delay in the projects. The primary scope of this paper is to enforce the importance of adopting the methodology of creating a Temporary Working Platform in the deep foundation industry. This will create a safe working environment, thereby increasing work productivity.

2 Need for Temporary Working Platform

As stated earlier, there lies greater responsibility for ensuring safety on construction sites. It is hard to accept the fact that, on average, 38 workers involved in construction activities die every single day in India, which includes accidents involving the failure of plants and machinery engaged in various construction works (Patel and Jha 2016). Building Research Establishment (BRE) study states that one-third of accidents involving plants and machinery in the construction industry are due to the bearing failure of the ground supporting the crawls and outrigger of such cranes and piling rigs (BR470 2004). A typical incident with a near-miss of crane toppling due to poor ground condition is shown in (Fig. 1)

This implies the need for a firm base to support and effectively disperse the loads from the cranes and piling rigs. These issues can only be solved by proper planning, designing and construction of a temporary working platform at the site. A temporary working platform not only provides the firm base for the operation of plant and machinery but also increases the productivity and safety of workers involved in such activities throughout. Despite the vital prerequisites of such working platforms, there are no such specific guidelines or regulations that persist in the Indian safety standards. Even the codes such as IS 5121 (1969) "Safety code for Piling and other Deep foundation" and SP 70 (2001) "Handbook on Construction safety practices"

Fig. 1 A typical incident of poor platform failure

neither insist on providing a temporary working platform despite engaging heavy plant and machinery nor provide the technical support in designing such working platform. Occasionally, the industry follows a few conventional methods such as granular blanketing and ground improvement through compaction, lime/cement/fly ash treatment (Pardakhe and Ban 2019). Although the above-stated conventional methods were implemented in limited project sites, no detailed guidelines or technical support exists for adopting such methods. The present study puts forth the available guideline documents and recommendations on the design and implementation of the temporary working platform by international associations and technical forums. These documents can be referred to formulate the implementation strategy for the temporary working platform in the Indian deep foundation industry. Also, this study briefly discusses the case study on such effective design and implementation of a geosynthetic reinforced temporary working platform in one of the Indian construction sites.

3 Existing Design Methods

3.1 *Temporary Working Forum—TWf 2019:02*

Temporary Working forum (TWf), which belongs to the Institution of Civil Engineers, London, released a guideline titled Working Platforms: Design of granular working platforms for construction plant—A guide to practice, which was referred to as TWf 2019:02. It provides guidelines for designing granular working platforms; however, it is limited to only unreinforced working platforms. For reinforced working platforms, it recommends adopting geosynthetic manufacturer's alternative method, which depends on product design and development. TWf 2019:02 follows the conventional projected area method, which employs load spread angle to estimate the load spread area. The reduced pressure on the subgrade due to the provision of a granular platform is then calculated using the estimated load spread area. Also, it is to be noted that the load spread angle is restricted to a maximum of 26.6°. TWf 2019:02 is not a design code by itself, and it is drafted in concurrence with Eurocode and British standards. TWf 2019:02 makes use of BS 8004 (2015) and EN 1997-1 (2004) EuroCode7 (EC7) for estimating the bearing capacity and settlement of both subgrade and platform. The overview of the design procedure of TWf 2019:02 in designing an unreinforced granular working platform is represented by the flowchart as given in (Fig. 2).

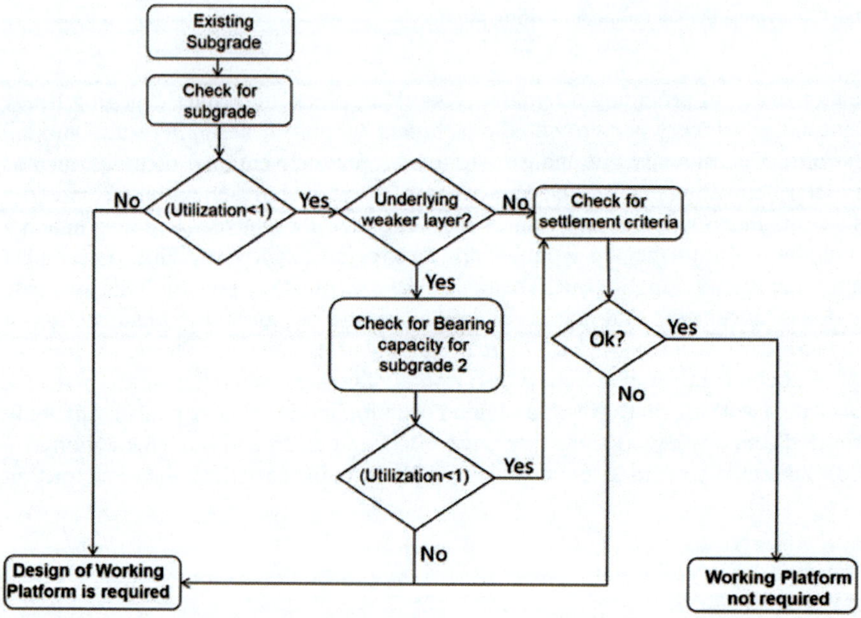

Fig. 2 Flowchart of TWf 2019:02 design process

Table 1 Details of different load cases (after BRE)

Load case 1	Crane/rig operator is unlikely to be able to support recovery, which includes Crane travelling standing handling casing and reinforcement cages
Load case 2	Crane/rig operator is able to control the load, which includes Drilling Installing/ extracting casing Rig travelling

The design of the granular platform starts with the calculation of the design load applied to the platform, which is derived from stress blocks beneath the footing of the plant or machinery. TWf 2019:02 recommends the usage of rectangular stress blocks in case of crane or piling rig with crawl mounted or outriggers. In the case of trapezoidal or triangular stress block, it recommends converting to an equivalent rectangular stress block, and for wheel loads, SP 123 (1996) is recommended. The design is carried out for two design load case, which depends on the operation mode of the plant or machinery, as referred to in (Table 1). According to the load cases, the partial safety factor is applied. The TWf 2019:02 guides well even in the presence of multiple subgrade soils.

3.2 *Br470*

Working platform for tracked plants is a technical guide published by Building Research Establishment Ltd. (BR470 2004), which is mainly referred to as BR470. The BR470 guideline is developed along with the Federation of Piling Specialists (FPS) as an initiative toward the safe working of plants and machinery at the ground engineering sites. BR470 provides various technical guidelines to be followed in activities involving heavy plants and machineries like cranes and piling rigs. It clearly recommends providing a working platform in all cases where the bearing capacity of the existing subgrade fails to support such heavy machinery. BR470 is recognized by BS 8006-1 (2010) as an acceptable technical guide for the design of both unreinforced and geogrid-reinforced working platforms. It recommends the use of EN 791 (1996) Drill rigs-safety and EN 996 (1996) Piling Equipment-safety standards in estimating the design load on the platform. Similarly, TWf 2019:02, BR470 also recommends rectangular stress block conditions. In the case of triangular or trapezoidal stress block existence due to asymmetry orientation of machinery during operation, it suggests Meyerhof's (1953) approach to convert into equivalent rectangular stress block. BR470 uses the analytical method developed by Meyerhof (1974) to access the bearing capacity of the platform and subgrade, which is based on the punching shear failure mechanism. While estimating the bearing capacity of a geosynthetic reinforced platform, BR470 includes an additional term $2T_d/W_d$ in Meyerhof's (1974) formula, which is the resistance provided by the 2D planar geosynthetic reinforcement. The T_d is the design tensile strength of the reinforcement, and W_d is the design track width of the plant. It also recommends a minimum platform thickness of 300 mm or half of the track width, whichever is less, in case of an unreinforced platform. In the case of a geosynthetic reinforced platform, a minimum cover of 300 mm to reinforcement is recommended. The calculated thickness of the geosynthetic reinforced platform should be in such a way that, when the same thickness is provided without reinforcement, the bearing resistance should be at least 25% greater than the design load case 1 and 5% greater than the design load case 2.

4 Upgradation with Technical Advancements

Although the above-discussed methods provide basic guidelines for the design of a temporary working platform, they have their own shortcomings. The design and usage of geosynthetic material such as geocell or geo-composite material for the formation of a temporary working platform are not elaborated on in the previously developed guidelines. As far as Indian codes and standards are considered, as mentioned earlier, there are no recommendations for the temporary working platform. The Federation of Piling Specialists (FPS) made an initiative to offer Working Platform Certificate, a mandatory certificate to be possessed by contractors when engaging heavy plant and machinery for construction activities. Such an initiative is highly appreciated, and the same kind of initiative needs to be taken in India. Also, in addition to the requirement of thorough examination and certification of the crane as per Annex C of SP 70 (2001), this working platform certificate should also be included in it. Another major problem faced by contractors in executing the working platform is that; no contract documents include the provision of providing a working platform. However, the contractors provide such working platforms following their company's safety policies without any additional bill of quantities for payments.

5 Case Study on Geosynthetic Reinforced Working Platform

The existing subsoil at one of the piling job sites predominantly consists of black cotton soil or clayey soil at the top. This subgrade soil with low bearing strength and CBR value tends to become slushy even with a slight drizzle. A typical view of the site soil condition without any working platform is shown in (Fig. 3). The unreinforced platform with the fill of granular material was found to be defunct to serve for the safe movement of plants, as it gets mixed with soft clay after 2 or 3 passes. The condition of the unreinforced granular platform immediately after formation and the functional failure of the platform after a few passes of vehicles are shown in (Fig. 4a, b) respectively.

To facilitate the safe movement of heavy equipment such as cranes, piling rigs, transit mixers, and other heavy machinery, at the piling site, a properly designed

Fig. 3 Site soil condition without the working platform

Fig. 4 **a** Formed Unreinforced platform. **b** Unreinforced platform failure

temporary working platform is required. Hence, it was proposed to design the temporary platform as per the guidelines of the Temporary works forum (TWf 2019:02) and BR470 (2004). The soil condition and load due to construction equipment such as heavy crawler cranes and piling rigs were considered for the design. Accordingly, for the present site condition, the thickness of unreinforced granular platform required as per TWf 2019:02 and BR470 was found to be 1100 mm and 900 mm, respectively. The geogrid-reinforced granular platform requires 650 mm as per BR470 method. To optimize the design thickness of platform and cost, geocell option was chosen. With the consideration of granular infilled geocell layer in the design, the thickness of temporary working platform was reduced to 475 mm. Consequently, it was recommended to construct the temporary working platform with the inclusion of the geosynthetic reinforcement for a stable operating platform over soft soil and to expedite the construction process. The geosynthetic materials as follows were recommended to improve the performance of the temporary working platform and road.

Fig. 5 Construction of geocell reinforced working platform

(a) A separation layer with non-woven geotextile is proposed to lay between the existing soft clay layer and the granular fill layer. This is essential as the mixing of clay and granular fill during the site operation reduces the strength of the working platform.
(b) A synthetic Geocell reinforcement layer is proposed to reduce the overall thickness of the temporary working platform fill material and cater to heavy loading. This will provide a three-dimensional confinement effect (stiffened load transfer mat) and enhances strength.

The design of a temporary working platform with the geocell reinforcement was made concerning the guidelines of the Temporary works forum (TWf 2019:02), BR470 (2004) and Geocell design philosophy. A typical view of constructing a geocell-reinforced temporary working platform is shown in (Fig. 5). The safe movement of the Piling rig, Crawler crane and safe construction operation over the geocell reinforced working platform is shown in (Fig. 6). The performance of the working platform was found to be a satisfactory engineering solution to expediting the piling process in difficult ground conditions that too in the heavy rainy season. The temporary working platform at the piling site satisfies the intended purpose with advantages such as (i) safe movement of the Piling rig and crane with ease, (ii) positioning of piling rig with respect to pile point without any delay and (iii) multiple movements of crane/piling rig without damage/failure of the working platform.

6 Conclusion and Recommendation

The Indian construction industry contributes to one-fourth of the country's total occupational fatal accidents, which is due to the lack of awareness among workers and an unfavourable workplace environment. One of the unfavourable workplace environments shall be mitigated by providing checks for the requirement of temporary working platform for the safe and effective operation of plants and machinery

Fig. 6 Safe pile construction practice with the reinforced working platform

in the deep foundation industry. As discussed in the paper, a comprehensive guideline or recommendation on the safe ground engineering practice through working platform construction and certification needs to be developed by considering Indian geological conditions and construction practices. The same needs to arrive with the integration of all the latest construction technologies and materials for a competent and cost-effective design solution. With the formation of guideline documents on a temporary working platform and by imposing the statutory requirement as part of the contract will ensure the ground engineering contractors and general contractor can practice safe and effective construction practices.

Acknowledgements The authors acknowledge the support and encouragement provided by the Chief Executive, Heads of all departments and Staff members from Headquarters and project sites of L&T GeoStructure Private Limited.

References

BR470 (2004) Working platforms for tracked plant. Building Research Establishment

BS 8004 (2015) Code of practice for foundations. British Standards Institution

BS 8006-1 (2010) Code of practice for strengthened/reinforced soils and other fills. British Standards Institution

EN 1997-1 (2004) Eurocode 7: geotechnical design—Part 1: general rules. European Committee for Standardization

EN 791 (1996) Drill rigs—Safety. British Standards Institution

EN 996 (1996) Piling equipment—Safety requirements. British Standards Institution

IS 5121 (1969) Safety code for piling and other deep foundation. Bureau of Indian Standards

Meyerhof GG (1953) The bearing capacity of foundations under eccentric and inclined loads. In: Proceedings of 3rd international conference on soil mechanics and foundation engineering, vol 1, Zurich, pp 440–445

Meyerhof GG (1974) Ultimate bearing capacity of footings on sand layer overlying clay. Can Geotech J 11(2):223–229

Patel DA, Jha K (2016) An estimate of fatal accidents in Indian construction. In: Proceedings of the 32nd annual ARCOM conference, vol 1. Association of Researchers in Construction Management, Manchester, UK, pp 539–548

Pardakhe P, Ban P (2019) Soil stabilization by using waste material—Brick dust. Int Res J Eng Technol 6(9):1935–1941

SP 123 (1996) Soil reinforcement with geotextiles. CIRIA Special Publication

SP 70 (2001) Handbook on construction safety practices. Bureau of Indian Standard

The Business Research Company. Global Construction Market Report (2022), website, https://www.thebusinessresearchcompany.com/press-release/construction-market-2022#:~:text=4%20Jan%2C%202022&text=The%20construction%20market%20is%20expected,at%20a%20CAGR%20of%2010.8%25

Ground Improvement Techniques for Mass Stabilization

Vibro Stone Columns with Preloading—An Effective Foundation Solution for Large Storage Structures

C. Ramanathan and P. V. S. R. Prasad

Abstract The most critical challenge for any large size storage structures to be constructed on soft compressible soils is bearing capacity and consolidation settlements. Consolidation is the time-dependent phenomena, as expulsion of water from soil voids on application of external load will take longer period. It is mainly influenced by soil consistency, permeability, thickness of compressible stratum and presence of free-draining layers. This study deals with the foundation solution of a large size Mounded Storage Vessel (MSV) to be constructed on thick compressible stratum whose loading intensity is around 155 kPa, and the estimated settlements are greater than 400 mm. Hence, the foundation solution using vibro stone columns is adopted as an effective technique to address the challenges of excess settlements and inadequate bearing capacity. Further to keep settlements within the permissible limit of 50 mm, preloading technique with 60% of actual loading intensity will be utilized. The consolidation analysis of stone column-reinforced foundations is carried out based on the method given by Han and Ye (2001). The quality of the stone columns is ensured by real-time quality control measures, and the performance of the stone columns shall be ensured by vertical load tests following IS 15284-Part 1:2003. Ground settlements were closely monitored with proper instrumentations and the actual degree of consolidation has been verified by Asaoka (1978) method.

Keywords Consolidation · Vibro stone columns · Preloading · Soft soils

C. Ramanathan (✉)
Sr. Geotechnical Engineer, Keller Ground Engineering India Pvt. Ltd., Chennai, India
e-mail: ramanathan@kellerindia.com

P. V. S. R. Prasad
DGM, Keller Ground Engineering India Pvt. Ltd., Chennai, Tamil Nadu, India
e-mail: prasad@kellerindia.com

A. M. Krishna et al. (eds.), *Deep Foundations for Infrastructure Development in India*, Lecture Notes in Civil Engineering 373, https://doi.org/10.1007/978-3-031-37256-8_13

1 Introduction

India being a country with world's second-highest population, the demand for oil and gas products increases rapidly which led to the surge in demand for related storage facilities to satisfy the requirement of supply chain. The increasing demand for storage facilities and scarce in availability of suitable construction land, steers the practicing engineers Valluri et al. (2015) and Gupta et al. (2017) towards the adoption of ground improvement techniques to convert the available land into constructible ground.

Mounded bullets are horizontal pressure vessels that are intended for the pressurized storage of Liquefied Petroleum Gas (LPG) under ambient temperature. In mounded storage facility, a mound of earth or suitable inert material is provided to cover the bullet, and this is kept above ground. Mounded bullets are installed on sand bed foundations, which allow the load to be transferred uniformly to the underlying sand. The diameter of these bullet tanks ranges between 4.0 and 8.0 m whose length commonly varies from 20.0 to 40.0 m. Rapid installation process relatively low installation and maintenance cost with high safety make these mounded type of storage vessels to be widely used.

This paper deals with the application of vibro stone column as an effective foundation solution for the construction of mounded storage vessel on compressible soils in the coastal area of North Chennai.

2 Structure Details and Performance Criteria

The LPG storage facility comprises of three numbers of 450 MT capacity mounded-type pressure vessels (bullets). The diameter and length of the bullets are 6.0 m and 37.0 m, respectively (see Fig. 1) with mound footprint area of around 1350 (30 × 45) m^2. The height of the storage mound is about 9.5 m, and loading intensity of the storage vessel is 155 kPa. The corresponding long-term (Post-hydrotest) settlements of the storage mound founded on improved ground should not be more than 50 mm with differential settlements less than 1:2500 of length of vessel.

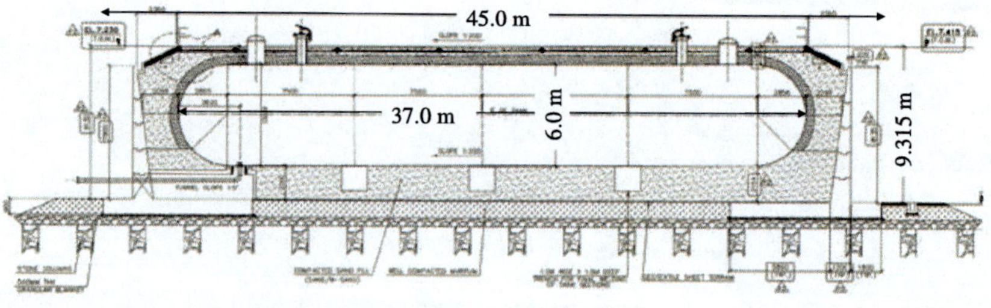

Fig. 1 Typical longitudinal section of storage vessel

3 Subsoil Conditions

To understand the subsoil conditions at site, 5 numbers of Electric Cone Penetration Tests (ECPTs) were conducted within the footprint area of mound. The average Ground Water Table (GWT) is observed at shallow depth (i.e., 1.0–1.5 m below NGL).

It could be inferred from the ECPT results that subsoil comprises of 2 m thick stiff to very stiff clay (qc ~ 1.1–2.5 MPa) at top, followed by 7 m thick, soft to medium stiff clay (qc ~ 0.5–0.6 MPa). These layers are underlain by loose to medium dense sand for 5.5 m thick before being encountered by hard clayey silt layer at 14.5 m deep. The hard clayey silt layer continued till the depth of exploration, i.e., 22.0 m (see Fig. 2) below Natural Ground Level. The subsoil conditions based on ECPT results are summarized in Table 1.

The presence of 7.0 m thick, soft to medium stiff silty clay encountered between 2.0 and 9.0 m results in inadequate bearing capacity and induces large settlements upon loading. When the external load is applied on top of these saturated clay layer, the porewater pressure will increase as the hydraulic conductivity of clay is very small. Significant time would be required to dissipate this excess porewater pressure and to transfer the increased stress to soil skeleton. This phenomenon of increase in effective stress in the clay layer upon loading, causing settlements over a period of time is consolidation. The estimated settlements of the insitu ground for the given loading intensity were in excess of 400 mm.

Further, the site falls under seismic zone III (0.16 g) with earthquake magnitude of 6.5, which also makes the loose silty sand layer below the cohesive layers more susceptible to liquefaction (IS 1893-Part 1:2016) during the occurrence of earthquake.

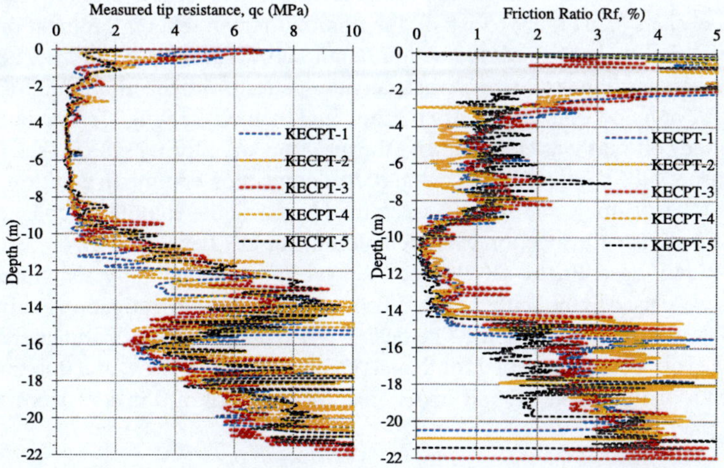

Fig. 2 ECPT PLOTS (Depth Vs. qc & Rf)

Table 1 Subsoil condition

Serial number	Depths (m)	Soil type	Average q_c (MPa)	Y, kN/m^3	C_u, kPa	Φ, deg	Constrained modulus D_s, kPa
1	0–1	Very stiff silty clay	2.5	17	125		37,500
2	1–2	Stiff silty clay	1.1	16.5	55		16,500
3	2–7	Soft silty clay	0.5	15.5	22.5		2,700
4	7–9	Soft to Medium stiff silty clay	0.6	16	30		6,000
5	9–12	Loose silty sand	2.5	16.5			10,500
6	12–14.5	Medium dense silty sand	6.4	18		27	31,500
7	14.5–22	Hard clayey silt	5.0	18	250	32	75,000

4 Foundation Solution

Based on the subsoil information and required performance criteria, ground improvement using vibro stone columns had been used as an effective foundation solution to address the challenges of bearing capacity, settlements and liquefaction. The treatment was provided under the mound footprint and also extended by three additional rows on all sides (see Fig. 3) to satisfy the minimum treatment requirement of liquefying soils (i.e., half of the liquefaction depth below foundation level). The stone columns were installed at 2.0 m c/c in triangular pattern with the minimum treatment depth of 12 m below NGL using dry bottom feed method (see Fig. 4). The quality of the installed columns was confirmed by the real-time monitoring system (see Fig. 5), which ensured the depth of treatment and the compaction energy. In addition, diameter formation of stone columns was examined based on the volumetric consumption of stones and visual inspection at stone column Cut Off Level (COL).

The widely used Priebe's (1995 & 1998) design methodology for vibro stone columns were used in the settlement analysis and liquefaction mitigation analysis for the improved ground conditions. The design (stiffness and shear) parameters of the composite soil were established for the adopted treatment scheme, and the estimated settlements of improved ground under loading intensity of 155 kPa were around 210 mm.

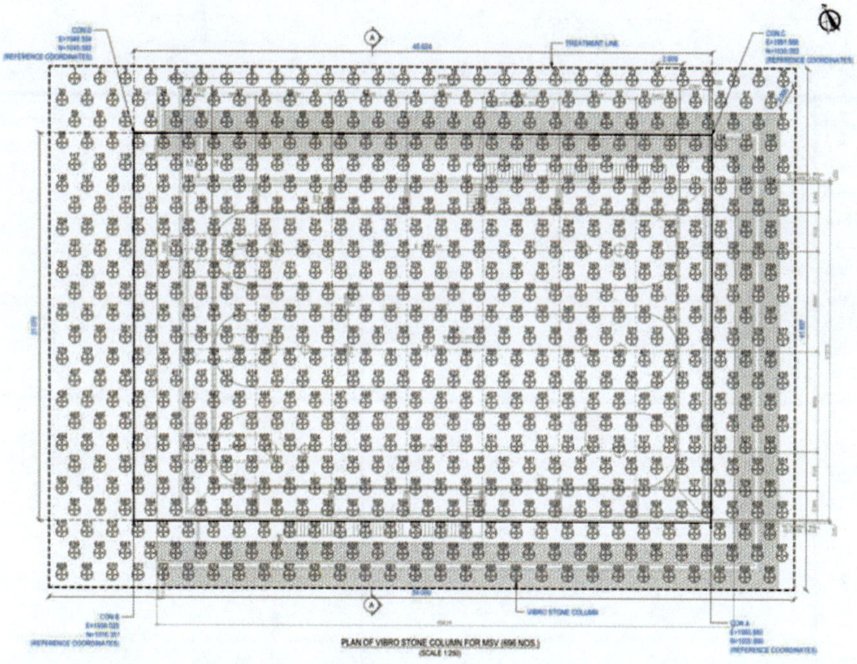

Fig. 3 Ground improvement treatment scheme

Fig. 4 Installation of vibro stone columns using dry bottom feed method

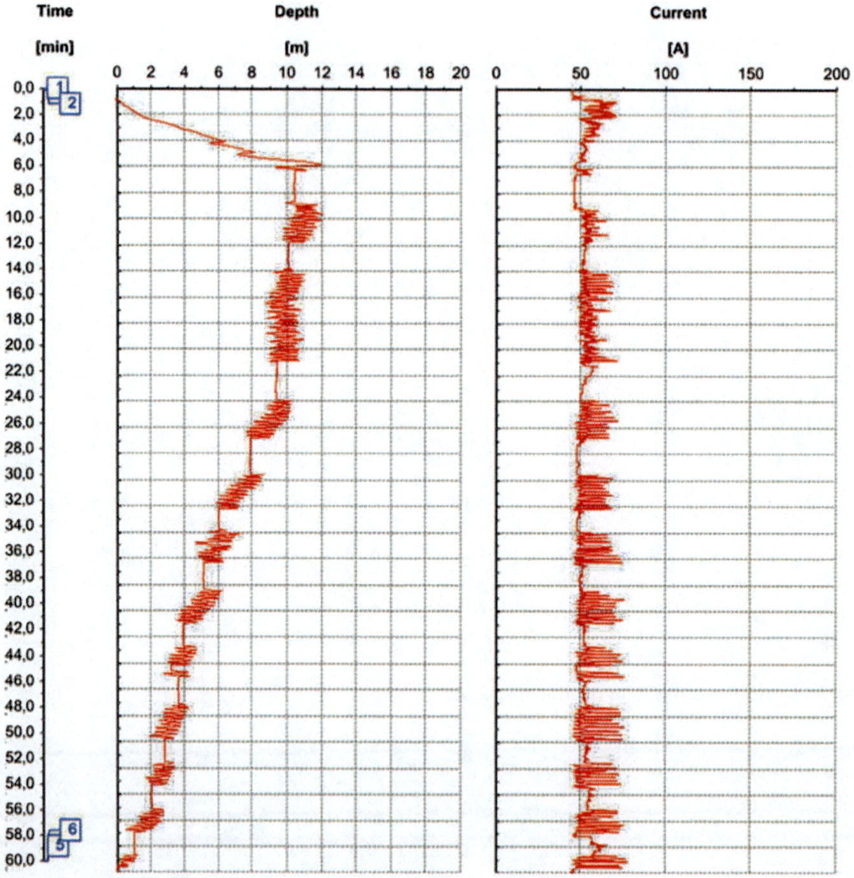

Fig. 5 Typical real-time quality monitoring record (M5 record)

5 Requirement for Preloading

As stated in the previous section, the estimated post-treatment settlement was around 210 mm. In order to control the settlements within the permissible range (i.e., Post construction long term settlements $\leq$50 mm), preloading technique was required. Preloading in combination with vibro stone columns (see Fig. 6) reduces the consolidation settlements and increases the rate of consolidation.

The consolidation settlement analysis of the treated ground had been carried out based on the method given by Han and Ye (2001) whose method of estimation is similar to that of Barron (1947) solution for drain wells in fine-grained soil. However, with modified coefficients of consolidation, it demonstrates the stress transfer from soils to stone columns, dissipation of excess porewater pressure due to drainage and vertical stress reduction during consolidation.

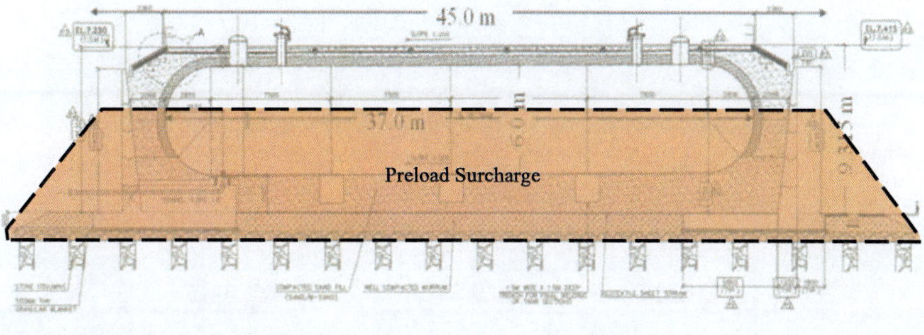

Fig. 6 Preloading of ground reinforced with stone columns method

Based on the above method of analysis, it was determined to use 60% of total loading intensity as preload, which is equivalent to 5.3 m of soil filling. The slow rate of surcharging (1 m/week) is considered to avoid the buildup of excess porewater pressure. Here, the rate of radial consolidation C_r is considered as 3.5 m^2/year and the duration of preloading is considered for 3.0 months (1.5 months of loading and 1.5 months of holding period). Post preloading, the residual settlements will be brought down, i.e., the soil would have completed 90% degree of consolidation for 60% of total loading intensity. Further, the construction of mound and hydrotest will have additional settlements which will be monitored, thus ensuring the post-construction long-term uniform settlements of storage mound will be less than the permissible range.

6 Settlement Monitoring and Its Inferences

Field monitoring was carried out through settlement gages established at nine points covering Mound footprint. Along with that four numbers of inverted settlement plates were placed on top of surcharge surface. Layout showing location of settlement monitoring plates is shown in Fig. 7. These settlement monitoring rods are placed at the stone column Cut off Level (COL) prior to the application of surcharges as shown in Fig. 8.

The Surcharge had been filled in layers up to the desired height of ~5.3 m (~100 kPa) above finished ground level (FGL). Monitoring of settlement plates took place during the layered filling of surcharge and during the holding period. Figure 9 shows the completion of preload surcharge of 5.3 m soil fill.

After attaining the final fill height, the surcharge had been maintained for a duration of 45 days. During the holding period, the settlements had been monitored closely and time rate of settlement results is shown in Fig. 10.

The degree of consolidation during holding period was estimated by Asaoka (1978) method (an observational approach). Asaoka method is a simple graphical procedure that uses field monitoring data at regular interval to determine the degree

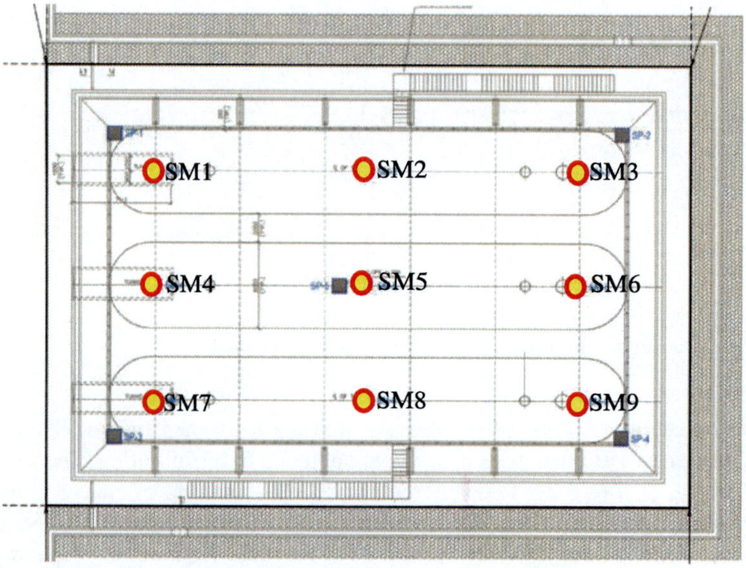

Fig. 7 Plan of settlement monitoring plates

Fig. 8 Location of settlement monitoring plates at COL

Fig. 9 View of completed surcharge load

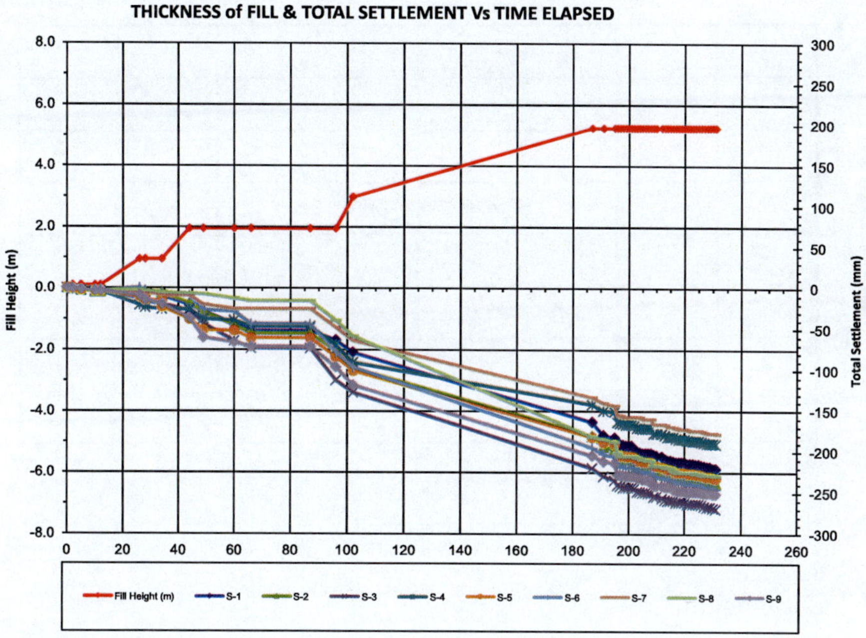

Fig. 10 Consolidation plot (Time Vs. Settlement) of preloading process period

of consolidation and predict final settlements of treated ground. As per the settlement monitoring data, the final settlements for 60% loading are around 230 mm (average settlements of monitoring plates), which is higher than the estimated settlements. The surcharge fill was unloaded after ensuring around 90% degree of consolidation by Asaoka's method as shown in Fig. 11. The settlements were governed by the soft silty clay layer. The initial estimation of settlements was based on the consideration that soil stiffness of soft silty clay would be around 100 * C_u. However, based on the field settlement data, the soil stiffness was back-calculated resulting ~55 * C_u – 80 * C_u soft silty clay.

At present, the unloading of surcharge activities is in progress at site. Further, the construction of storage mound will take place in stages followed by the hydrotest activities. Later the holding period of hydrotest load will be extended till achieving 90% degree of consolidation, thus ensuring the post-hydrotest long-term settlements is within the permissible range.

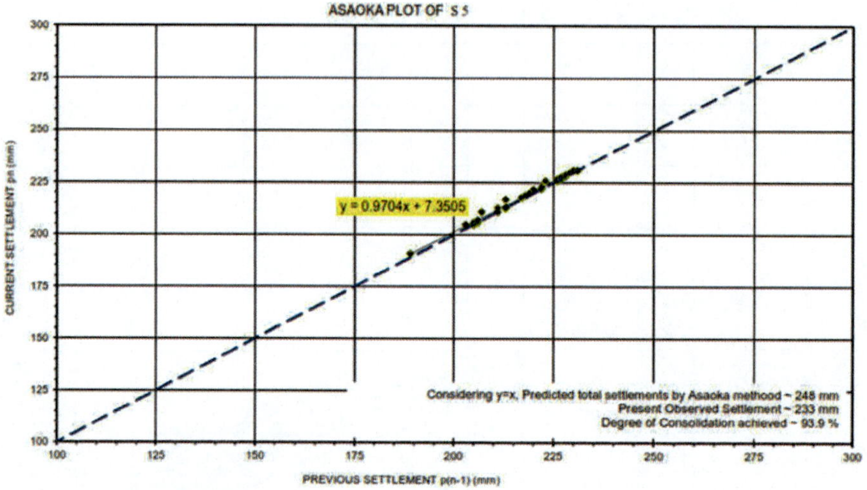

Fig. 11 Settlement prediction by Asaoka's method period

7 Conclusion

Consolidation settlement is the critical challenge for the construction of large storage structures on soft/compressible stratum, which can be effectively addressed by ground improvement using vibro stone columns. Further, the feasibility of vibro stone columns in such conditions could be very well enhanced by the adoption of preloading technique by utilizing the drainage characteristics of stone columns.

In this current study, the subsoil conditions were explored by Electric Cone Penetration Tests and ground improvement using vibro stone columns was adopted as an effective foundation solution for the given subsoil condition and loading intensity. Based on the consolidation analysis of treated ground, 60% of total loading intensity is considered as the preload. The consolidation settlement was closely monitored during the holding period, and the unloading was scheduled after ensuing 90% degree of consolidation for the surcharge load using Asaoka's method (observation approach). Further, the design considerations were back-calculated based on the actual settlement data. In addition to the contribution for settlement reduction, vibro stone columns also help to mitigate the liquefaction potential of the loose saturated sands present below the compressible stratum.

Thus, the combination of vibro stone columns with preloading will make an effective foundation solution for large storage structures like mounded vessels and tanks farm.

References

Asaoka A (1978) Observation procedure of settlement prediction. Soils and foundations

Barron RA (1947) Consolidation of fine grained soils by drain wells. ASCE 73(6):811–835

Gupta A, Sridhar V, Annam MK (2017) Optimal foundation solution for storage Terminal in Mangalore. In: Indian Geotechnical Conference

Han J, Ye S-L (2001) Simplified method for consolidation rate of stone column Reinforced foundations. J Geotech Environ Eng

IS 15284-Part 1:2003. Design and construction for ground improvement—Guidelines, Bureau of Indian Standards, New Delhi

IS 1893-Part 1:2016. Criteria for earthquake resistant design of structures. Bureau of Indian Standards, New Delhi

Priebe HJ (1995) The design of vibro replacement. Ground Engineering

Priebe HJ (1998) Vibro replacement to prevent earthquake induced liquefaction. Ground Engineering

Valluri S, Raj D, Raj D, Shukla V (2015) Vibro stone columns to support large oil storage tank farm on west coast of India. Adv Soft Ground Eng

Application of Deep Mixing and Stone Column Techniques for Stabilization of Large Earth Cum Rock Fill (ECRF) Dam for an Irrigation Project

Vidyaranya Bandi, Cemal Can Maslak, Karsten Beckhaus, and Vijaya Bhushan Rao

Abstract A Large Earth Cum Rock Fill, ECRF Dam, is under construction at Polavaram in the state of Andhra Pradesh in India as part of an irrigation project. The Embankment crest width of the dam is 12.5 m and the base width is approximately 136.75 m with a stepped sloping height of 27 m on both upstream/downstream sides. Once the 1.2 m Thick Plastic concrete cutoff wall is constructed below the dam along the longitudinal central axis to avoid seepage, this magnificent structure will be supported on a series of Deep Mixing Columns and Stone Columns that are evenly arranged under the footprint of ECRF Dam forming a smooth transition of gradually reducing stiffness moduli of treated ground from the centre of the embankment, where the foundation stresses are high, towards the edges in order to be able to progressively limit the settlements under the dam to a specified allowable limit. As such, the required post-treatment stiffness in the case of stone columns is 2 times the in situ stiffness of foundation soil, whereas it is expected to be 5 times the improvement for Deep Mixing as per project specifications. Laboratory testing and Field Trials were undertaken successfully to verify the design.

Keywords Earth cum rock fill dam · Deep mixing · Stone columns · Laboratory tests · Field trials

---

V. Bandi (✉)
Bauer Engineering India Pvt. Ltd, Chennai, India
e-mail: vidyaranya.bandi@bauer-india.com

C. C. Maslak · K. Beckhaus · V. B. Rao
Bauer Spezialtiefbau GmbH, Schrobenhausen, Germany
e-mail: Cemal-Can.Maslak@bauer.de

K. Beckhaus
e-mail: Karsten.Beckhaus@bauer.de

V. B. Rao
e-mail: Vijay.Bhushan@baueruae.ae

A. M. Krishna et al. (eds.), *Deep Foundations for Infrastructure Development in India*, Lecture Notes in Civil Engineering 373, https://doi.org/10.1007/978-3-031-37256-8_14

Fig. 1 Location of ECRF DAM at Polavaram

1 Introduction

The Government of Andhra Pradesh Irrigation and Command Area Development (I&CAD) proposed to construct a multipurpose dam across the river Godavari near Polavaram village, Andhra Pradesh, India (Fig. 1). Bauer is the subcontractor of the project to build the Earth Cum Rock Fill (ECRF) dam for a length of 1750 m across the Godavari River. Ground improvement using Deep soil mixing (DSM) is proposed by Bauer along the existing plastic concrete diaphragm wall (upstream side and downstream side) to ensure the overall stability of the ECRF dam by improving the stiffness of soils at GAP I and GAP II locations. Deep Soil Mixing (DSM) is introduced for the first time by Bauer in India at this project and it's a very sustainable solution replacing the traditional ground improvement techniques.

2 Need for the Installation of the DSM Column Below the ECRF Dam

The DSM columns will form a transition treatment between the existing Cut-off Wall and the proposed ground improvement by stone columns below the footprint for ECRF Dam. The DSM columns act as the transition of the deformation modulus of the ground from the centre of the dam to the toe of the dam embankment. The soil improvement works consist of Single Column Deep Soil Mixing (DSM) and shall be carried out along the installed Plastic Concrete Diaphragm Wall/Cut-off Wall in the clay sections as shown in Fig. 2. As the stiffness of the Diaphragm Wall is higher than the stone columns, this DSM treatment will provide intermediate stiffness by transition. Installation of DSM columns shall confirm the improvement of the initial soil stiffness by 5 times.

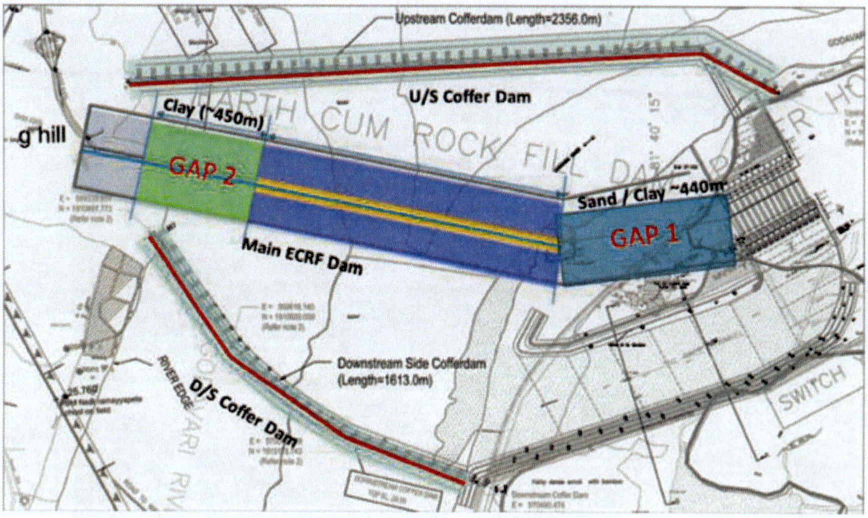

Fig. 2 Plan of deep soil mixing works at GAP I and GAP II

3 Designing the DSM Columns for Protection Works

For soil treatment beneath the rock fill embankment, Stone Columns (Deep Vibro Floatation Method) and Rigid Columns (DSM method) are used for ground improvement. Field trials are conducted near the GAP I area to establish the design parameters and mix the design of the DSM columns to be installed at the site. The data obtained from the trials shall be used for the execution of DSM column production at GAP I and GAP II.

4 Summary of the Geotechnical Data at the Job Site

Based on the available geotechnical information for the site, the summary of the grain size distribution at the locations of improvement by Deep soil Mixing (DSM) is summarized below in Table 1.

Table 1 Summary of the soil profile of the site

Location	Chainage, m	Clay %	Silt %	Sand %	Gravel %
GAP 2	1365–1195	34.3	49.8	7.5	8.4
GAP 2	1195–1025	24.4	46.1	22.2	7.3
GAP 2	1025–910	<50	<50	>50	>50
GAP 1	250–425	16.7	48.6	30.3	4.5

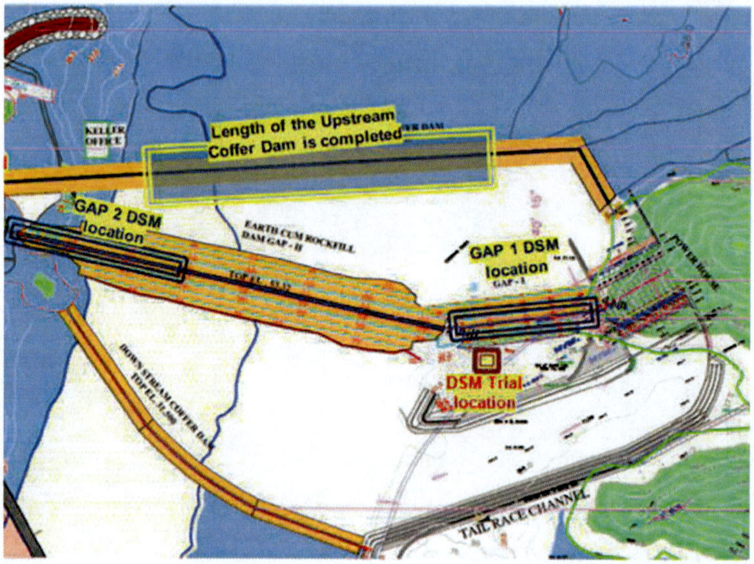

Fig. 3 Tentative location of the DSM works and proposed trial locations

4.1 Methodology for Trial Works for Deep Soil Mixing

Field trials are conducted prior to the commencement of the main works to ensure the performance of the proposed ground improvement scheme for both Vibro Stone Columns and Deep Soil Mixing columns. The tentative trial locations at the GAP I and GAP II are shown in Fig. 3. The results of field trials establish the correctness of the assumed material properties (elastic modulus) which were considered in the numerical modelling for the case of post-ground improvement scenario, in dam foundation design. The anticipated post-treatment stiffness in the case of stone columns is 2 times the in situ stiffness, whereas it is expected to be 5 times the improvement for deep soil mixing.

4.2 Trial Columns

The field trial test section includes the installation of 18 nos of DSM columns. The layout of the DSM columns is shown in Fig. 4. For a treatment area ratio of 40%, with a column diameter of $D_{\text{col}} = 1$ m and spacing between the columns in a squared grid is considered at $a = 1$ m. However, the plan is to instal trial columns with a 1.3 m spacing as per the tender specified for production. The quality control measures for the installation of the trial columns are specified in Table 2.

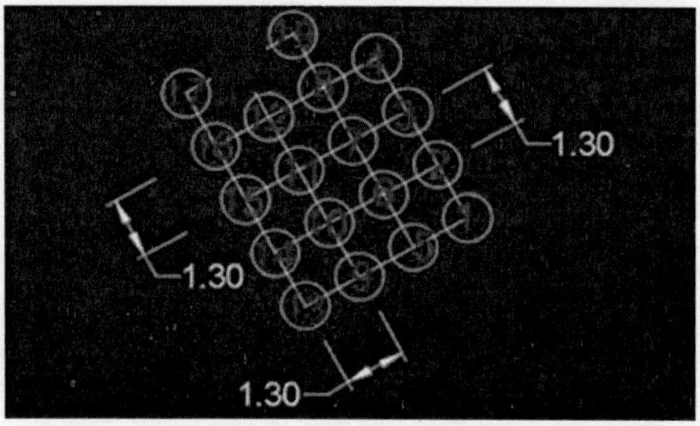

Fig. 4 DSM field trial columns setup

Table 2 DSM quality control requirements

ITEM	Description
Grout mix (W/C ratio)	1:1.2
Design stiffness	150 kPa
Target UCS of DSM columns	750 kPa
Amount of binding	150–300 kg/m^3 of Soil
Methods of collection of samples (at various depths)	1. Wet grab sampling method 2. 3 samples at various depths 3. Backflow sampling method
Tests conducted on DSM samples	1. UCS test 2. Stiffness measurement
Curing period of DSM samples	7, 14 and 28 days

4.3 Deep Soil Mixing Technology and Work Sequence at Field Trial Works and Main Works

The Deep Soil Mixing technology implies mixing of existing soil with cement or compound binder (Lime, Fly Ash, Slag, etc.,) and aims to achieve significant improvement of mechanical properties of the soil. The improvement becomes possible by strengthening with cement, bonding of soil particles and/or filling of voids by chemical reaction products. As a result of this mixing, soil-cement grout columns with higher strength than the natural soil are created. Along with improved strength, due to the known constitutive mechanical equation for cement-based materials, the DSM columns will have an improved stiffness (FHWA-HRT-13-046).

Regarding the homogeneity, it shall be noted that soil particles of fist to cobble size will remain unmixed in the DSM material which is subject to mechanical mixing with the DSM mixing tool but is not deemed to compromise the column stiffness significantly. Major advantage of Deep Soil Mixing technology is reduced transportation of the construction material and use of in-situ soil for improvement. The mass of excavated material and spoil that must be treated, transported, hauled, and

Fig. 5 Equipment for deep soil mixing technology

dumped off site is also significantly reduced. This makes the method environmentally friendly as well. The complete process of the Deep Soil Mixing Technology is depicted in Fig. 5 (Kitazume and Terashi 2012). The deep soil mixing works are executed by a single auger known as Single Columns Mixing (SCM). Installed DSM trial columns will not be removed from the site after testing.

The wet method of Deep Soil Mixing is carried out by setting up a drill rod with a DSM mixing tool at the column position. The DSM tool penetrates to the termination depth of treatment disaggreagating the soil and blending the soil with water and binder material. The process is continued till the auger reaches the final depth of treatment. The binder slurry is injected into the soil during withdrawal; optionally by additional downward and upwards strokes, where the mixing tool rotates in the horizontal plane and mixes the soil and the binder. The final column is formed in the ground as the auger is retrieved. The construction sequence of the DSM Column (FHWA-HRT-13-046) installation is shown in Fig. 6.

4.4 Methodology of Collection of Samples at Various Depths

The samples of soil mixing will be collected using both the wet grab sampling method and the backflow sampling method for better quality. The wet samples are collected using Wet Grab Sampling Box, typically attached as extension to the to the excavator, piling rig or crane. The sampler is pushed into the freshly mixed columns to collect in-situ blended soil sample for sample collection at different depths. The wet sample is cast in standard cylindrical moulds in ratio of 2:1 (L:D) atleast 3 samples at same depth. It is suggested to collect the samples from every column at 3 different depths and it covers every 1 m depth among all columns. The max depth of sample collection by wet sampler is limited upto 12m from the ground level. Figure 7 shows the typical

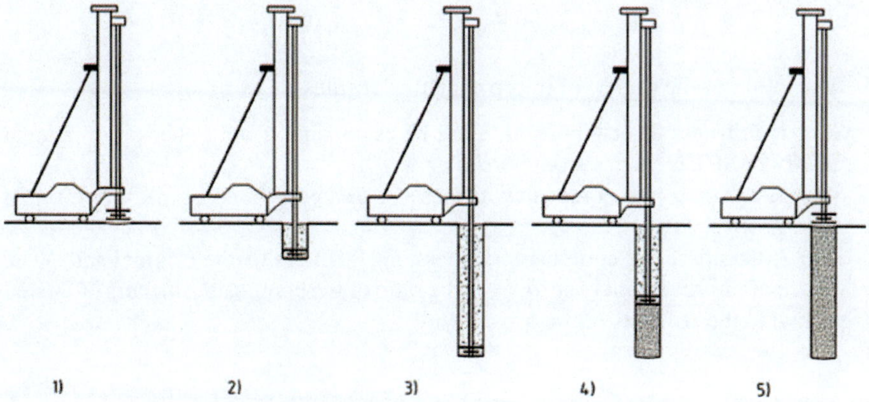

Fig. 6 Construction sequence of DSM columns

Fig. 7 Typical pictures of wet grab sampler

setup of a wet grab sampler. In any condition, wet grab samples are not able to collect from the desired depth, whereas backflow samples shall be collected from the surface for testing purposes. Backflow samples refer to the sludge generated during the installation of the wet mix column. Samples are collected from the sludge, and cast in standard mould (cube or cylinder). At least 3 nos of samples shall be collected from all columns after installation of DSM columns for necessary stiffness testing.

4.5 Assessment of Improvement

The improved composite stiffness after treatment shall be calculated after achieving the actual DSM column stiffness values. Improved composite stiffness of treated ground was established using the below set of formulas (Kitazume and Terashi 2012).

$$E_{comb} = A_r E_{50,col} + (1 - A_r) E_{50,soil} \quad (1)$$

Estimation of the degree of improvement in stiffness is as follows:

Step 1: Stiffness of original soils shall be established from either pre-treatment ECPTs or SPT N values from boreholes.
Step 2: Estimate the composite stiffness of treated ground as per the formula shown above.
Step 3: Compare the composite stiffness and stiffness of the original soils. The degree of improvement in stiffness is the ratio of the composite stiffness of treated ground to the stiffness of the original soils.

4.6 Main Works

As a result of the performed field trials, a variation of mixing slurry and other construction parameters are examined, which cover the construction parameters necessary for the construction work, i.e., Need to allow observing the dependence of the construction procedure and mechanical properties of installed columns. At main works, installing numerous DSM elements with previously established specific mixing parameters will be used to achieve the specified mechanical properties over the column depth and cross section. For this, slurry mix designs and mixing parameters obtained from the trial field shall be used.

In main works, the same equipment proposed for the trial field shall be used for the installation of the DSM Columns. Based on the field trial results, the mix design and construction procedure shall be defined to achieve treated soil meeting the project-specific acceptance criteria.

4.7 Post-performance Criteria for DSM Works

The whole DSM Field Trial programme is aimed to establish all necessary DSM parameters to achieve the following post-performance criteria for the DSM material. Based on the results of the trials field and considering aspects of required stiffness and compressive strength, the following requirements on DSM material were suggested by Project Authority. The required unconfined compressive strength (UCS) shall be proved as a characteristic value, tested at a cylindrical specimen with minimum dimensions of 100 mm, at the age of 28 days shall be within 0.5–1.0 MPa. The resulting average unconfined compressive strength shall correspond with an average design stiffness of the DSM columns respectively a deformation modulus of DSM material E within 40–135 MPa. This deformation modulus is supposed to be evaluated as a tangent modulus from the stress–strain graph from the UCS test. A correlation shall be assessed from a representative series of test samples taken from UCS samples.

5 Details on the Design and Execution Works

For the design study of Deep Soil Mixing (DSM) columns in the Polavaram project, the layout of columns in field for GAP I is shown in Fig. 10 for Main Works and Transistion Zone, respectively. The goal was to achieve the desired characteristic compressive strength f_{cm} of the DSM material in a defined grid. The general idea was to achieve a composite modulus for a lower bound of 50 MPa and an upper bound of 150 MPa. A transition zone that allows for a smooth and gradual transition of settlement between the areas treated with DSM columns and treated with stone columns is also implemented along with the lower and upper bound. To allow a mutual understanding of the effect of the soil improvement, a numerical model of the dam and an idealized underground situation has been set up in Plaxis 2D. Theoretical settlements have been calculated for the situations with the following features: (1) Lower bound and (2) Upper bound of treated soils, considering the differently treated areas perpendicular to the dam (or cutoff wall) axis, i.e., across the DSM column zone, transition zone, and stone-column zone. Although the total settlement values can be read only qualitatively, the comparison performed may help to decide on the necessary transition between DSM and vibro stone columns to smooth the incremental settlements across the dam.

5.1 *Transition Zone*

To provide a smooth transition of settlements between the DSM columns and stone columns, a consistent or gradual decrease in oedometric modulus ($E_{oed,composite}$) must be maintained. The average values of the lower and the upper bounds between the

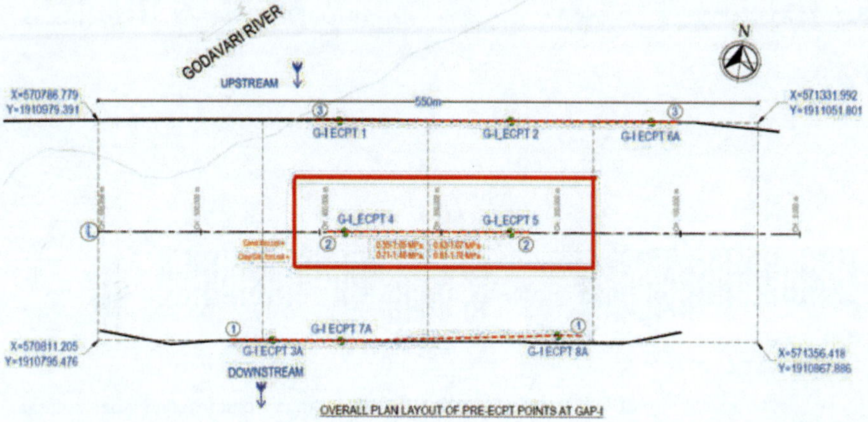

Fig. 8 Layout for GAP I with the range of f_{cm} values

oedometric modulus of stone columns and DSM columns results in the range of 32.5–82.5 MPa (Fig. 9).

The length of the Transition Zone is taken to be 10 m, which allows 5 rows of DSM columns between the stone columns. To utilize the transition zone to the optimal capacity and allow the stiffness modulus to shift along the profile, the first three layers of DSM columns in the transition zone are taken at their original depth of 12 m, whereas the depths of the last two layers are reduced to 6 m. The detailed summary of the DSM works in the main section and transition zone are provided in the Table 3 (Fig. 10).

Fig. 9 Transition zone visualization

Table 3 Overview of soil improvement dimensions

Particulars	DSM up to 10 m wide	Transition zone				
		1st	2nd	3rd	4th	5th
Spacing, m	1.3	2	2	2	2	2
Diameter, m	1	1	1	1	1	1
Length of the DSM column, m	12	12	12	12	6	6
Length of the VSC column, m	17	17	17	17	17	17
Area column, m	0.79	0.79	0.79	0.79	0.79	0.79
Ratio of replacement, %	46.47	19.63	19.63	19.63	19.63	19.63

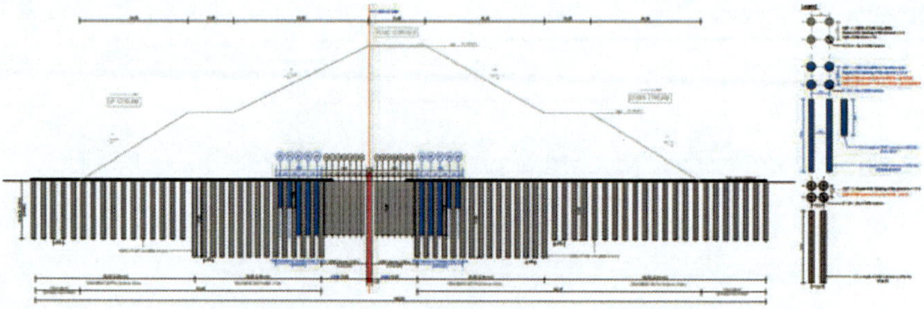

Fig. 10 Section Detailing of DSM, transition zone (DSM + Stone Columns) and stone columns for GAP I

Table 4 Soil parameters

GAP-I	Model	E_{50}^{ref} (KPa)	E_{oed}^{ref} (KPa)	E_{ur}^{ref} (KPa)	$C_{ref'}$ (KPa)	Φ' (phi)	Rinter
Silty clay 1	HS	7500	7500	22,500	10	10	0.67
Silty clay 2	HS	9000	9000	27,000	17.50	30	0.67
Rock bed	HS	200,000	200,000	600,000	50	30	0.67
Shell and core	Linear elastic	γunsaturated = 19 kN/m^3					0.67

5.2 *Introduction to Soil Stabilization Analysis*

The aim of this study is to analyze the settlement of the soil after treatment with DSM columns and stone columns. A Plaxis 2D model was created to analyze the settlement in the soil of GAP I. The settlements in treated soil, after construction of the dam without water filling and with water filling. This allows us to compare the effect of both cases. Four stages of construction that are considered in this analysis are initial soil, soil treatment, dam construction and water filling are considered in this analysis. The summary of the soil parameters considered for the design analysis is presented below in Table 4. For soil treatment, the DSM columns are studied with a lower bound E_{oed} value of 50 MPa and an upper bound E_{oed} value of 150 MPa. The composite E_{oed} Stone Columns have been given as 14.77 MPa.

5.3 *Soil Settlement*

The settlement along the cross section of the treated soil is checked for the understanding or influence of soil treatment. The study is conducted for two scenarios, considering the settlement of soil at a cross section with lower bound values and the settlement of soil at the same cross section with upper bound values. Scenario 1—Lower Bound—Figs. 11 and 12 show the settlements of treated soil before filling the water respectively in GAP I.

Scenario 2—Upper Bound—Figs. 13 and 14 show the settlements of treated soil before filling the water and after filling the water, respectively, in GAP I.

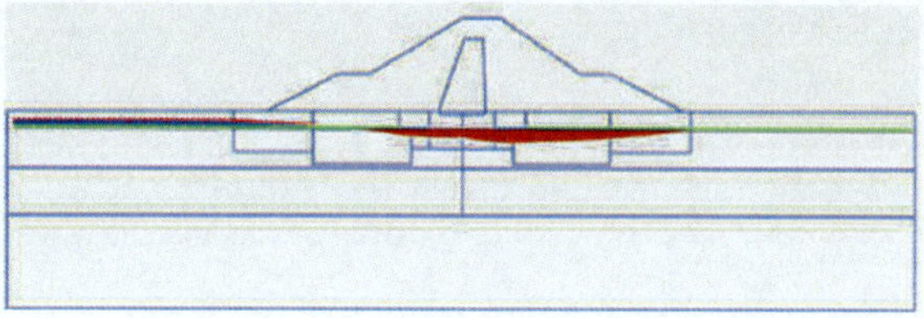

Fig. 11 Lower bound—maximum settlement before filling the water with the dam is 48 cm (indicative)

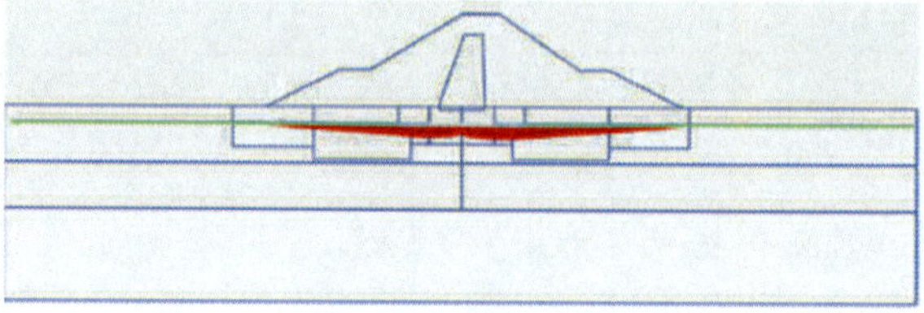

Fig. 12 Lower bound—maximum settlement after filling the water upstream is 50 cm (indicative)

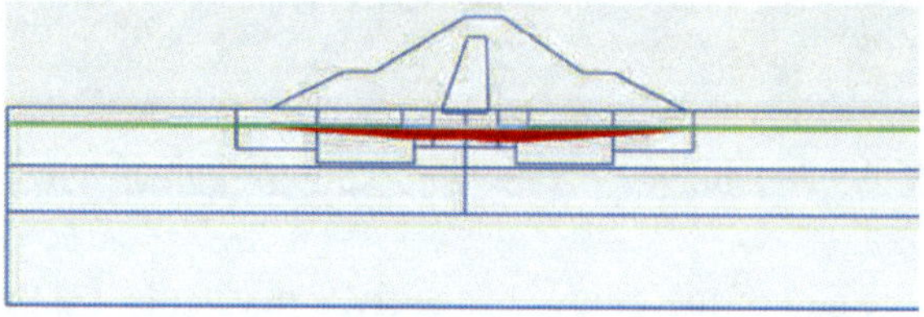

Fig. 13 Upper bound—maximum settlement before filling the water with the dam is 46 cm (indicative)

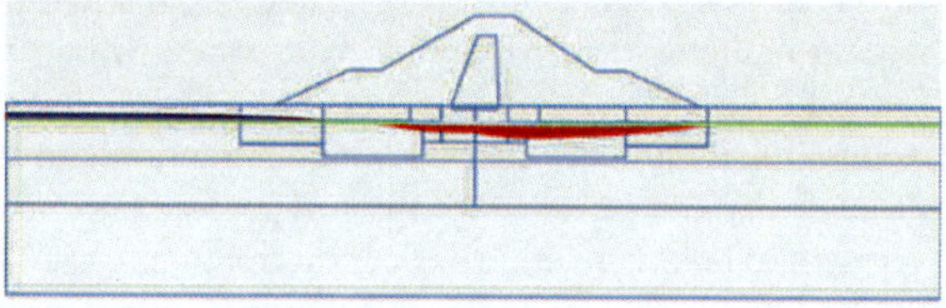

Fig. 14 Upper bound—maximum settlement after filling the water upstream is 48 cm (indicative)

6 Conclusion

In this paper, the use of the Deep Soil Mixing (DSM) technique for the ground improvement beneath the dam foundation is presented. Detailed design analysis is carried out to estimate the settlements under different conditions like the dam being empty and flooded to the high flood levels. DSM acts as a reinforcement element in providing protection to the plastic concrete cutoff wall along the axis of the dam, thus reducing the settlements and improving the deformation modulus of the ground. The paper describes the installation of the Deep Soil Mixing (DSM) methodology for ground improvement introduced for the first time in India. The ground improvement by DSM is a sustainable solution and offers an alternative solution to the traditional ground improvement solutions by stone columns and other techniques.

References

Federal Highway Administration Design Manual: Deep Mixing for Embankment and Foundation Support. Publicatoin No. FHWA-HRT-13-046

Kitazume M, Terashi M (2012) The deep mixing method. CRC Press, Tokyo, Japan

Ground Improvement By Controlled Modulus Columns—A West Africa Case Study

S. Vinothkumar and Sunil S. Basarkar

Abstract Ground Improvement Techniques serve to improve engineering properties for supporting various foundations. These may be undertaken either by conventional or advanced methods. The Controlled Modulus Column (CMC), also known as Rigid Inclusion, is a soil improvement technique to reduce the total and differential settlement by using semi-rigid soil reinforcement columns. The CMC technique can be executed by displacement or non-displacement methods. It does not aim to bypass the compressive layer or to instal a pile that will directly support the entire load from the structure. The principle of the technique is to introduce a load transfer platform between the head of the column and the structure, and transform a uniform load into the soil. Initially, the structural load transmitted into the soil generates a settlement greater than CMC piles. When the settlement between the soil and column became equal at the neutral point, the maximum structural load shall be balanced by the positive friction and tip resistance to stabilize the foundation without exceeding the allowable settlement limit. This paper describes the behaviour of soil reinforcement with CMC and their installation process with reference to a project at the Dry Bulk Terminal Port at Western coast of Africa. The methodology of CMC, requirements of Ground Improvement, arrangements and outcome of this method are presented in this paper.

Keywords Geotechnical investigations · Ground Improvement · Controlled modulus column · Pile load test

1 Introduction

Authors' organization (herein referred to as EPC Contractor) had been awarded a marine project at a port city along western coast of Africa for the construction and expansion of the dry bulk terminal at the existing port location. The port is located at 4°44′ 38″ N Latitude and 6°37′4″ W Longitude. The scope of the project mainly

S. Vinothkumar · S. S. Basarkar (✉)
Afcons Infrastructures Limited, Mumbai, India
e-mail: sunil.basarkar@afcons.com

A. M. Krishna et al. (eds.), *Deep Foundations for Infrastructure Development in India*, Lecture Notes in Civil Engineering 373, https://doi.org/10.1007/978-3-031-37256-8_15

comprised berth construction, reclamation and dredging works, construction of buildings, truck unloading/vehicle manoeuvring area stockpile and pavements. The project client provided a ground investigation report to all tenderers to invite their proposals. Post-award of the contract, the EPC Contractor carried out subsequent geotechnical investigations which indicated ground conditions more critical than those interpreted from the tender information. Fresh investigations indicated a significant presence of soft clays which warranted steps and measures to ensure the stability of reclamation and earthworks by ground Improvement works and achieve agreed acceptance criteria with respect to post-construction settlement for the earthworks.

2 Geotechnical Background

The tender stage and post-award stage investigations comprised a series of boreholes and a limited number of CPTs to delineate variations in ground conditions across the site and provide information to produce a suitable earthworks and structures design. The boreholes considered relevant to each critical earthwork are detailed in Table 1.

Table 1 Zones for ground improvement measures at the backyard area

Zone ID	Location	Pre-tender BH	Post award BH	Load kPa	Area m^2	Computed Pre-GI settlement, mm	Permitted settlement mm
3A	Truck parking—1	MBH-9	MBH-G	25	2160	675	50
3B	Stockpile	–	MBH-H	100	2200	741	50
3C	Truck parking—2	–	MBH-H	25	1020	645	50
3D	Workshop area	–	MBH-I	70	3523	1196	50
3E	Non-berth revetment	–	CPT 1,2,4&6	10	8163	529	50
3F	Rigid pavement on reclamation fill	MBH-4&6	CPT-9, 14, 16	25	12,895	645	50
3G	Berth-2 revetment	LBH-3	MBH-S	10	1167	698	50
3H	Other areas near workshop	–	MBH-I	10	613	1196	50
3 J	Truck parking	–	CPT-11	25	1200	790	50

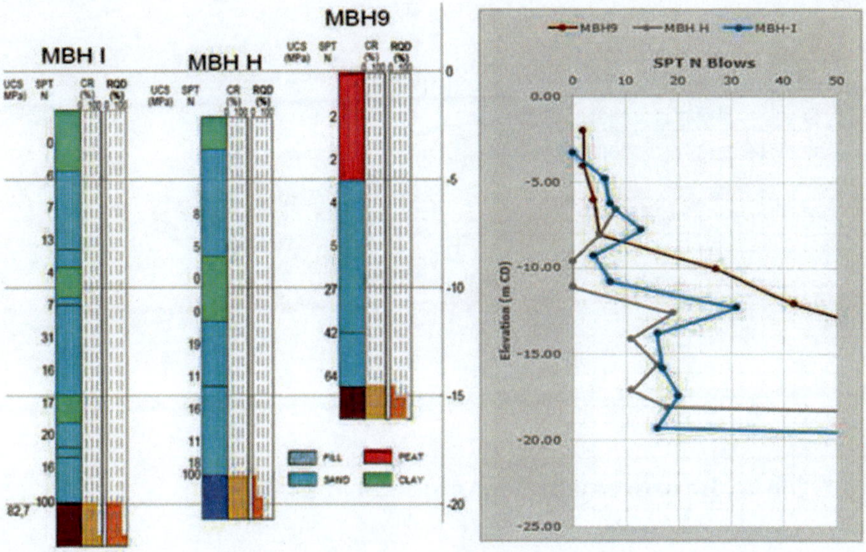

Fig. 1 Comparative bore logs of tender and post-award stage at 3A zone CMC location

Inadequate mapping of the sub-surface strata during the tender stage geotechnical investigations was evidenced as per the following information:

(a) The presence of soft clay and peat deposits were not completely captured in most of the tender stage boreholes (MBH9, Fig. 1) at various depths above the rock. These soft layers were very vivid in the confirmatory investigations (MBHI & MBHH, Fig. 1).
(b) The tender stage boreholes consistently measured higher SPTs over the top 5 to 10 m of the residual soil deposits compared to the confirmatory boreholes. Without a supplementary investigation method in the tender investigation, there was no way of confirming the authenticity of the SPTs.
(c) The complete absence of the measurement of consolidation parameters either using laboratory testing or in situ testing such as piezo-cones meant that the tenderers could not reliably estimate time dependant settlement which would occur during and post-construction.
(d) The limited scope of in situ stiffness measurement (no pressuremeter tests conducted within the planned area) meant that soil stiffness parameters could only be predicted using SPT and Index tests. These factors resulted in a significant under-prediction of settlements. Predicted stiffness parameters from SPT were observed to be high compared to the pressuremeter.

A review of ground investigation exercises after the award of work indicated that the information provided at the tender stage was inadequate to produce a safe design and the EPC contractor proposed ground improvement work on various zones as indicated in Fig. 2.

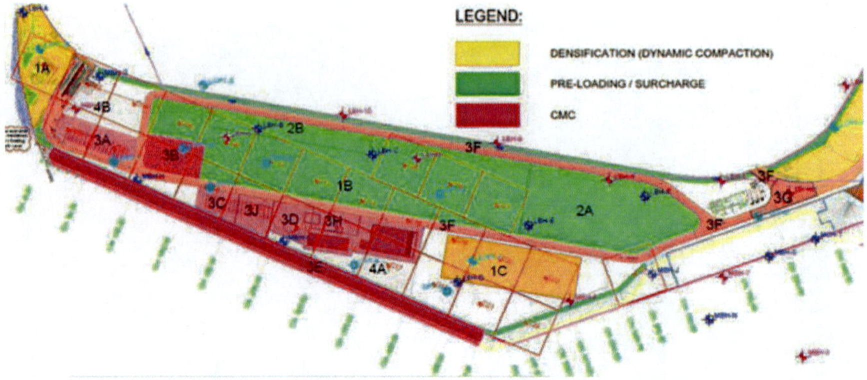

Fig. 2 Layout ground improvement works and associated ground investigation

3 Ground Improvement Technique Proposals

The project location witnessed the following sub-surface scenario requiring deep ground improvement and shallow compaction measures:

(a) ***Reclaimed fill improvement (coarse-grained material)***

The natural ground had to be reclaimed to +4.00 m CD with fill height ranging from 5 m to 8 m. Densification of the reclaimed fill thickness all over the project had to be ensured wherever reclamation is carried out. This kind of improvement was mandatory irrespective of ground conditions to ensure that reclaimed fill did not undergo settlement beyond acceptable limits (refer Table 1) under working loading. Reclamation fill was compacted by Vibro-rollers.

(b) ***Existing loose sand layer under loaded areas (coarse-grained soil)***

Densification of the identified sand layers was desired to address the settlements before the structural loads were applied. Such layers were improved by densification using shallow compaction and deep vibro-compaction. Alternative ground improvement approach also included pre-compression where the sub-surface layers were subjected to higher load through surcharging and maintaining the load till the settlement rate was within limits (Asaoka 1978).

(c) ***Soft compressible layers (fine-grained)***

Improvement of identified soft compressible layers was intended to limit the long-term settlements. This category of ground improvement was planned using Rigid Inclusions or what is referred to in this paper as Controlled Modulus Columns (CMC). CMC technique is a ground modification technique, planned for Zone 3, in which soft compressible, or unstable soils are treated in situ by means of grouted piles for strengthening (enhancing the physical and mechanical properties) the soil layers to arrest global and differential settlements by relieving the soil from excess load imposed by the structure within acceptable criteria.

4 Controlled Modulus Columns (CMC)

The CMC method was developed in France to support structures such as warehouses, industrial buildings, medium-weight housings, roads, railways, embankments, and storage tanks. CMC method is predominantly used for sites with soft cohesive soils, loose sand, chalk, organic soil and peat. The current practice is to instal columns with diameters between 280 and 500 mm. The principle of this soil Improvement system is to form a composite grout column surrounding soft soil material. More details about the CMCs method are the subject of publications (Plomteux and Liausu 2007; Racinais and Plomteux 2011; Racinais et al. 2016). The CMC technique is intended to globally improve the sub-soil performance and reduce its deformability by using semi-rigid soil reinforcement columns. The entire process is vibration free and provides for cleaner project sites.

A schematic of CMC is shown in Fig. 3. Under uniformly loaded structures such as embankments and slabs, load transfer is developed through a load transfer platform (LTP) installed between the head of the columns and the structure to be supported. LYP is made of well-compacted granular material and has a minimal thickness of 0.4 to 0.8 m depending on the type of structure and soil conditions.

Two methods of CMC execution are used, namely, drilled and displacement methods:

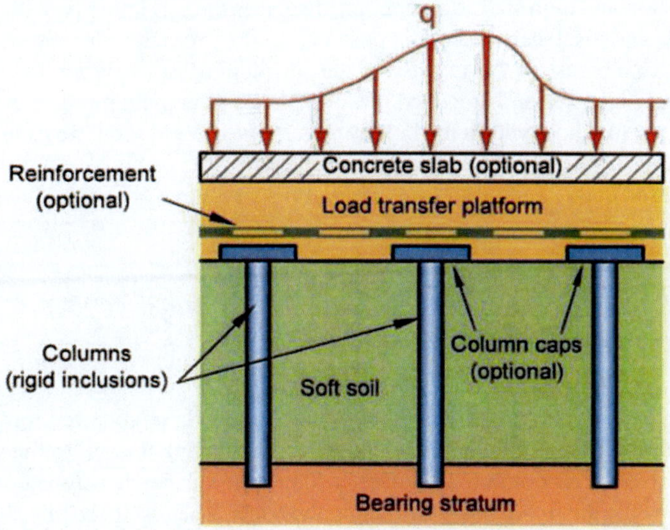

Fig. 3 Schematic principle of CMC

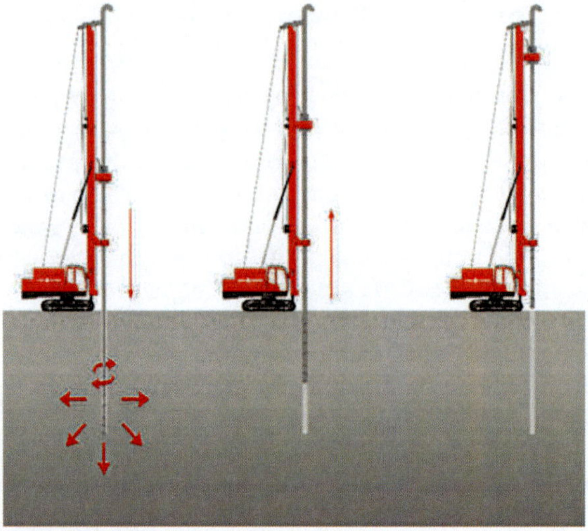

Fig. 4 Displacement CMC

4.1 Displacement CMC (Fig. 4)

A specially designed tool, powered by equipment with a large torque capacity of about 180 kN.m and high static down thrust, displaces the soil laterally, with no spoil or vibration. The displacement tool is pushed into the ground to the required depth resulting in an increase in the density of the surrounding soils. When the required depth or a refusal criterion is reached, a workable grout-cement mixture is pumped through the annular hole within the hollow auger. The cement grout flows under low pressure out of the auger base as it is retracted. The result is a high-capacity column that can be used even in close vicinity of sensitive structures.

4.2 Drilled CMC Using Continuous Flight Auger (CFA) Drilling Equipment (Fig. 5)

The system uses a CFA powered by a hydraulic piling rig of required torque. The auger is drilled into the soil while simultaneously extracting the soil to the working platform surface. The column is drilled to the required depth before concreting commences while simultaneously extracting the auger. The result is a pile shaft-type column that is effectively bonded to the surrounding soil.

Fig. 5 Drilled CMC using CFA rig

5 Design and Modelling Philosophy of CMC

The design of the CMC under static loading requires the use of the long-term modulus for concrete, calculated according to the following:

$$E_Y = 3700 \cdot f_{ck}^{1/3} \quad \text{in MPa} \tag{1}$$

To calculate the maximum allowable compressive stress, based on Eurocode 2 norm (NF EN 1992-1-1, Design of concrete structures—Part 1–1), the allowable Ultimate Limit State design compression strength f_{cd} of CMC plain concrete/mortar is calculated using the following relationship:

$$f_{cd} = a_{cc}\left(f_{ck}/\gamma_c\right) \quad \text{in MPa} \tag{2}$$

where

f_{ck}: the characteristic compression strength of the CMC material (concrete or grout),
a_{cc}: the coefficient taking into account long-term effect on the compression strength and unfavourable effects, depending on the way the load is applied for unreinforced concrete (acc = 0.8)
γ_c : the partial factor for the CMC concrete ($\gamma_c = 1.5$).

The mechanical characteristics of the CMC adopted at the present location are presented in Table 2.

Table 2 CMC properties for design

Diameter	f_{ck}	EY	N	acc	
M	(MPa)	(MPa)	(–)	(–)	
0.4	15	9100	0.2	0.8	

Two-dimensional finite element software PLAXIS was used to assess the effect of the proposed CMC ground improvement. Axi-symmetrical models centred on a column were carried out to determine CMCs spacings and to meet settlement criteria. Whereas plane strain models were implemented to check CMCs integrity.

6 CMC Working Mechanism

The working mechanism of the ground improvement by CMC is detailed in Fig. 6. The CMC and the soil are firstly mobilized by the loads applied on the top ($Q_p(0)$ and q_{soil} respectively). Over the upper part of depth ($h < h_c$), the load transmitted to the soil, q_{soil}, generates a settlement higher than those occurring in CMC. This effect takes the form of a negative friction load, F_N, providing an increase in the CMC compressive load until reaching the maximum load Q_{max} at the neutral point N, corresponding to a critical height $h = h_c$, in which the soil settlement and the CMC settlement are equal. Below the neutral plane, the CMC settlement becomes larger than that in the soil, leading to the mobilization of the positive friction, F_p, and tip resistance, Q_p. The maximum axial load in the CMC, Q_{max}, shall be balanced by the positive friction and the tip load, that is

$$Q_p(0) + F_N = Q_p(L) + F_p = Q_{max}$$

The anchor length or termination level of CMC was calculated based on the Geotechnical information of the site location. At the site, several Static cone penetration tests at Zone 3 were performed and logged (Figs. 7 and 8) to assess the presence of soft layers and determine the probable termination of CMC.

Layer-wise design parameters are summarized in Table 3.

The CPT cone resistance plot indicated the presence of a soft clay layer up to −9.0 m followed by dense Sandy strata. This geotechnical information was favourable for the CMC technique since a non-displacement rig comprising CFA of length 15–20 m would be suitably deployed.

Geotechnical parameters and their basis of determination are enumerated as follows:

The Pressuremeter modulus, E_M is obtained from the CPT data according to Cassan (1988) correlation:

$$E_M = \alpha\left(qc - \sigma'_{v0}\right) \tag{3}$$

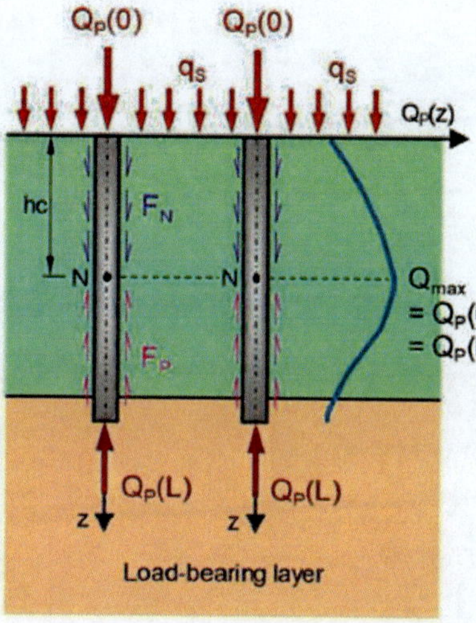

Fig. 6 CMC loading philosophy—positive and negative friction

where σ'_{v0} is the effective overburden pressure

α – 3–6 for the Clay layer, 3 for the Silt layer, and,

1–1.5 for the Sand layer.

The Young's modulus E_Y is determined from the pressuremeter modulus E_M using the correlation proposed by Menard (1975):

$$E_y = E_M / \alpha \tag{4}$$

where

α–2/3 for Clay layer, 1/2 for Silt layer and, 1/3 for Sand layer.

The relationship between oedometric modulus E_{oed} and the Young modulus E_Y is derived from principles of mechanics as

$$E_{oed} = (1 - v) \cdot E_y / [(1 - 2v)(1 + v)] \tag{5}$$

$$\text{where } v - \text{Poisson ratio, } v = 1/3 \tag{6}$$

With the above basis, layerwise values of geotechnical design parameters are determined and summarized in Table 3. The CMC pile was designed to terminate 0.5 m into the dense sandy strata.

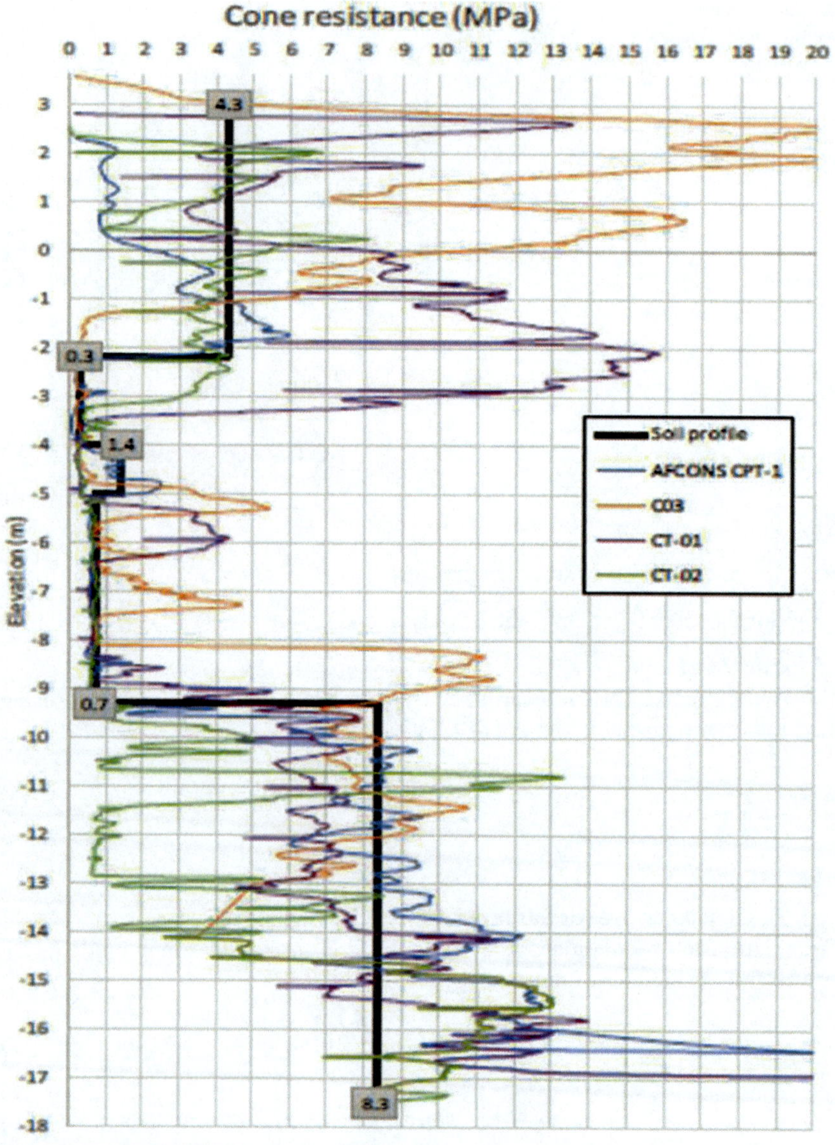

Fig. 7 CPT data (Zone 3A, CH 0 to CH 70)

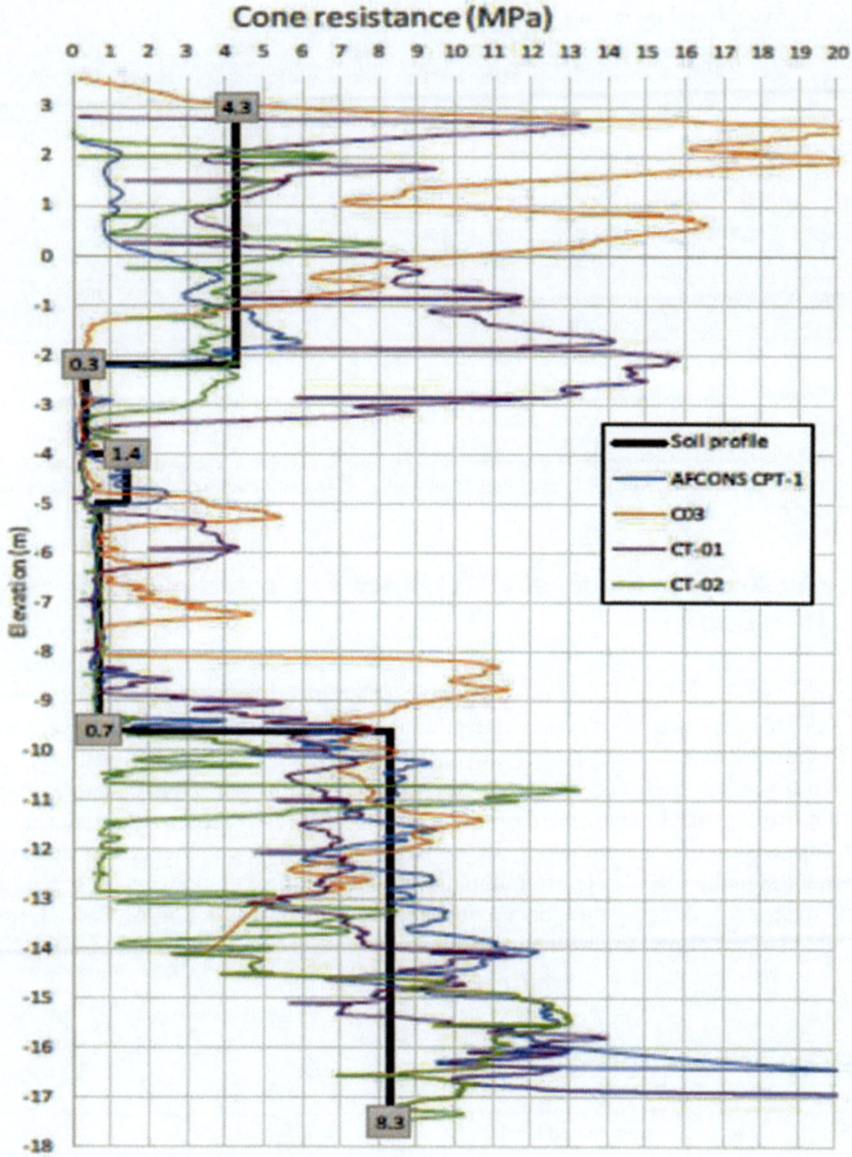

Fig. 8 CPT data (Zone 3A CH 70 to CH 100)

Table 3 Geotechnical parameters for zone 3A

Layer	Description	Depth m CD	q_t MPa	E_M MPa	E_y MPa	E_{oed} MPa	N –	γ kN/m^3	c' kPa	φ °
Layer 0	LTP		–	11.2	34	50	0.33	18	0	38
Layer 1	Reclamation fill	3.5 to –2.2	4.3	6.4	19.3	28.6	0.33	18	0	35
Layer 2	Soft clay	−2.2 to −4.3	0.3	1.4	2.1	3	0.33	17	3	15
Layer 3	Sandy silt	−4.3 to −5.3	1.4	2.9	4.3	6.3	0.33	20	0	24
Layer 4	Medium stiff clay	−5.3 to −9.6	0.7	3	9.1	13.4	0.33	17	8	20
Layer 5	Sands	−9.6 to −17.5	8.3	12.4	18.5	27.5	0.33	20	0	32

Notes q_t = tip resistance of CPT; γ = unit weight of soil; c' = drained cohesion; φ = angle of internal friction

7 Construction Stages of CMC Based on Geotechnical Investigations

Working stages, referred to as phases are presented in Table 4 and Fig. 9.

The CMC rig was positioned at the pile location and boring activity commenced with CFA till dense sand or stiff clay layer determined by CFA refusal (refer to Table 3) was reached. The concrete pump is connected by a flexible concrete hose to the drilling machine. After the arrival of the concrete RMC truck, and under the supervision of the pump operator, concrete pumping commenced. The drilling operator signalled to commence pumping of concrete through the annular space in the auger stem. All real-time data associated with the construction of CMC were recorded in the digital log which included the drilling rates, lifting rate of CFA, pile shaft formation, and concrete flow speed (Fig. 10).

Table 4 Stage construction

Stage #	Description	Ground level [m CD]
Phase 0	Existing ground level	−1.6
Phase 1	Reclamation works	3.5
Phase 2	Installation of a working platform	3.5
Phase 3	CMC works	3.5
Phase 4	LTP and Pavement construction	4.08
Phase 5	Live load—25 kPa	4.08

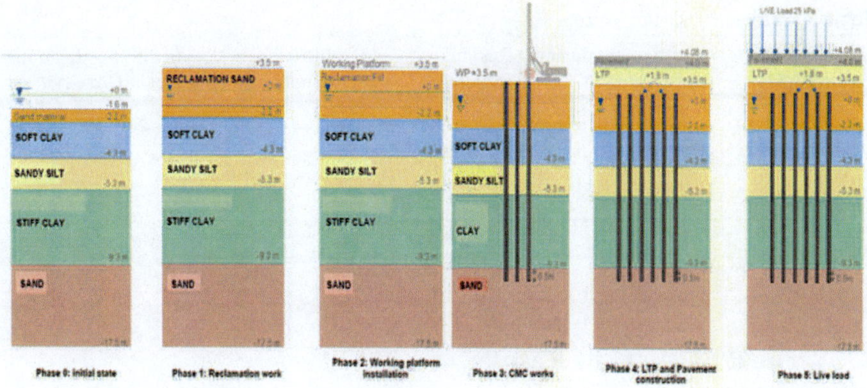

Fig. 9 Work sequence phases for zone 3A CH 0 to CH 70

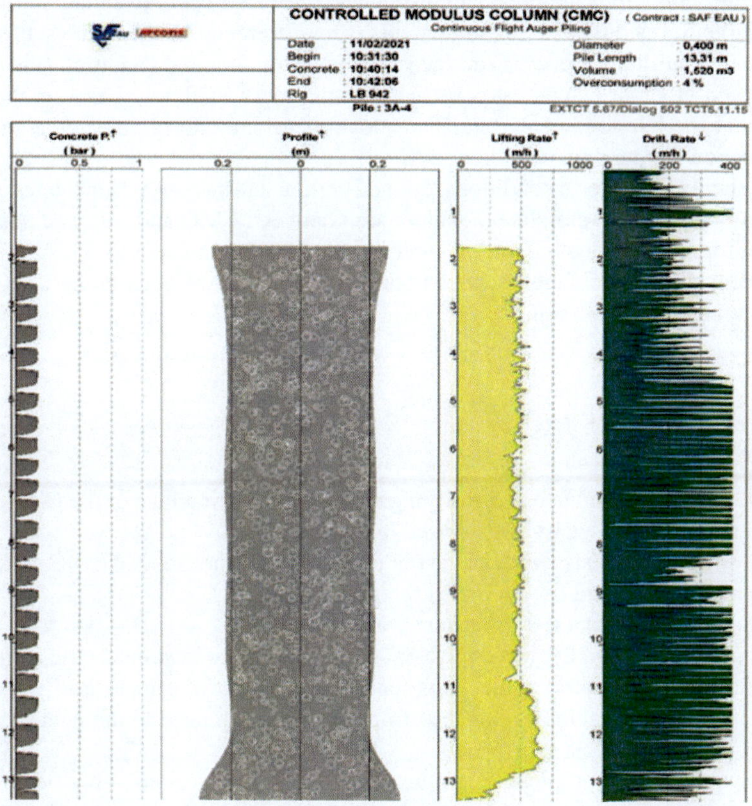

Fig. 10 CMC construction AMU display unit for 3A zone

Table 5 Settlement results for zone 3A

Zone	Chainage	Settlement during construction	Residual settlement (*) (1)	Residual settlement criteria (2)	Is the specification met? (1) < (2)?
		Undrained condition	Drained condition		
	CH (m)	mm	mm	mm	
3A	0–70	6.7	25.1	50	Yes
	70–100	6.6	25.0	50	Yes

8 Vertical Settlement Results

Table 5 presents the results of CMC-based ground improvement analyses as proposed by the EPC contractor based on soil profile and with applicable live load:

In Table 5, residual vertical settlement is the difference between the settlement in drained condition (Phase 5) and the settlement in undrained condition (Phase 4). As covered in Table 1, the post-construction settlement shall not be greater than 50 mm under the live load of 25 kPa. The above-mentioned design meets the project specification for the revetment zone 3A.

To establish long-term conditions, drained parameters have been considered in the calculation model. To simulate the interaction between CMC and soil, and possibly account for displacement effects, an interface element had been used. The use of the interface in FEM calculation creates node pair, one node belongs to the CMC and the other belongs to the soil.

9 Load Test on CMC

As part of quality assurance and verifying the design working load of the CMC, load tests were planned at each zone referred to in Table 1.

The vertical load tests were carried out on completed and isolated CMC columns. The maximum test load Q_T applied is reported in Table 6.

Following CMC load test procedure was adopted:

The design test load Q_T was equivalent to the maximum load inside the inclusion at the level of the neutral plane. This load is the sum of the inclusion head load (Q) and the negative skin friction (F_N) from the inclusion head to the neutral plane (Fig. 11).

The plate transmitting load to the test CMC was placed directly on the column head. A layer of fine sand about 1 cm thick was sandwiched between the head of CMC and the plate to ensure a uniform distribution of load. A primary structural steel girder was then positioned on the plate (Fig. 12) and the level of the beam was adjusted with leveling screws from beam supports. The metre (comparator) was positioned

Table 6 CMC load test details

CMC ID	Chainage	CMC execution date	Length, m	CMC Diameter, mm	Load test execution date	Test load, Q_T (MT)
CMC 3A-110	CH 0–70	10-Feb-21	11.8	400	3-Mar-21	49
CMC 3C-51	CH 150–200	18-Dec-20	7	400	18-Feb-21	60
CMC 3D.9b-8	CH 300–350	4-Feb-21	13.5	400	6-Mar-21	62
CMC 3E.1-2-8	CH 6.6–120	23-Jun-21	15.8	400	28-Aug-21	48
CMC 3F-2-202	CH 0–120	12-Mar-21	15.2	400	7-May-21	49
CMC 3 J-1	CH 150–200	19-Dec-20	7.4	400	12-Feb-21	54

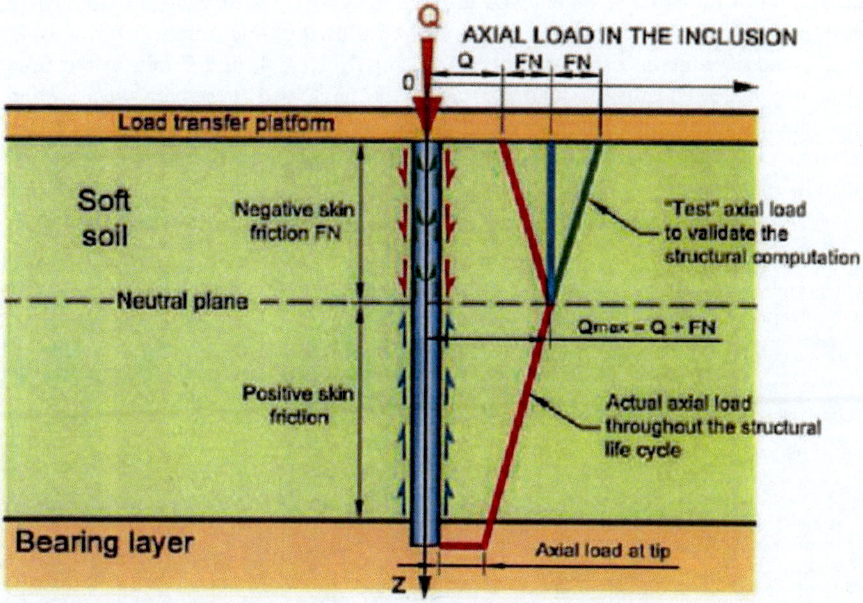

Fig. 11 Schematic of load distribution along CMC (ASIRI 2013)

at the second end of the beam. It was ensured that the rod of the comparator was in contact with the moving part of the beam. The cylinder was finally positioned on the plate and connected to the pump.

Fig. 12 Load test on CMC using Structural Girders

A first load cycle was equivalent to 10% of the maximum test load and maintained for a period of 15 minutes. This cycle is called "preloading". After unloading, the maximum test load that was achieved in increments of 25% of the test load, in five increments. Settlements were measured at the following time durations: 1, 2, 4, 8, 15, 30, and 60 mins at each loading stage; and 1, 2, 3, 4, and 5 min at the final unloading stage. During the loading stages, the next load increment was applied when the displacement was observed to be less than 0.02 mm/min on at least two consecutive measurements, subject to a minimum time of 15 mins. Otherwise, the entire loading was maintained for one hour (Fig. 13).

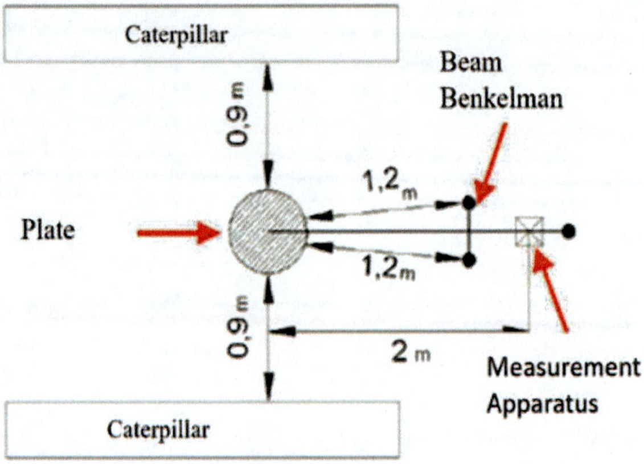

Fig. 13 Schematic of load test on single CMC using Caterpillar loader

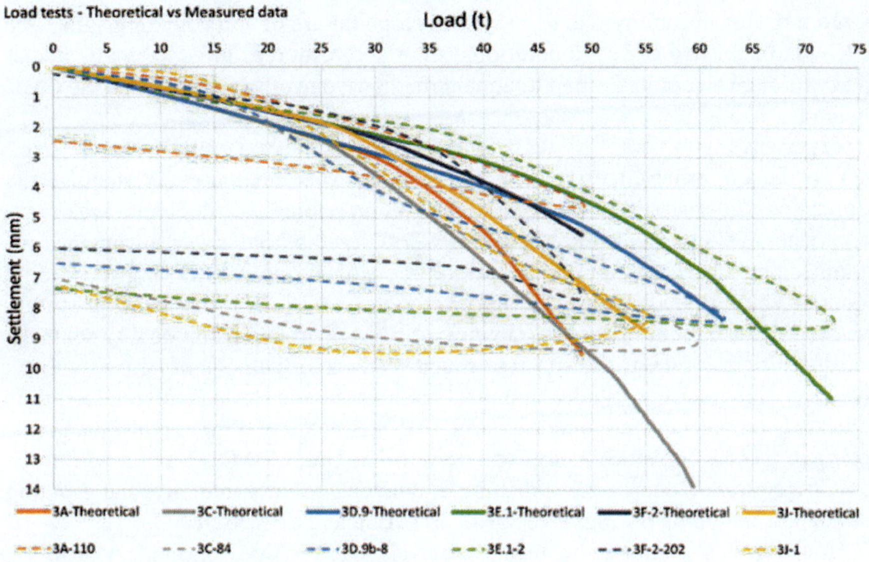

Fig. 14 CMC load test results at various zones

10 Load Test Results

Acceptance of the design criteria was verified through Finite Element Method (FEM) modelling using Plaxis 2D. The settlement result obtained from Plaxis 2D was compared to the field Plate Load Test results. From the geotechnical parameters, theoretical settlements of CMC 3A were compared to field values under different loads.

The pile load test was performed with 100% of the load on the top of CMC to confirm design criteria. LTP settlement was observed to be very low.

Maximum test load Q_T, corresponding to 100% of the design column load has been applied without ground or concrete failure. As per design provisions, the theoretically expected settlements under SLS uniform load were about 14 mm (Fig. 14). These test results thus validated the performance of the CMC to support embankment on weak ground.

11 Summary, Concluding Remarks and Lessons Learnt

In the foregoing case study, the inadequacy of tender stage geotechnical investigations in mapping complete sub-surface features resulted in missing out on weaker layers of soil within the stratigraphy. This under-estimated earthwork settlements and time expected for pore pressure dissipation during post construction. This, therefore, posed a risk of instability due to shear and slope failure of soft founding soils. The EPC contractor had to spend additional time and resources to address this inadequacy through elaborate ground treatment methods, one of which included the CMC method.

Experience with CMC installation indicated this method could meet the stringent settlement criteria to support embankments. This method could stabilize the foundation under stratified soil where the compressible layer thickness was more than 5 m. At the current project site, the CMC method has given advantages for commencing structural foundation works immediately after achieving the design concrete strength and vertical capacity through CMC load tests. CMC execution method afforded monitoring of parameters and hence offered better quality controls; and the progress rate at the site was much faster than its contemporary methods like Deep Soil mixing or stone columns. Quality assurance tests involving load tests indicated settlements were under control. There is better and more real-time monitoring of the column since Automatic Monitoring Units (AMU) housed within the CMC rigs present records of all drilling, alignment, column integrity, and grouting parameters including boring rates during lowering and extraction.

However, it is cautioned to rely excessively on the AMU on matters of piling levels. CMC cut-off levels (CoL), as in the present project, should be measured physically by sounding, since in several of the CMC piles, actual termination was seen to be much below the CoL. This omission resulted in unnecessary excavation works below the working platform and building up the concrete up to the CoL. This needless exercise incurred not only additional costs but also retarded the progress resulting in the extension of handing-over time.

There is at present, a single case of use of CMC technique in India. Experience at this project site of West Africa by author's organization indicates that this method could be fruitfully deployed under similar or suitable sub-soil conditions in India.

Acknowledgements The authors are thankful to the M/s AFCONS Overseas Singapore PTE Ltd. and Terminal Industrial Polyvalent De San Pedro for providing learning opportunities and experiences in this project. The authors are also grateful to their project associates COWI, AECOM and MENARD for technical assistance throughout the project execution of Ground improvement works.

References

ASIRI National Project (2013) Recommendations for the design construction and control of rigid inclusion ground improvement. Accessed from Presses des Ponts, Paris, France

Asaoka A (1978) Observational procedure of settlement prediction. Soils Found 18(4)

NF EN 1992–1–1 (2005) AFNOR–" Eurocode 2–Design of concrete structures Part 1–1: General rules for buildings

Cassan (1988) Les essais in situ en mécaniques des sols 1. Réalisation et interpretation

Menard (1975) The interpretation and application of Pressuremeter test results to foundation design Saulx-les-Chartreus (Essonne), France

Plomteux C, Liausu P (2007) Construction of flat glass float factory on loess deposits using controlled modulus columns in seismic condition. Accessed from 16th SEAGC

Racinais J, Plomteux C (2011) Design of slab-on-grades supported with soil reinforced by rigid inclusions. Accessed from EYGEC 2011. Rotterdam, The Netherlands

Racinais J, Thomas B, Ong R (2016) Twenty years of CMC successful application. Accessed from 19th SEAGC & 2nd AGSSEA. Kuala Lumpur, Malaysia

Stability Analysis of Micropile Reinforced Idukki Soil Slopes

C. P. Anjana and K. Ranga Swamy

Abstract The hilly regions of Kerala are severely affected by excess rainfall, causing tremendous damage due to landslides and slope failures. It is necessary to strengthen the existing slope profiles to prevent such disasters and avoid their risk to human lives and civil infrastructures. This paper studies the effect of micropile to reinforce the slopes in Idukki district of Kerala, where slope failures are frequent. Three-dimensional slope stability analysis is carried out on slopes from Idukki to obtain safety factor values using the MIDAS GTS NX program. The slope is fully saturated herein by taking the model's piezometric surface at the ground level to simulate high rainfall intensity. The critical slopes are then modelled with micropiles to analyze their effectiveness in improving stability under static conditions. Different micropile arrangements are considered, and a series of slope stability analyses have been carried out to obtain the optimum configuration with the required safety factor. The FOS values increased by more than 50% with micropiles for all the slopes. The optimum micropile arrangement obtained for each slope shows that micropile distribution between the toe and middle of the slope is an effective location for reinforcing slopes.

Keywords Stability · Micropile · GTS NX · Rainfall intensity · Piezometric surface

1 Introduction

Slope failures frequently occur in hilly regions of Kerala, especially during rainfall. In order to decrease the hazards of slope failures on human life and property, geotechnical engineers need to make the necessary plans. Understanding the slope

C. P. Anjana (✉) · K. Ranga Swamy
National Institute of Technology Calicut, Calicut, Kerala, India
e-mail: anjanalakshminarayanan94@gmail.com

K. Ranga Swamy
e-mail: ranga@nitc.ac.in

A. M. Krishna et al. (eds.), *Deep Foundations for Infrastructure Development in India*, Lecture Notes in Civil Engineering 373, https://doi.org/10.1007/978-3-031-37256-8_16

failure mechanisms is essential, and this may be accomplished by conducting a slope stability study and implementing the appropriate slope stabilizing techniques. Griffiths and Lane (1999) initiated the study on the finite element method (FEM) as a powerful alternate methodology for slope stability analysis compared to the conventional limit equilibrium approaches. The shear strength reduction method for finite element slope stability analysis was verified through field test data published by Matsui and San (1992). According to their findings, the FEM portrays failure mechanisms more accurately than the Limit Equilibrium Method. The primary cause of slope failures is the loss of matric suction in unsaturated soil caused by rainfall infiltration (Li et al. 2006). The exact cause was identified by Aslam and Sreekumar (2016) in their research on mechanisms of recent slope failures in hilly regions of Kerala.

Piles are usually used for slope stabilizations. Several researchers (Cai and Ugai 2000; Raouf 2019) have conducted experimental and numerical analyses on pile-reinforced slopes. Their findings demonstrate that piles significantly raise the slope's factor of safety (FOS). The centre of the slope is the best location for piles, and increasing pile diameter and reducing pile spacing increases slope stability. Micropiles are now preferred over traditional anti-slide piles because they are simpler to place without causing damage to the surrounding structures or the soil mass. Less than 300 mm in diameter, these small-diameter piles are typically built by drilling, placement of steel reinforcing, and grouting. Two methods of installing micropiles on slopes are described in the FHWA (2005) manual. One is to install the battered micropiles far enough down the slope, past the slope's possible slip surface, to a stable soil layer. They mobilize axial, bending, and shear resistance to offer resistance to soil movement. The second method involves installing a group of micropiles closely spaced apart so that they act as a retaining wall. This group provides resistance by acting as a single unit while the individual piles take only small loads. Numerous studies are being conducted to examine the performance of micropiles on slopes and embankments.

Esmaeili et al. (2013) conducted an experimental study on micropile reinforced embankments and then validated the results by numerical analysis using PLAXIS. The best location for micropiles was found at the toe of the embankment soil and concluded that the micropile inclination should not exceed 45°. Sun et al. (2014) analyzed the coupled effect between a group of micropiles and the slope from the failure surfaces formed at different micropile positions. Micropiles stabilize slopes under static as well as dynamic conditions. The behaviour of inclined micropiles under dynamic loading conditions was studied by Ismail and Wardany (2017). Increased micropile density and micropile inclination result in a reduction in the horizontal displacement and velocity amplification of soil particles under dynamic loads. It is possible to analyze the seismic response of slopes containing micropiles using shaking table model tests (Li et al. 2018). Micropiles are effective at reducing seismic vibrations, delaying sliding time, decreasing the sliding speed, and reducing sliding distance.

Anirudhan and Niranjana (2017) investigated the use of micropiles in stabilizing the slopes in a section of Kerala's western Ghats. It demonstrates that micropiles may also be installed on Kerala's soil slopes to raise FOS. The present study investigates the effectiveness of micropiles in Kerala's Idukki area, where slope failures are common. To perform stability analysis, the Idukki soil slopes are modelled in MIDAS GTS NX. The strength reduction approach is used by the program to execute the finite element analysis. Micropiles with different arrangements are incorporated into the critical slopes to analyze the stability of micropile reinforced slopes. The optimum micropile arrangement for each slope is then determined.

2 Stability Analysis of Idukki Soil Slopes

2.1 *Properties of Soil Slopes*

Idukki district of Kerala is mostly a hilly area that faces slope stability issues and is highly susceptible to slope failures. The occurrence of soil slope failures is frequent in the Idukki district, mainly during the rainy season, which causes severe damage, loss of lives and property, disruption of traffic, etc. Following the 2018 heavy rain and the occurrence of a series of landslides and slope failures, a detailed study of the stability of slopes has become necessary. Therefore, soil slopes in the Idukki district of Kerala have been selected for the present stability analysis study. The location details, geometry, and soil properties of selected slopes in Idukki have been collected from the Geotechnical laboratory, NIT Calicut, for modelling and tabulated in Tables 1 and 2.

2.2 *Numerical Modelling and Analysis of Soil Slopes*

The slopes were modelled using the geometrical data and soil properties in MIDAS GTS NX software. Three-dimensional slope geometry was developed using the slope height and angle values. The soil was modelled using the Mohr–Coulomb model, and meshing was done using the hybrid meshes option to get accurate results. In the present study, slope stability analysis using the strength reduction method was carried out for two cases: (1) Dry state and (2) Rainfall (saturated) state. The ground surface of the slope was taken as the piezometric level to simulate the fully saturated status of slopes.

The safety factor of all the slopes obtained from analysis under dry and saturated conditions is tabulated in Table 3. Among those, four soil slopes, namely, St. Maries Chapel, Munnar Govt. Arts College compound, Velliyamattam Panchayath, and Kallar landslide top sides completely failed under the saturated condition as the safety factor is less than 1. During the rainfall (saturated state) period, the safety factor of slopes decreases and the soil slopes are prone to failure. The soil slope at Aadit, which was safe (FOS > 1.5) in the dry condition, is found unsafe (FOS <

Table 1 Location and geometry of soil slopes

Sl. No	Location	Latitude	Longitude	Slope height (m)	Slope angle (°)
1	St. Maries Chapel, Ninevu Nagar	N10°1′2.7″	E76°59′20.6″	15	25
2	Lourdes Church, Karadippara	N10°2′25.0″	E77°0′19.5″	10	25
3	Spice country resort, Shankupadi	N10°1′31.1″	E77°2′26.9″	6	17
4	Aadit, Shankupadi	N10°1′15″	E77°2′48.8″	5	25
5	Pathuvalavu Pachadi	N9°50′54.6″	E77°06′59.3″	9	30
6	200 m from Pachadi Nedumkandam	N9°50′51.00″	E77°06′56.3″	10	35
7	Munnar Govt Arts College Compound	N10°04′57.5"	E77°04′23.8″	21	40
8	Velliyamattam Panchayath, Thadiyana Colony	N9°49′42.3″	E76°51′59.0″	12	25
9	Kallar landslide top side	N10°02′25.3″	E77°00′19.9″	8	20
10	2nd mile top side	N1″	E77°02′17.6″	12	26

1.5) during the saturated state. The slopes of Lourdes Church, Spice country resort, Pathuvalavu Pachadi, 200 m from Pachadi Nedumkandam, and 2nd mile top sides are entirely safe. Drastic reduction of stability of slopes under fully saturated conditions shows that it is the worst state experienced by the slopes.

3 Stability Analysis of Micropile Reinforced Soil Slopes

Four slopes that failed in the stability analysis (FOS $<$ 1.5) are modelled with micropiles to analyze the effect of micropile reinforcement in stabilizing the slopes. Different arrangements of micropiles are considered. Then a series of stability analyses are carried out on each soil slope to obtain an optimum micropile arrangement with a more significant improvement in FOS against slope failure.

Table 2 Geotechnical properties of soil slopes

Sl. No	Location	Unit weight (kN/m^3)	Water content (%)	Shear strength parameters	
				c (kPa)	φ (°)
1	St. Maries Chapel, Ninevu Nagar	18	22.2	10	32.3
2	Lourdes Church, Karadippara	13	31.9	30	31.7
3	Spice country resort, Shankupadi	19	27.8	0	46
4	Aadit, Shankupadi	17	10.2	40	41.2
5	Pathuvalavu Pachadi	18	25.9	50	22.8
6	200 m from Pachadi Nedumkandam	13.5	13	30	22
7	Munnar Govt Arts College Compound	15.9	15	3	29.7
8	Velliyamattam Panchayath, Thadiyana Colony	12	25.9	10	15.9
9	Kallar landslide top side	15	11	50	24.8
10	2nd mile top side	14.4	16	30	24.5

Table 3 Factor of safety values of existing Idukki soil slopes

Sl. No	Location	Factor of safety (Dry state)	Factor of safety (Saturated state)
1	St. Maries Chapel, Ninevu Nagar	1.99	0.99
2	Lourdes Church, Karadippara	3.29	1.65
3	Spice country resort, Shankupadi	3.78	1.65
4	Aadit, Shankupadi	2.66	1.28
5	Pathuvalavu Pachadi	3.08	2.36
6	200 m From Pachadi Nedumkandam	3.37	2.40
7	Munnar Govt Arts College Compound	1.21	0.34
8	Velliyamattam Panchayath, Thadiyana Colony	1.43	0.62
9	Kallar landslide top side	2.04	0.82
10	2nd mile top side	2.56	1.60

Table 4 Material properties of micropile (Esmaeili et al. 2013)

Parameter	Value
Elastic modulus of cement grout (MPa)	31,000
Compressive strength of cement grout (MPa)	34.5
Elastic modulus of reinforcing wire (MPa)	200,000

3.1 Material Properties of Micropile

According to FHWA guidelines, standard reinforcing steel conforming to ASTM A615/AASHTO M31 is used to reinforce the micropile. The grout comprises a cement mix with a w/c ratio between 0.45 and 0.6 by weight. The elastic modulus of steel and cement grout was taken according to Esmaeili et al. (2013) in this study given in Table 4.

3.2 Geometrical Parameters of Micropile

The different geometric parameters of micropile considered are:

- Length of the micropile (L): The length of the micropile depends on the depth of the sliding surface of unreinforced slopes obtained from stability analysis. For improved stability, micropiles are extended beyond the sliding surface.
- Diameter of the micropile (D): Maximum diameter, i.e., 30 cm, is considered to reduce the number of micropiles.
- The angle of the micropile concerning the vertical axis (θ): Vertical, inclined, and combination of both have been considered for different trials to get the required stability. The upslope micropile angle is selected between 5° and 10°, while the downslope angle is varied between 15° and 30°. It depends on the length of the micropiles such that they cross the sliding surface.
- Micropile spacing: In the longitudinal direction (S_{long}), a micropile spacing of S = 2D and 3D, where D is the diameter of the micropile, is chosen. Trial and error are used to determine the lateral spacing, or spacing along the slope length (S_{lat}) until the necessary safety factor is attained.

3.3 Cap Beam

The steel-reinforced concrete cap beams are constructed on slopes to anchor the micropile array at the ground surface. According to FHWA guidelines, the cap beam should have sufficient height and width to embed at least two to three micropile diameters into the cap beam and provide fixity of the upslope and downslope micropile. The material properties of the concrete cap are tabulated in Table 5.

Table 5 Material properties of cap beam

Parameter	Value
Grade of concrete	M_{20}
Elastic modulus (MPa)	22,361
Poisson's ratio	0.15
Unit weight (kN/m^3)	25

3.4 Modelling of Reinforced Soil Slopes

The soil is modelled using the Mohr–Coulomb model and micropiles as a linear elastic material. The soil-pile interface characteristics are automatically taken into account by the interface wizard in MIDAS GTS NX. The cap beam that anchors the micropile array at the top is also modelled as a linear elastic material. It is a rectangular concrete beam with a size of 2 m × 1 m. Different arrangements of micropiles have been modelled by varying the geometrical parameters, and stability analysis was carried out with each micropile arrangement. The micropile arrangements adopted in the model are tabulated in Table 6.

4 Results and Discussion

The present study considers four locations of soil slopes to reinforce the soil slopes with different micropile arrangements. Figures 1, 2, 3 and 4 show the total displacement contours with FOS values of both unreinforced and reinforced slopes with optimum micropile arrangement. The FOS of each soil slope obtained from the analysis for each micropile arrangement is shown in Table 7.

Tables show that for slope 1 (St Maries Chapel), micropile arrangement no. 5, where the micropiles are installed in the lower half of the slope, has given the satisfied high FOS (1.59 > 1.5). Installing micropiles in the bottom part of the slope enhances stability against slope failure since the unreinforced slope's toe experienced the maximum deformation. Hence, arrangement no. 5 is the optimum arrangement for soil slope 1. However, the small lateral spacing between piles makes it challenging to install in the field.

Slope 2 (Aadit, Shankupadi) is a relatively stable slope with less height than the other three slopes. Installing the micropiles inclined at the toe of the slope itself increases the FOS of the slope to 1.47. A combination of micropiles at the slope's middle, toe, and crest (arrangement no.8) gives the required FOS > 1.5 against failure.

Slope 3 (Kallar landslide) and Slope 4 (Velliyamattam Panchayath) are the two most critical slopes with low FOS and deep failure surface. Their FOS against stability could improve only by providing battered micropiles on both upslope and downslope of soil slopes. The FOS of slope 3 increased to 1.29 by installing battered micropiles along the entire length of the slope with a spacing of 6 m (arrangement no. 12). In the

Table 6 Micropile arrangements implemented in the model

Arrangement no	Position	D (m)	θ (degree)	L (m)	S_{long} (m)	S_{lat} (m)
Slope 1: St Maries Chapel, (slope height:15 m, slope length: 35 m)						
1	Middle of slope	0.3	0	8	0.6	0
2	Middle, toe, and crest of the slope	0.3	0	8	0.6	0
3	Middle, toe, and crest	0.3	10° upslope, 30° downslope	8–9.5	0.6	0
4	Full length of the slope	0.3	10° upslope, 30° downslope	8–9.5	0.6	9
5	½ length	0.3	0° and 15° downslope	8.5–9	0.6	4
Slope 2: Aadit, Shankupadi, (slope height: 5 m, slope length: 9.4 m)						
6	Middle of slope	0.3	0	4	0.9	0
7	Toe of slope	0.3	15° downslope	4	0.9	0
8	Middle, toe, and crest	0.3	0	3.5	0.9	0
9	Middle, toe, and crest	0.3	0° at toe and crest, 0° and 15° downslope at middle	3.5–4	0.9	0
Slope 3: Kallar landslide, (slope height: 8 m, slope length: 23 m)						
10	½ length	0.3	0° and 15° downslope	5–5.5	0.6	4
11	Full length	0.3	0° and 15° downslope	6–6.5	0.6	4
12	Full length	0.3	5° upslope and 20° downslope	8–8.5	0.6	6
Slope 4: Velliyamattam Panchayath, (slope height: 12 m, slope length: 29 m)						
13	Middle of the slope, upper and lower middle	0.3	0° at middle, 10° upslope and 25° downslope at upper and lower middle	12–13	0.6	7
14	Full length	0.3	0	10	0.6	6
15	Full length	0.3	0° and 20° downslope	11–12	0.6	6
16	½ length	0.3	0° and 15° downslope	10–10.5	0.6	3

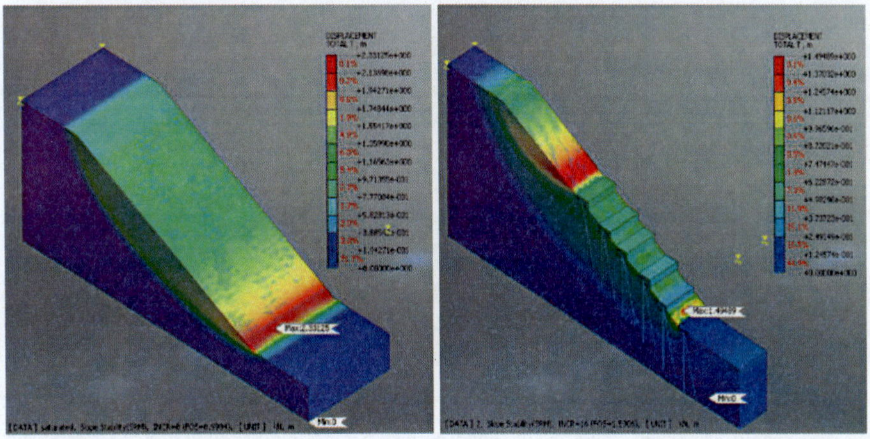

(a) (b)

Fig. 1 Total displacement contours for slope 1 (St. Maries Chapel): **a** unreinforced slope, and **b** reinforced slope with micropiles arrangement no. 5

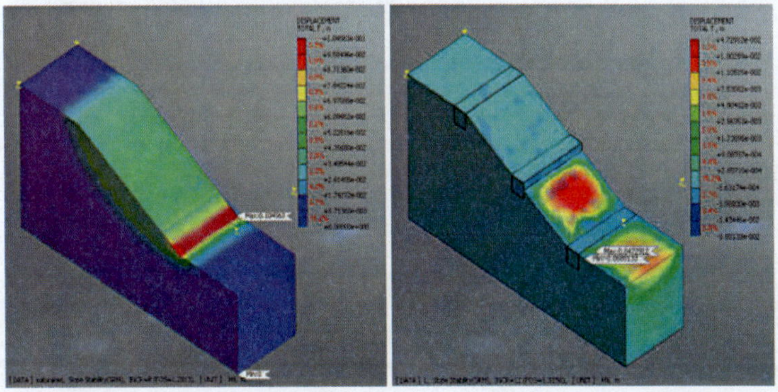

(a) (b)

Fig. 2 Total displacement contours for slope 2 (Aadit): **a** unreinforced slope, and **b** reinforced slope with micropiles arrangement no. 8

case of slope 4, driving vertical and inclined micropiles into the downslope along the lower half-length gives a FOS of 1.23 (arrangement no. 16). With micropile reinforcement, the increased FOS of Kallar and Velliyamattam Panchayath soil slopes are less than the soil slope at St. Maries Chapel due to less frictional resistance between the soil particles. Therefore, these soil slopes require micropile reinforcement heavily to bring them into a stable condition with a high factor of safety.

In summary, comparing the FOS values of micropile reinforced soil slopes with unreinforced soil slopes, more than a 50% increase in FOS against slope failure is observed. Since micropile systems with battered micropiles have provided more FOS

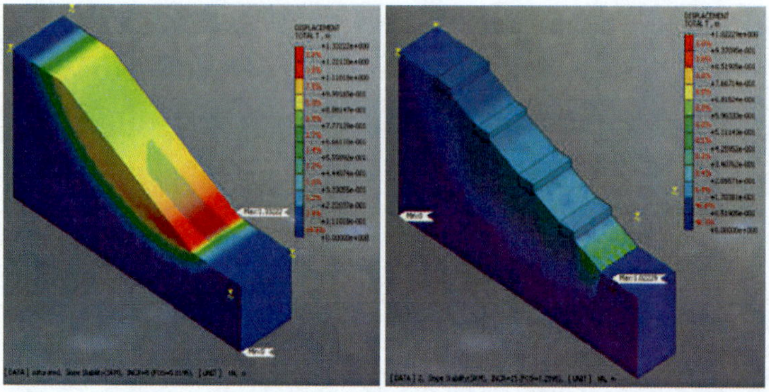

(a) (b)

Fig. 3 Total displacement contours for slope 3 (Kallar landslide): **a** unreinforced slope, and **b** reinforced slope with micropiles arrangement no. 12

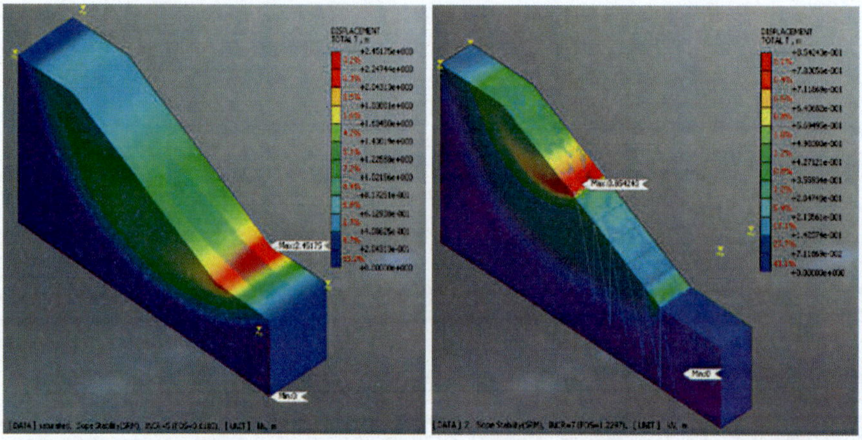

(a) (b)

Fig. 4 Total displacement contours for slope 4 (Velliyamattam Panchayath): **a** unreinforced slope, and **b** reinforced slope with micropiles arrangement no. 16

against slope failure, they offer more excellent resistance than vertical micropiles. For most unstable soil slopes with low internal friction angles, such as Velliyamattam and Kallar slides, installing micropiles precisely in the middle or toe of the soil slope is ineffective as failure occurs in the upper region. Micropiles should be installed for such soil slopes either in full length or half-length in slope with reduced spacing. It shows how crucial it is to use this technique to stabilize slopes that are particularly steep and unstable. Such slopes fail unless they are heavily reinforced with micropiles that reach extremely deep into a stable layer which is not practical from the application point of view. Maximum displacement values obtained from the displacement

Table 7 Factor of safety obtained with various micropile reinforced arrangements

Arrangement no	Factor of safety
Slope 1: Initial FOS 0.99	
1	1.06
2	1.1
3	1.13
4	1.39
5	1.59
Slope 2: Initial FOS 1.28	
6	1.42
7	1.47
8	1.51
9	1.64
Slope 3: Initial FOS 0.82	
10	1.14
11	1.16
12	1.29
Slope 4: Initial FOS 0.62	
13	0.73
14	0.75
15	1.02
16	1.23

contours of the slopes show that slope displacements of all four slopes decreased by providing a micropile system compared to unreinforced soil slopes. After analyzing the optimum arrangement obtained for each soil slope, it is concluded that micropile distribution between the toe and middle of the soil slope effectively reinforces the soil slopes.

5 Conclusions

The existing soil slopes of Idukki were characterized after carrying out the stability analysis. Ten slopes were modelled in MIDAS GTS NX software, and slope stability analysis using the strength reduction method was carried out to get the FOS of existing soil slopes against slope failure. The FOS against the stability of soil slopes decreases during the rainfall (fully saturated) period, and hence the slopes are prone to failure. Among the ten soil slopes modelled, four soil slopes entirely failed, and FOS obtained in the analysis was less than one under saturated state. The soil slope at the Munnar Govt Arts College compound shows a minimal FOS value than other soil slopes. Four slopes that were found unsafe were selected and modelled with micropile reinforcements to analyze and increase the FOS against slope stability. The optimum arrangement of micropiles for the slopes has been determined. The significant conclusions obtained from the present study are:

- FOS of slopes against slope stability increased by more than 50% by reinforcing it with micropiles.
- FOS of 1.5 against slope failure has not been achieved for all soil slopes because micropiles need to be embedded deep into a stable layer to provide good fixity and utilize total micropile capacity for weak and unstable soil slopes. But providing such long and non-standard lengths of micropiles with less spacing is not practical as it is challenging to install on soil slope sites.
- This technique of installing micropiles as reinforcements to stabilize the slopes is most critical for slopes that are extremely high and unstable.
- Battered piles are more effective than providing vertical piles alone.
- Increasing the length of micropiles well beyond the failure surface and reducing the micropile spacing (both lateral and longitudinal) increases FOS against soil slope failure.
- Micropile distribution between the toe and middle of the slope effectively reinforces soil slopes.

References

Abd El Raouf ME (2019) Improvement stability of slopes using anti-slide piles. Eng Res J 1(42):104–108

Anirudhan KM, Niranjana K (2017) Design of micro pile walls for slope stability improvement. Int J Recent Eng Res Dev 2(5):60–66

Aslam A, Sreekumar S (2016) Characteristics of slope failures in tropical hill slopes of South West India. Int J Sci Res 5(3):1584–1588

Cai F, Ugai K (2000) Numerical analysis of the stability of a slope reinforced with piles. Soils Found 40(1):73–84

Esmaeili M, Gharouni Nik M, Khayyer M (2013) Experimental and numerical study of micro piles to reinforce high railway embankments. Int J Geomech 13(6):729–744

FHWA (2005) Micro pile design and construction. Publication No. FHWA NHI-05-039, U.S. Department of Transportation, p 456

Griffiths DV, Lane PA (1999) Slope stability analysis by finite elements. Geotechnique 49(3):387–403

Ismail S, Wardany RA (2017) A finite difference study on the use of inclined micro piles for soil slope stabilization under seismic loading. In: Geo Ottawa conference

Li R, Yu Y, Deng L, Li G (2006) Stability analysis of unsaturated soil slope by 3-D strength reduction FEM. In: Advances in unsaturated soil, seepage and environmental geotechniques (GSP), vol 148, pp 62–69

Li N, Men Y, Gao O, Liu X (2018) Seismic response of landslide with micro piles. In: IOP conference series: materials science and engineering. IOP Publishing, pp 1–7

Matsui T, San KC (1992) Finite element slope stability analysis by shear strength reduction technique. Soils Found 32(17):59–70

Sun S, Wang W, Zhao F (2014) Three-dimensional stability analysis of a homogeneous slope reinforced with micro piles. In: Mathematical problems in engineering. Hindawi Publishing Corporation, p 11

Performance of Stone Columns in Multi-Layered Soils Under the Action of Seismic Loads

P Kiruthika and A Murali Krishna

Abstract Soil liquefaction is a dangerous phenomenon that involves huge damages during strong earthquake ground motion events. Stone columns (SCs) are widely used as a liquefaction mitigation element which is expected to perform improvements against liquefaction hazard by densifying, increasing the stiffness and rapid dissipation of excess pore water pressure in the surrounding soil. Although in recent times, few studies on the behaviour of stone columns in the seismic zone have evolved, the performance of SCs in multi-layered soils is not been extensively examined. This paper presents the numerical investigations on the performance of SCs installed in multi-layered soils as a liquefaction mitigation measure against seismic loading. The acceleration time histories, Excess Porewater Pressure (EPP), vertical displacement and stress ratio were evaluated along different depths using PLAXIS 2D. With the installation of a stone column, there is a considerable reduction in the acceleration amplitude, which in turn reduces the EPP and vertical displacement at different locations and the amplitude of reduction depends on the soil stiffness levels at the depth.

Keywords Stone column Multi-layered soil · Liquefaction · Excess porewater pressure · Stress ratio

1 Introduction

Liquefaction is a phenomenon in which the loss of strength or stiffness of the ground occurs resulting in the settlement under the footings, failure of slopes and retaining structures due to the induced irregular/earthquake motion. Various ground improvement methods have been extensively used to mitigate liquefaction and the associated

P. Kiruthika (✉) · A. M. Krishna
Indian Institute of Technology Tirupati, Chindepalle, Andhra Pradesh, India
e-mail: ce20d502@iittp.ac.in

A. M. Krishna
e-mail: amk@iittp.ac.in

A. M. Krishna et al. (eds.), *Deep Foundations for Infrastructure Development in India*, Lecture Notes in Civil Engineering 373, https://doi.org/10.1007/978-3-031-37256-8_17

hazards. Earlier studies have shown that the stone columns are widely used as a liquefaction (induced due to seismic loading) mitigation element which is expected to perform the following improvements including densifying the surrounding soil increases the Cyclic Resistance Ratio, higher stiffness of the column reduces the Cyclic Stress Ratio, and the high permeability of the column leads to the rapid dissipation of excess pore water pressure (Priebe 1989, 1991). The higher modulus of the granular column will reduce the stress transferred onto the surrounding weak soil by acting as a reinforcing element (Baez and Martin 1992; Goughnour and Pestana 1998; Adalier and Aydingun 2003; Olgun and Martin 2008; Rayamajhi et al. 2014; Salem et al. 2017; Badanagki et al. 2018; and Tiznado et al. 2020). The densification and drainage mechanisms of the granular column as a liquefaction mitigation were studied by Krishna et al. (2006). Many physical (centrifuge and small scale 1–g shaking table tests) and numerical modelling studies have been performed on the granular columns (Brennan and Madabhushi 2006; Kamai and Boulanger 2013; Dinesh et al. 2021). However, the study on the performance of the mitigation mechanism of granular columns installed in the layered ground is not well explored.

In this paper, the performance of the granular columns in the multi-layered soils studied by the authors (Kiruthika et al., 2022) is extended to predict the behaviour under the seismic excitation using the Hardening soil model with small-strain stiffness (HS small). The outcomes of these investigations are aimed to provide insight into the magnitude of acceleration at different depths; Excess Porewater Pressure (EPP) generation with and without stone columns; vertical deformation for different cases; and stress ratio distribution between column and the surrounding soil during the seismic event.

2 Soil Profile and Input Motion Characteristics

The present study is conducted on the soil profile located on the East coast of India. The soil stratification consists of three soil layers: the top dense sand layer (0–13 m from Ground Level, GL); the middle soft clay layer (13–26 m from GL); and the bottom stiff clay layer (26–35 m from GL). To enhance the behaviour of the weak soil, improvement using a stone column of 0.8 m in diameter (d) with a spacing (s) of 1.7 m arranged in a triangular pattern was employed. Stone columns extending up to a depth of 27 m and an area replacement ratio (A_r) of 20% for the effective design was determined. The SPT N value and the corresponding shear wave velocity profile of the site soil are presented in Fig. 1.

The horizontal earthquake time history recorded from the Imperial Valley USA earthquake, 1979 with a frequency range of 0.1 to 40 Hz and predominant frequency of 7.14 Hz (Recording station: USGS Station 5115) was given as an input at the rigid bottom of the model at a depth of 35 m from the ground level. The total duration of the input signal is 40 s with a time step of 0.01 s as shown in Fig. 2.

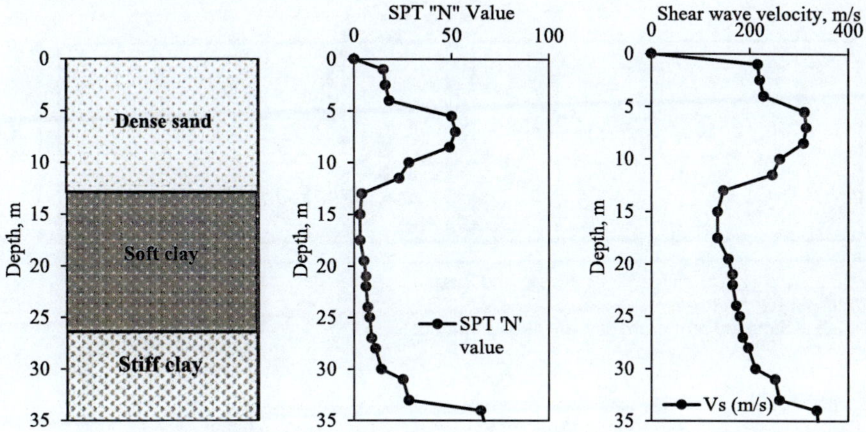

Fig. 1 Site soil stratification, SPT blow count and Shear wave velocity profile

Fig. 2 Input acceleration time history

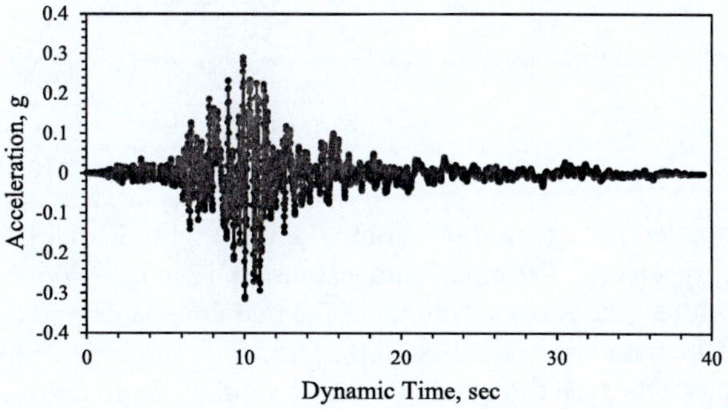

3 Numerical Modelling

To examine the effectiveness of the installation of the stone column against liquefaction, a two-dimensional numerical study was conducted using PLAXIS finite element software. Due to the computational constraints, a two-dimensional analysis was performed. The present study is divided into four different cases with and without improvement (1) Homogenous dense sand (DS), (2) Homogenous soft clay (SC), (3) Homogenous stiff clay (STC) and (4) Multi-layered site stratification (SS).

3.1 Validation Study

At the outset, a validation study is performed for which the large-scale shaking table results of Sasaki and Taniguchi (1982) were considered. The sinusoidal loading was

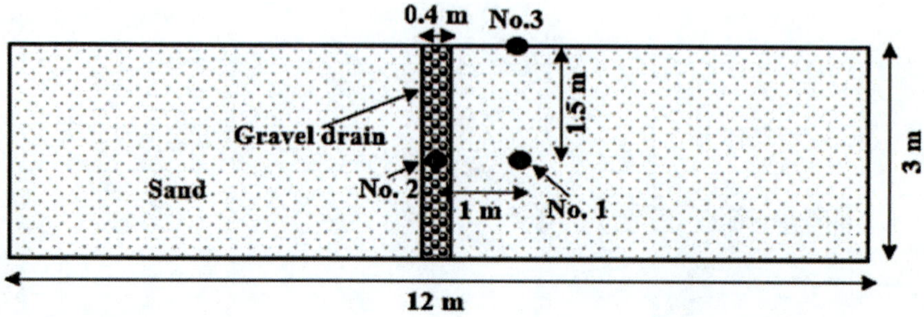

Fig. 3 Schematic representation of validation model

Table 1 Material properties for validation model

Properties	Sand
Relative density, %	41.9
Dry density (g/cm^3)	1.46
Wet density (g/cm^3)	1.91
Void ratio, e	0.893
K_v (cm/s)	1.57 × 10^{-2}
K_h (cm/s)	9.02 × 10^{-3}

applied for a period of 1 min with an amplitude and frequency of 2 m/s^2 and 5 Hz, respectively. The model dimensions are 12 m in length, 2 m in width and 3 m in height with a central stone column of 0.4 m diameter, extending to the model depth and the schematic representation of the validation model is displayed in Fig. 3. The material properties used for the numerical validation are considered from the experimental study and are provided in the Table 1. From the analysis, the EPP examined at two locations No.1 (in the soil) and No. 2 (in the stone column) are presented in Fig. 4. Although the peak EPP at location No. 2 is in good agreement with the experimental values, the EPP reduction is not attained with the dynamic time that may be due to the undrained behaviour exhibited by the stone column in the numerical analysis.

3.2 Numerical Modelling for the Present Study

An initial static analysis was performed on the multi-layered soils with stone columns using the Hardening soil model. To take into account the dynamic soil parameters, HS Small model is employed as the material model that considers two parameters G_0 and $\gamma_{0.7}$ to define the variation of stiffness with strain, and the properties are reported in Table 2. Other stiffness parameters $E_{50}{}^{ref}$, $E_{oed}{}^{ref}$ and $E_{ur}{}^{ref}$ are adopted similarly to the Hardening soil model by Kiruthika et al. (2022). The meshed model displayed in Fig. 5 consists of 15 noded triangular elements with an average element size

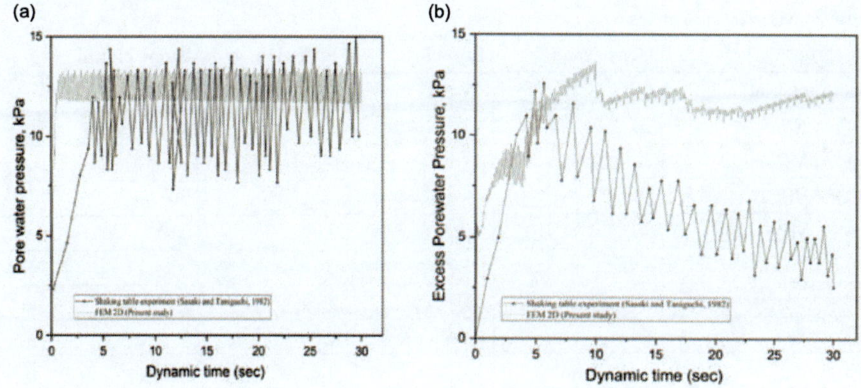

Fig. 4 Effective porewater pressure distribution at the location **a** No. 1 **b** No. 2

calculated as $AES \leq \frac{V_{s,\min}}{8 f_{\max}} = 0.45\,\text{m} \cdot V_{s,\min}$ of the soft clay soil is calculated using correlation with SPT N value $95.64N^{0.301}$ reported by Maheshwari et al. (2010). The earthquake motion is applied at the bottom of the model as prescribed displacement and the load is given as a dynamic multiplier in terms of acceleration time history. Due to the high rate of loading by the earthquake accelerations, undrained analysis is performed. The bottom boundary of the model is set a fixed based, fully reflective boundary. The two vertical boundaries are set as tied degrees of freedom. The proportionality damping coefficients, α and β are determined by keeping a target damping ratio of the soil, fundamental frequency as $\frac{V_s}{4H}$. The additional dynamic parameters are calculated as given in the following Eqs. 1 and 2.

$$G_0^{ref}(MPa) = 33\frac{(2.97 - e^2)}{1 + e}(\text{Hardin and Black 1969}) \tag{1}$$

$$\gamma_{0.7} = \frac{1}{9G_0}\left[2C'(1 + \cos(2\varphi')) - \sigma_1'(1 + K_0)\sin(2\varphi')\right] \tag{2}$$

4 Results and Discussions

The four different cases of soils profiles were modelled as unit cells and the results in terms of excess porewater pressure, acceleration time history, vertical deformation, and shear stress ratio along different depths at a distance of 0.5d from column edge for the case with and without stone columns are presented.

Table 2 Material properties

Properties	Dense sand	Soft clay	Stiff clay	Stone column
Depth interval, m	0–13	13–26	26–35	0–27
γ_{bulk}, kN/m^3	20	17	18	20.5
$E_{50}{}^{ref}$, MPa	30	–	–	70
$E_{oed}{}^{ref}$, MPa	30	–	–	70
$E_{ur}{}^{ref}$, MPa	90	–	–	210
C_C	–	0.37	0.46	–
I_p	–	55%	59%	–
C_S	–	0.04	0.05	–
m	0.5	1	0.6	0.5
$e_{initial}$	0.21	1.39	1.28	–
c_u, kPa	0	20	150	0
φ', °	38	0	0	42
G_0^{ref}, MPa	202	35	42	280
$\gamma_{0.7}$	1.8×10^{-4}	3×10^{-4}	9×10^{-4}	5×10^{-4}

4.1 Excess Pore Water Pressure

The variation of EPP to investigate the effectiveness of stone columns in different soil layers has been examined. The graphical representation of the variation of EPP at different locations and time for 4 cases with and without stone columns are presented in Fig. 6. At the surface level in the DS (Homogeneous Dense Sand) case, the EPP generated is 17 kPa (negative) without improvement and 6 kPa with improvement. This indicates that the dense sand dilates during the application of dynamic load. For other layers in the DS and (Homogeneous Stiff Clay) STC cases, the reduction in EPP is not prominent. A significant reduction in the EPP is seen in the SC (Homogeneous Soft Clay) and SS (multi-layered Site Stratification) profiles, where the EPP reduced from 50, 40 and 26 kPa to about 50 kPa (in the negative direction) at the depths 13, 19 and 27 m, respectively. The reduction in the EPP is attributed to the reduction in the length of the drainage path due to the higher permeable material. It can also be noted that the EPP is reducing with depth, due to lesser stress distribution at the deeper soil layers.

4.2 Acceleration Time History

The horizontal time histories of acceleration at different locations for all the cases without a stone column and with a stone column are presented in Fig. 7. The plot shows that the higher acceleration is observed at the top surface for all the cases.

Fig. 5 Meshed model

However, in the SS case up to 13 m from GL where the dense sand layer occurs, the acceleration amplitude is the same for with and without improvement case, i.e., about 1 m/s^2. A slight reduction in the level of acceleration can be seen at a depth of 19 m (in the mid-level of the soft clay layer) for the improved case (3.2 m/s^2) than the unimproved case (3.9 m/s^2). This depicts that the presence of a stone column plays a very little role in reducing the acceleration amplitude on the surrounding weak soil, therefore, more cases need to be analysed to firmly conclude its efficacy.

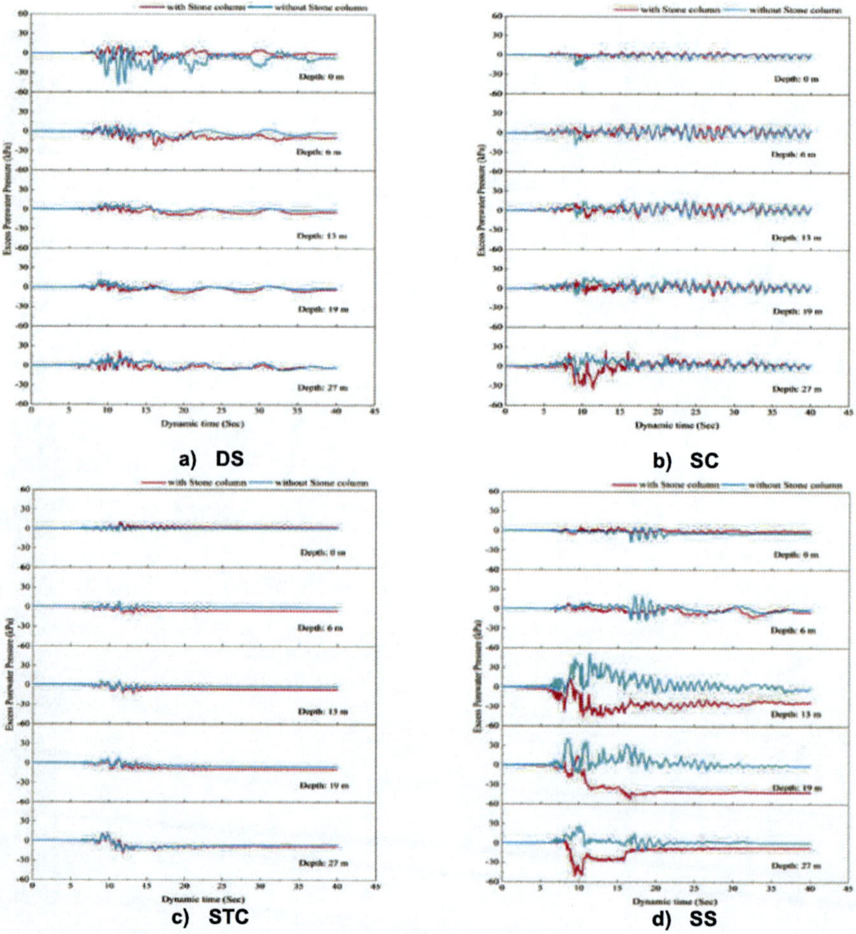

Fig. 6 Excess pore water pressure variation during the seismic event

4.3 Vertical Deformation

The total settlement of ground can be reduced with stone column and the vertical deformation of the soil recorded at different locations during the seismic event are presented in Fig. 8. The following are the maximum vertical deformation predicted at the top layer in all four cases: DS—0.15 m; SC—0.76 m; STC—0.73 m; and SS—0.34 m with no stone column and DS—0.10 m; SC—0.35 m; STC—0.15 m; and SS—0.09 m with stone column. The maximum deformation was attained after the 20 s of random loading after which the settlement remained constant. The observed settlement is higher in the case of unimproved ground than that in the improved case. The presence of stiffer stone columns aids in the reduction of settlement.

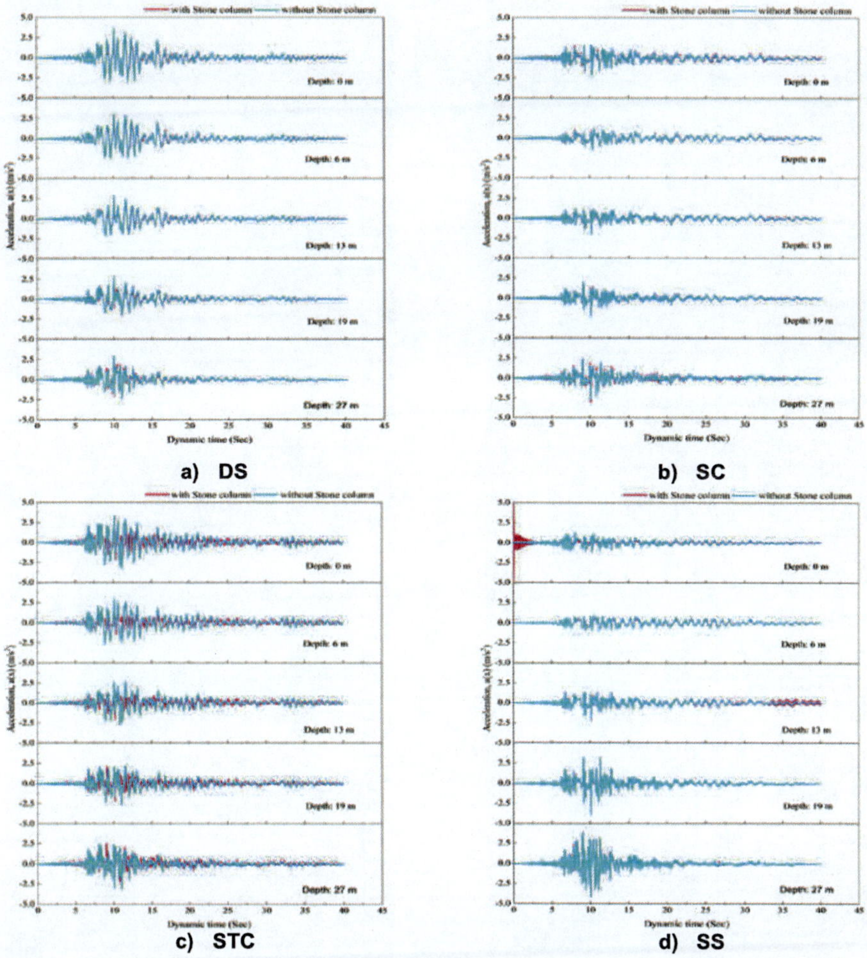

Fig. 7 Acceleration time history at different locations during the seismic event

4.4 Cyclic Stress Ratio

The installation of a stone column aids in reducing the stress carried by the surrounding soil due to the higher stiffness. The variation of the maximum cyclic stress ratio along the depth of the soil in the case of SS soil profile without improvement, with improvement at two locations at the centre of the stone columns and at 0.5d distance from the column edge (in the soil) is presented in Fig. 9. The stress ratio for the topmost dense sand layer varies between 0.42 and 0.3 on both with and without improvement case. However, below this layer, the stress ratio suddenly drops down to 0.1 in the case of soft clay and 0.18–0.3 on the stone column within 13–26 m

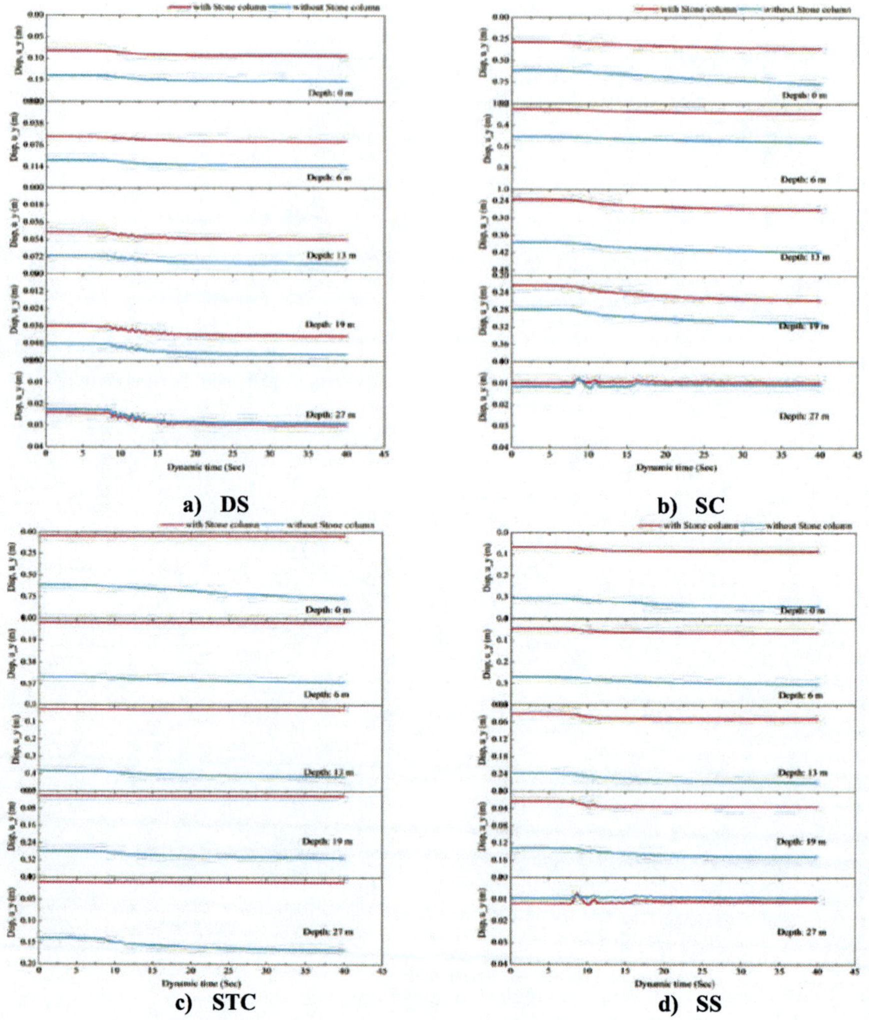

Fig. 8 Vertical deformation of different locations during the seismic event

depth. No significant change in the shear stress ratio is seen in the no improvement and improvement case. Nevertheless, the cyclic shear stress carried by the column is higher than the surrounding weak soil.

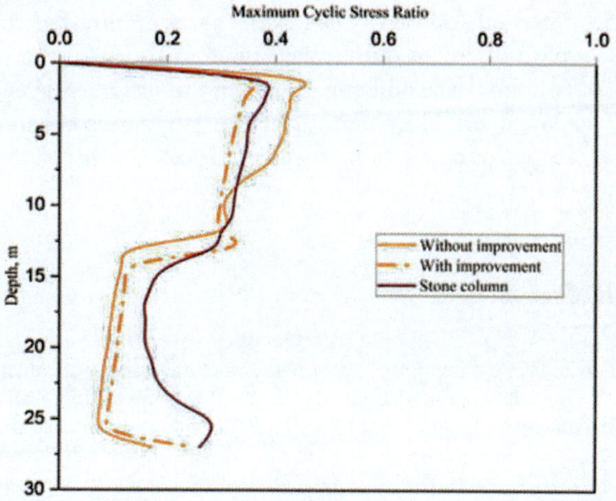

Fig. 9 Maximum cyclic stress ratio along the depth

5 Conclusion

The Finite Element analysis is performed to comprehend the behaviour of the stone column in the multi-layered soil profile under the action of seismic load. Three different soil layers, homogeneous dense sand, soft clay and stiff clay layer, and a site stratification layer were considered. The results in terms of EPP, acceleration amplitude, vertical deformation and cyclic stress ratio were studied at different locations. The significant outcomes are summarised below:

1. With the presence of a stone column, the treated ground responded in a stiffer manner exhibiting the lesser EPP generation. The reduction in the EPP in case of dense sand and stiff clay layer is minimal. However, the installation of the high permeable stone column reduces the drainage path and EPP to a greater degree in the soft clay layer.
2. Without the stone column, the higher EPP and settlement were observed at a duration of 8–12 s where the magnitude of the acceleration is higher. The maximum acceleration is experienced by the soft clay layer is about 3.2 m/s^2 with improvement.
3. The vertical deformation was observed maximum at the top layer, where the acceleration and EPP values are also higher. The reduction in the maximum vertical settlement in the site soil layer is 53% with the granular column. With the depth of the stone column extending into the soil layer, the settlement is also reduced.
4. The cyclic stress ratio of the site soil with and without improvement has negligible differences. The drastic reduction in the maximum cyclic stress ratio of the soft clay layer is clearly observed. Nevertheless, the installation of a stone column carries higher shear stress than the surrounding soft clay layer.

5. The study obviously states that stone columns are beneficial in reducing the EPP and settlement due to the higher stiffness of the reinforced ground. The stone column has promising effects in the soft clay layer than the dense sand and stiff clay layers. However, further investigations need to be done to strongly validate its performance in multi-layered soils under seismic loads.

References

Adalier K, Aydingun O (2003) Numerical analysis of seismically induced liquefaction in earth embankment foundations. Part II. Appl Remedial Meas Can Geotech J 40(4):766–779
Badanagki M, Dashti S, Kirkwood P (2018) Influence of dense granular columns on the performance of level and gently sloping liquefiable sites. J Geotech Geoenvironmental Eng 144(9):1–14
Baez JI, Martin GR (1992) Quantitative evaluation of stone column techniques for earthquake liquefaction mitigation. In: Proceedings of 10th world conference on earthquake engineering. Madrid, Spain
Brennan AJ, Madabhushi SPG (2006) Liquefaction remediation by vertical drains with varying penetration depths. Soil Dyn Earthq Eng 26:469–475
Dinesh N, Banerjee S, Rajagopal K (2021) Numerical analysis of shallow foundations resting on Granular Columns embedded into the liquefiable deposit. J Earthq Tsunami 15(3)
Goughnour RR, Pestana JM (1998) Mechanical behavior of stone columns under seismic loading. In: Proceedings of 2nd international conference in ground 186 improvement techniques. CI-Premier, Singapore, pp 157–162
Hardin BO, Black WL (1969) Closure to vibration modulus of normally consolidated clays. In: Proc ASCE: J Soil Mech Found Div 95(SM6):1531–1537
Kamai R, Boulanger RW (2013) Simulations of a centrifuge test with lateral spreading and void redistribution effects. J Geotech Geoenvironmental Eng 139(8):1250–1261
Kiruthika P, Banerjee S, Krishna AM, Boominathan A (2022) Performance of stone columns in multi-layered soils system. In: Geo-Congress 2022, ASCE. North Carolina, USA, pp 50–60
Krishna AM, Madhav MR, Latha GM (2006) Liquefaction mitigation of ground treated with granular piles: densification effect. ISET J Earthq Technol 43(4):105–120
Maheswari RU, Boominathan A, Dodagoudar G (2010) Use of surface waves in statistical correlations of shear wave velocity and penetration resistance of Chennai soils. Geotech Geol Eng 28(2):119–137
Olgun CG, Martin JR (2008) Numerical modelling of the seismic response of columnar reinforced ground. In: Geotechnical earthquake engineering and soil dynamics IV GSP 181, ASCE. Reston, VA
Priebe HJ (1989) The prevention of liquefaction by vibro-replacement. In: Proceedings of earthquake resistance construction and design. Berlin, Germany
Priebe HJ (1991) Vibro-replacement-design criteria and quality control. In: Esrig B (eds) Deep foundation improvements: design, construction, and testing, ASTM STP, vol 1089. Philadelphia, pp 62–72
Rayamajhi D, Nguyen TV, Ashford SA, Boulanger RW, Lu J, Elgamal AW, Shao L (2014) Numerical study of shear stress distribution for discrete columns in liquefiable soils. J Geotech Geoenvironmental Eng 140(3):1–9
Salem ZB, Frikha W, Bouassida M (2017) Effects of densification and stiffening on liquefaction risk of reinforced soil by stone columns. J Geotech Geoenvironmental Eng 143(10)
Sasaki Y, Taniguchi E (1982) Shaking table tests on gravel drains to prevent liquefaction of sand deposits. Soil Found 22(3):1–14

Tiznado JC, Dashti S, Ledezma C, Wham BP, Badanagki M (2020) Performance of embankments on liquefiable soils improved with dense granular columns: observations from case histories and centrifuge experiments 146(9)

Geotechnics for Marine, Near-Shore, and Coastal Construction

Geotechnical Assessment of Highly Saline Soil Stabilized with Fly Ash–Cement Admixtures

Jahnavi Parmar, P. S. Prasad, Saloni Pandya, and Pranav R. T. Peddinti

Abstract Coastal cities are in desperate need of suitable land to meet the ever-increasing demand for urban infrastructure to support commercial, residential, tourism, and off-shore activities. Countries with a lengthy coastal line, such as India, possess significant areas of widespread highly saline soils subjected to periodic seawater intrusion. One such soil stratum was encountered in the Gulf of Cambay, Bhavnagar district, Gujarat, India. Due to the high volume of river runoff, the Gulf has a positive water balance. The relative humidity ranges from 65 to 86%, making the climate semi-arid to sub-humid. The high salinity, mineral content, basaltic origin, and deeper water table offered the impetus to investigate the applicability of various soil treatment options for such saline soils. Admixture stabilization techniques have been shown to help with problematic soil features. The Current study aims to determine the effectiveness of the addition of locally available lignite fly ash (10–30% by weight) and cement (6–9% by weight) to Bhavnagar saline soil in respect of strength, electrical conductivity, and CBR. In this assessment, a comparison of stabilized and unstabilized saline soil mixtures is expected to highlight the need of understanding the influence of treatment on the instinctive performance of stabilized saline soils, as well as assist practitioners in efficacious treatment of extreme saline soils for diverse geotechnical and construction applications.

Keywords Saline soil · Strength · Electrical conductivity · CBR · Pavement

J. Parmar (✉)
Transportation Engineering, Pandit Deendayal Energy University, Gandhinagar, Gujarat, India
e-mail: parmarjahnavi38@gmail.com

P. S. Prasad
CSIR-CRRI Central Road Research Institute, New Delhi, India

S. Pandya
Civil Engineering, Navrachana University, Vadodara, Gujarat, India
e-mail: salonip@nuv.ac.in

P. R. T. Peddinti
Civil Engineering, Pandit Deendayal Energy University, Gandhinagar, Gujarat, India

A. M. Krishna et al. (eds.), *Deep Foundations for Infrastructure Development in India*,
Lecture Notes in Civil Engineering 373, https://doi.org/10.1007/978-3-031-37256-8_18

1 Introduction

Owing to the increase in urbanization, soil stabilization has been rendered as a frequent necessity in the construction of various civil projects (road pavements, runways, embankments, dams, and foundations) in order to utilize ground possessing poor quality soil, problematic soils, and reclaimed lands. The type of soil stabilization technique to be employed in a project is to be determined partially by the structure's design and by the geotechnical, physical, and chemical parameters of the soils involved (Akpokodje 1985). Coastal areas are becoming increasingly popular for both individual and commercial real estate developments along with the infrastructure that supports them. However, coastal and off-shore ground conditions are vulnerable to seawater intrusion. The main dissolved salts in seawater include sodium and chloride which constitute 85% of dissolved salts. The presence of a variety of salts that possess abundant ions such as Cl^-, Na^+, $SO^2{}_4{}^-$, Mg^{2+}, Ca^{2+}, and K^+ has been found to be unfavourable to civil engineering structures (Jakhrani et al. 2012). In the coastal regions, due to the presence of shallow groundwater tables and significant evaporation in the Spring and Summer season, the salt accumulates and crystallizes within the soil skeleton providing additional bonds among the soil particles (Wei et al. 2020). However, during the Monsoon season, salt undergoes dissolution resulting in sudden loss of the additional binding, which results in sudden loss of the mechanical strength of soil (Moncmanová 2007). Moreover, water absorption of common salt also subsides instinctive characteristics of saline soil (Onyejekwe and Ghataora 2015; He et al. 2014; Wei et al. 2020; Messad and Moussai 2016). Hence, the permanent presence of seawater or seawater intrusion severely affects the strength and stability of internal soil fabric, which causes instability to the civil engineering infrastructure constructed on saline soils. Physical and chemical treatment processes, strengthening with quality materials, and stabilizing with stabilizers are often used to enhance the strength and other physiognomies of saline soils during water logging conditions (Rahmat and Kinuthia 2011; Naveena and Reddy 2015).

Fly ash (FA) is one such stabilizer that is produced as a by-product (waste substance) from thermal power plant in vast quantities all over the world. Owing to the rise in urbanization, the generation of thermal power energy produces a large amount of fly ash. Waste management of fly ash is difficult in developing countries such as India. Various usage options have been investigated in order to overcome the issues related to fly ash management. Along with the use of FA in the fabrication of construction materials, land development, mining gap filling, and road construction, greater potential has been seen in utilizing fly ash as a soil stabilizer for ground improvement of weak and problematic soils. Combustion of lignite-based coal produces a Class C FA (self-cementing property), which is widely used in various civil engineering works to augment the instinctive properties of soils (Senol et al. 2002). Conventional cement has also been widely used as a stabilizer for the treatment of soft marine soils as it creates permanent and long-lasting cementation within the soil particles, which increases the strength and compressibility characteristics of such problem soils.

Despite the fact that the technology is widely established, research on fly ash stabilization along with cement for extremely saline soils has received little attention. Given the presence of extremely saline soils in India and the largest coastline in Gujarat state, the current study examines the applicability of fly ash–cement stabilization for such saline soils found in the Bhavnagar district of Gujarat, India.

The present study aims to determine the effectiveness of employing Class C FA–cement mixes for the treatment of saline soil in terms of strength, CBR, stiffness, electrical conductivity, and durability in order to examine the efficacy of treatment on instinctive behaviour of stabilized saline soils, as well to illustrate the application of the treatment for stabilization of extreme saline soils in a variety of geotechnical and construction applications.

2 Material Properties

2.1 Soil

In the current study, soil was collected 1–1.5 m below ground level in the vicinity of salt production works located in Bhavnagar (Gujarat, India). The Gulf of Khambhat, which connects the Saurashtra peninsula to mainland Gujarat in India, is one such major coastline route where road building is in high demand. Bhavnagar soil's geological origin has been traced to volcanic activity (Mehta and Sachan 2017). Denudation was induced by volcanic eruptions and lava dispersal, resulting in the creation of basaltic topography. From the physical examination, the soil was found to be dark in colour, smooth in texture, and fine-grained. Detailed laboratory testing and particle size distribution (GSD) of saline soil have been performed and are represented in Table 1. The saline soil was classified as ML (Low plasticity Silt) according to IS classification (IS 2720 Part-4 1985). The electrical conductivity (EC) of the soil was used to determine its salinity (USDA 2014). As per laboratory investigation, the EC of Bhavnagar soil was obtained to be 21,500 μS/cm, which indicated high saline levels (USDA 2014).

2.2 Fly Ash (FA)

Locally available FA (class C) was collected from the Surat Lignite Thermal Power Station in Gujarat, India to stabilize saline soil. Class C FA is widely reused as an engineering material to improve its qualities by utilizing its pozzolanic properties (Cockrell and Leonard 1970). When no additional activators or stabilizers are required, this type of fly ash can be used, making it a more cost-effective alternative for a wide range of stabilizing needs (Senol et al. 2002).

Table 1 Properties of saline soil

Test		Saline soil	Codal provision
Modified proctor	MDD	1.86 (g/cc)	I.S. 2720 (Part 7): 1980
Test	OMC	15.4 (%)	
GSD	Gravel	0.00 (%)	I.S. 2720 (Part 4): 1985
	Sand	0.40 (%)	
	Silt	82.45 (%)	
	Clay	17.14 (%)	
FSI (FSI)		10 (%)	I.S. 2720 (Part 40): 1977
Liquid Limit (LL)		33.33 (%)	I.S. 2720 (Part 5): 1985
Plastic Limit (PL)		24.20 (%)	
Plasticity Index (PI)		9.13 (%)	
Electrical Conductivity (EC)		21,500 (μS/cm)	I.S. 14,767: 2000

2.3 Cement

OPC (Ordinary Portland cement) grade 43 was used to stabilize the saline soil.

3 Experimental Programme

To evaluate the efficiency of treatment by locally available Class C FA and cement mixes on strength, stiffness, salinity, and durability properties of Bhavnagar Saline soil; UCS, CBR, EC, and Durability tests were performed, respectively. The tests were performed as per the I.S code procedures at 10, 20, and 30% fly ash dosages and 6 and 9% cement by weight. EC tests were also performed at varying percentages of FA and cement to assess the significance of stabilization on the salinity levels in the stabilized mixes. Details of fly ash–cement mixes considered for the experimental programme have been represented in Table 2.

Table 2 Nomenclature for stabilized saline soil mixes

Sample no.	Details of bhavnagar soil + Fly Ash (FA) and Cement (C) mix
S0	Untreated bhavnagar soil
S1	Soil + 10% FA + 6% C
S2	Soil + 20% FA + 6% C
S3	Soil + 30% FA + 6% C
S4	Soil + 10% FA + 9% C
S5	Soil + 20% FA + 9% C
S6	Soil + 30% FA + 9% C

3.1 Unconfined Compressive Strength (UCS)

UCS tests were performed on S0, S1, S2, S3, S4, S5, and S6 soil mixes by weight in accordance with IS 2720 Part-10. Specimens of 38 mm diameter and 76 mm height were prepared at MDD and OMC (Table 1) of Bhavnagar Saline soil. These samples were kept in a desiccator and tested at 7 and 28 days.

3.2 California Bearing Ratio

The CBR value is extensively utilized and deliberated in the design of pavement construction. The CBR value is critical as an indicator of the strength and bearing capacity of the soil. CBR tests were performed on samples (Table 2) in accordance with IS 2720 Part-16 (1987) under soaked condition. The resilient modulus (M_R) of a pavement material is also one of the most essential design parameters. The resilient modulus defines how a pavement material reacts to repeated loads. The pavement is described as a three-layer system in this paper. The critical pavement responses due to wheel loads were analysed using the mechanistic empirical software IIT-PAVE. The CBR value of the subgrade soils and the embankment layer underneath the subgrade might differ significantly. Alternatively, the 500 mm thick subgrade could be divided into two layers, with the CBR value of each layer varying. In these circumstances, the effective modulus/CBR value of a single-layer subgrade, which is similar to the combined subgrade and embankment layer, should be used.

3.3 Electrical Conductivity

EC tests (IS 14767 2000) were carried out on S1, S2, S3, S4, S5, and S6 to assess the effect of stabilization on salinity levels of the Bhavnagar soil using Electrical Conductivity Meter. The EC value of the untreated (S0) sample has been represented in Table 1.

3.4 Durability

Durability tests were conducted in accordance with IS 4332-4; 1968 on soil mixes as indicated in Table 2. After 7 days of curing, specimens were placed in water for 5 h and then kept in an oven at 70 °C for 42 h. The procedure was repeated continuously for 12 cycles for each specimen. The weight of the specimens was taken before the water submersion and after oven drying during each cycle to evaluate weight loss.

4 Results and Discussion

4.1 Influence of FA + Cement Treatment on CBR of Bhavnagar Saline Soil

CBR tests were done on native and stabilized Bhavnagar saline soil to determine the efficiency of fly ash–cement stabilization in enhancing subgrade stiffness. The CBR value increased from 1.62 for untreated soil to 37.18, 53.85, and 117.26 for S1, S2, and S3, respectively (Fig. 1). The CBR value increased from 1.62 for untreated soil to 68.62 and 100 for S4 and S5 respectively with the addition of 9% cement, respectively (Fig. 2). The Mr was evaluated based on Eqs. (1) and (2).

$$\mathrm{Mr} = 10 * \mathrm{CBR}\,\mathrm{for}\,\mathrm{CBR} < 5 \quad (1)$$

$$\mathrm{Mr} = 1.76(\mathrm{CBR})^{0.64}\,\mathrm{for}\,\mathrm{CBR} > 5 \quad (2)$$

The obtained values of Mr and effective CBR are represented in Table 3. It was observed that both M_R and effective CBR also improved with an increase in FA and cement content in the mix.

The CBR of a FA–cement stabilized soil mix would be attributed to proportional input of frictional resistance from FA and enhanced cohesive resistance from the cement (Pandian et al. 2002). The addition of FA–cement mix in varying proportions would lead to densification of microstructure in treated soils owing to the formation of CSH gel (Teerawattanasuk and Voottipruex 2019). This would lead to improved particle contact, and as a result, an increase in the CBR value of treated soil.

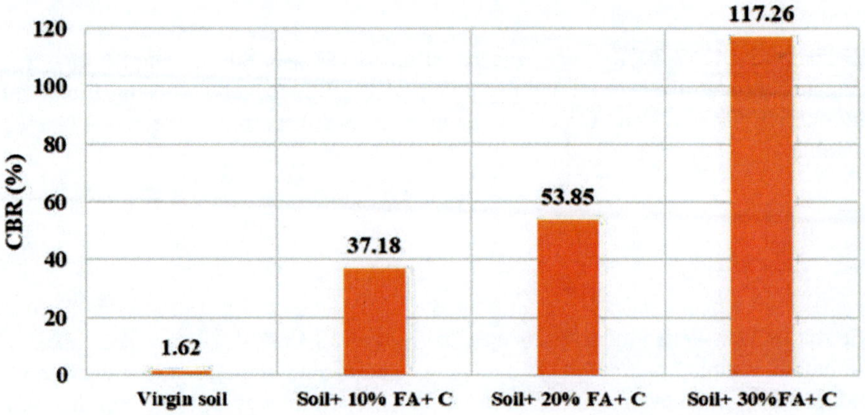

Fig. 1 Influence of varying FA content and 6% C mix on CBR of Bhavnagar saline soil

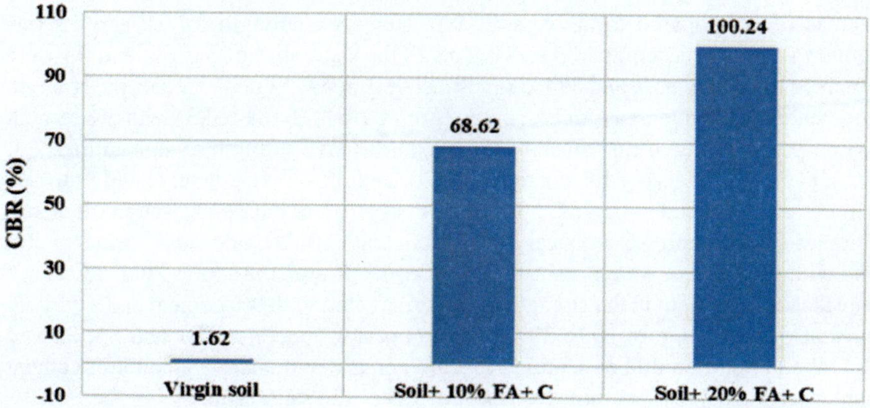

Fig. 2 Influence of varying FA content and 9% C mix on CBR of bhavnagar saline soil

Table 3 Effective CBR and M_R for stabilized subgrade

Soil Mix	CBR	Resilient modulus	Effective CBR
S0	1.62	16.2	1.62
S1	37.18	178	10.58
S2	53.85	225.6	12.88
S3	117.26	371.33	18.75
S4	68.62	263.53	13.26
S5	100.24	335.87	15.8

The effective modulus/CBR value was evaluated in accordance with IRC 37-2018 using IIT-PAVE software as shown in Table 2. According to IRC 37-2018, the effective subgrade CBR for roads carrying more than 450 commercial vehicles per day (CVPD) (two-way) shall be greater than 5%. Thus, the virgin subgrade was found to be not suitable for pavement construction. However, it could be identified from Table 3 that the effective CBR values of treated samples were found to be greater than 5%, and hence treated soil samples would be suitable as subgrade for pavements.

4.2 Influence of FA + Cement Treatment on UCS Tests of Bhavnagar Saline Soil

UCS tests were performed to assess the strength growth and permanence in the treated mixes. UCS (q_u) of native soil and soil-treated samples for 7 days and 28 days of curing has been demonstrated in Figs. 3 and 4, respectively. It was observed that the 7 days strength for specimens treated with FA and 6% cement (S1, S2, and S3) increased as compared to native soil (S0). However, strength for S2 and S3 was found to decrease as compared to S1. The 28 days strength of S1, S2, and S3 were obtained to be 826, 859, and 894 kPa, respectively at 6% cement. Hence, the strength augmented with an increase in FA content from 10 to 30% (S1 to S3) with 6% cement as compared to the virgin soil sample (S0). The 7 days strength of soil samples S4, S5, and S6 with varying FA content (10, 20, and 30%, respectively) and with 9% cement was obtained to be 487, 562, and 640 kPa. Also, the 28 days strength of soil samples S4, S5, and S6 with varying FA content (10, 20, and 30%, respectively) and with 9% cement were obtained to be 906, 1537, and 1588 kPa. This indicates a substantial increment in the strength of the soil treated with 9% cement and a varying percentage of FA (from 10 to 30%). On comparing specimens S1 and S4; S2 and S5; S3 and S6, it could be identified that an increase in the small amount of cement content led to substantial improvement in UCS of the saline soil.

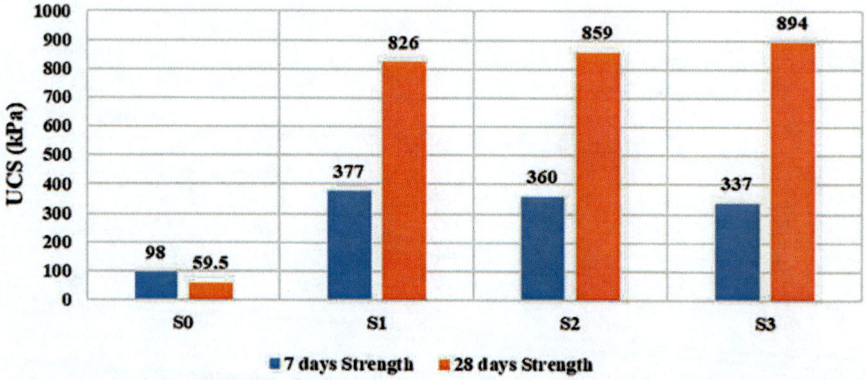

Fig. 3 Influence of varying FA content and 6% cement mix on UCS of Bhavnagar saline soil

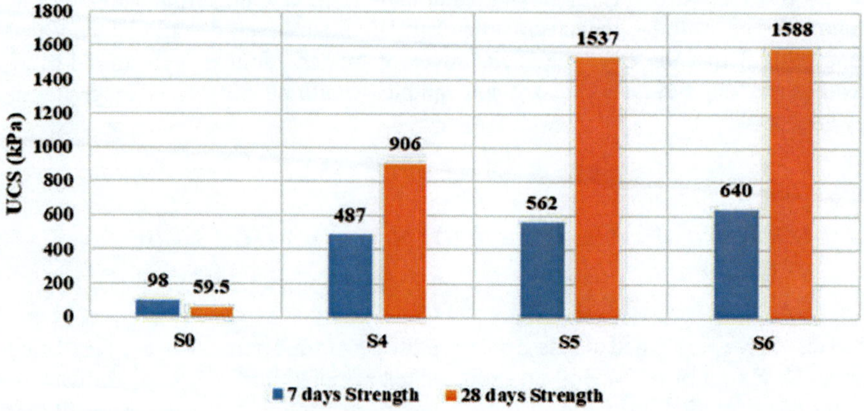

Fig. 4 Influence of varying FA content and 9% cement mix on the UCS of Bhavnagar saline soil

Substantial enhancement in the strength of treated samples would be accredited to the permanent cementation caused due to acclimatized amalgamation of FA and cement. The inclusion of FA and cement leads to the generation of pozzolanic reactions between FA and $Ca(OH)_2$ which is produced due to cement hydration. This results in the creation of CSH gel within the pore spaces of individual soil particles. The formation of thick CSH gel would induce higher binding within the soil skeleton, which would lead to the growth of improved cohesion amid soil particles (Cocka 2001; Senol et al. 2002). The inclusion of a higher percentage of FA and cement led to the development of a larger amount of CSH gel which might besiege the soil particles from all sides. As a result of the increased attraction, soil–FA–cement particles are held together, resulting in a sizeable increase in the strength and stiffness of the treated Bhavnagar saline soil. The decrease in the 7th day strength for samples S2 and S3 as compared to S1 would be because of partial development of CSH gel between soil particles, increased amount of FA (20 and 30%), and 6% cement content. It usually takes one or more weeks for Pozzolanic reactions to develop to an appreciable extent (Ma and Brown 1997).

From the field application perspective, as per IRC: SP:89 (Part II)-2018, if the stabilized soil has to be used for the base layer, a minimum strength of 4.5–7 MPa in 7/28 days is required. The obtained strength for all stabilized mixes does not fulfil the criteria and hence needs further investigation to improve strength.

As per IRC: SP-89 (Part II)-2018, recommended stabilized sub-base layer strength shall be in the range of 0.75–1.5 MPa. Thus, soil mixes of S4, S5, and S6 fulfils the criteria for the stabilized soil to be used in the sub-bass layer (Fig. 4).

4.3 Effect of FA–Cement Mix Treatment on Durability Characteristics of Bhavnagar Saline Soil

The durability test of stabilized soil samples was performed according to IS 4332-4 (1968). The tests were performed to judge the degree of permanence of the FA–cement mix. The resistance to weathering; evaluation of erosion due to traffic, speed of vehicles, and construction; and volume change due to soil–water chemistry are various examples of areas where durability research could be highly beneficial. In this study, Bhavnagar saline soil was stabilized with the combination of Class C fly ash (10, 20, and 30%) and Ordinary Portland Cement (6 and 9%) by weight. From the durability tests, it was found that the soil samples S1, S2, and S3 failed in the first cycle of wetting after a curing period of 7 days. Figure 5 shows the failed samples at all fly ash content (10, 20, and 30%) with 6% cement. However, soil mixes S4, S5, and S6 sustained 12 cycles of durability. The average weight loss at all fly ash content is shown in Fig. 6. The durability test with class C FA along with 9% cement can be used as construction material for its use as sub-base and base course of pavements. However, according to IRC: SP:89 (Part II)-2018, for ML type of soil classification

Maximum Allowable Weight Loss is 10%. Thus, combination S6 (30%FA + 9%C) satisfies the criteria of average weight loss for stabilized pavement construction.

Fig. 5 Failed samples (S1, S2, and S3) from durability test on Bhavnagar saline soil

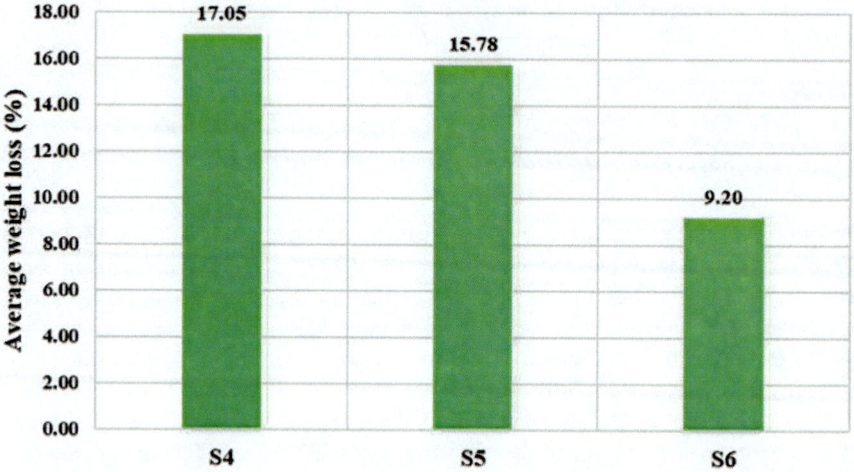

Fig. 6 Average weight loss from durability test at various FA dosages with 9% cement

Table 4 Influence of varying FA content and cement mix on the salinity of bhavnagar soil

Soil mix	Electrical conductivity (μS/cm)
S0	21,500
S1	8340
S2	8020
S3	7270
S4	8730
S5	7820
S6	7770

4.4 Effect of FA–Cement Mix Treatment on Salinity Characteristics of Bhavnagar Saline Soil

The extent of salinity in the untreated and treated mixtures was determined using the EC tests of the Bhavnagar saline soil and results are presented in Table 4. With the addition and increase in the fly ash–cement mixes, the EC was found to be reduced. This indicated a considerable reduction in soil salinity. With the addition of FA, water present in the soil mass might get embedded in the hollow spherical microscopic structure of fly ash particles (Hui et al. 2014). In addition to this, positive charges would be available due to free CaO present in the FA which would counterbalance the negative charges (Senol et al. 2002). Also, due to the formation of CSH gel on the inclusion of cement, solidification of the salt solutes could take place, which would have occasioned the diminution of EC of the treated soil.

5 Conclusions

The UCS and CBR of Bhavnagar saline soil is very low and EC is very high. The addition of FA (class C)–cement considerably improved the instinctive properties such as UCS and CBR of the Bhavnagar Saline soil. Treatment with 30% FA and 9% cement (S6) demonstrated the highest strength and stiffness. The effective modulus/ CBR value of stabilized saline soil mixes is noticeably higher than the virgin soil and capable of withstanding higher traffic loads. The EC of stabilized saline soil was found to decrease as salinity in the soil decreased. The durability for 9% cement with varying amounts of FA was found to show improved resistance to weathering, erosion, and environmental volume change. Thus, it could be concluded from the present study that the amalgamation of 30% FA and 9% cement mix (S6) was the most effective proportion in stabilizing saline soil as it enriched the strength, stiffness, salinity, and durability physiognomies of saline soil. Congress (2018, 2015) are refered as IRC in the text which stands for Indian Road Congress which is India's national body for laying down and designing standards for roads and highway construction and provides a stage for exchanging expertise and latest research development. Efficacious treatment and easy obtainability of FA (class C) and conventional cement make it an effective stabilization technique for saline soil in geotechnical and pavement applications.

References

Akpokodje EG (1985) The stabilization of some arid zone soils with cement and lime. Q J Eng GeolHydrogeol 18(2):173–180

Cockrell CF, Leonard JW (1970) Characterization and utilization studies on limestone modified flyash, vol 60. Coal Research Bureau, West Virginia University

Cokca E (2001) Use of class c fly ashes for the stabilization of an expansive soil. J Geotech Geoenviron Eng 127(7):568–573

Congress IR (2018) Guidelines for the design of flexible pavements. Indian code of practice, IRC, 37

Congress IR (2015) Guidelines for the design of Flexible Pavements for Low volume rural roads. Indian code of practice, IRC, SP 72

He B, Cai YL, Ran WR et al (2014) Spatial and seasonal variations of soil salinity following vegetation restoration in coastal saline land in eastern China. CATENA 118:147–153

IS 14767 (2000) Methods of test for soils-determination of the specific electrical conductivity of soils. Bureau of Indian Standards, New Delhi

IS 2720 part 16 (1987). Methods of test for soils: Part 16 Laboratory determination of CBR. Bureau of Indian Standards, New Delhi

IS 2720-Part 10 (1991) Methods of test for soils-Determination of unconfined compressive strength. Bureau of Indian Standards, New Delhi

IS 2720-Part 4 (1985) Methods of test for soils: Part 4 Grain size analysis. Bureau of Indian Standards, New Delhi

IS 2720-Part 5 (1985) Methods of test for soils: Part 5: Determination of liquid and plastic limit. Bureau of Indian Standards, New Delhi

Jakhrani AQ, Samo SR, Sobuz HR, Uddin MA, Ahsan MJ, Hasan NMS (2012) Assessment of dissolved salts concentration of seawater in the vicinity of Karachi. Int J Struct Civ Eng 1(2):61–69

Kaniraj SR, Havanagi VG (1999) Compressive strength of cement stabilized fly ash-soil mixtures. Cem Concr Res 29(5):673–677

Ma W, Brown PW (1997) Hydrothermal reactions of fly ash with $Ca(OH)_2$ and $CaSO_4$ $2H_2O$. Cem Concr Res 27(8):1237–1248

Mehta B, Sachan A (2017) Effect of mineralogical properties of expansive soil on its mechanical behavior. Geotech Geol Eng 35(6):2923–2934

Messad A, Moussai B (2016) Effect of water salinity on Atterberg limits of El-Hodna sabkha soil. Bull Eng Geol Environ 75(1):301–309

Moncmanová A (2007) Environmental factors that influence the deterioration of materials. Environ Deterior Mater 21:1

Murmu AL, Dhole N, Patel A (2020) Stabilisation of black cotton soil for subgrade application using fly ash geopolymer. Road Mater Pavement Des 21(3):867–885

Naveena S, Reddy GS (2015) Strength characteristics of expansive soils using eco-friendly Xanthan Gum. Int J Sci Res 6(6):2439–2442

Notman CF (2011) Durability testing of fine grained stabilised soils. Doctoral dissertation, University of Nottingham

Onyejekwe S, Ghataora GS (2015) Soil stabilization using proprietary liquid chemical stabilizers: sulphonated oil and a polymer. Bull Eng Geol Environ 74(2):651–665

Pandian NS, Krishna KC, Leelavathamma B (2002) Effect of fly ash on the CBR behaviour of soils. In: Indian geotechnical conference, vol 1. Allahabad, pp 183–186

Rahmat MN, Kinuthia JM (2011) Compaction of fills involving stabilisation of expansive soils. Proc Inst Civ Eng-Geotech Eng 164(2):113–126

Senol A, Bin-Shafique MS, Edil TB, Benson CH (2002) Use of class C fly ash for stabilization of soft subgrade. In: Fifth international congress on advances in civil engineering. Istanbul Technical University, Istanbul, Turkey, pp 25–27

Teerawattanasuk C, Voottipruex P (2019) Comparison between cement and fly ash geopolymer for stabilized marginal lateritic soil as road material. Int J Pavement Eng 20(11):1264–1274

USDA. United States Department of Agriculture (2014) https://www.nrcs.usda.gov. Accessed 20 Jan 2022

Wei L, Chai S, Guo Q, Wang P, Li F (2020) Mechanical properties and stabilizing mechanism of stabilized saline soils with four stabilizers. Bull Eng Geol Env 79(10):5341–5354

Recycled Polymer Application During Pile Boring Operations: A Review

Hitendrabhai S. Dewangan and Aashish A. Mehta

Abstract Pile foundation is widely used in the construction of marine structures. In offshore and marine piling work, drilling/boring supporting fluid is used to stabilize the borehole (excavation) prior to tremie concrete. The poly-fluid is a synthetic drilling granular polymer mainly used for boring operations of deep foundations. Disposal of polymers after using them in the sea pollutes the sea environment, which may be harmful to living organisms inside the sea. In this study, an attempt is made to review the methodology and tests conducted on recycled supporting fluid to provide economic and environmental solutions during pile boring operation. First, the selection of supporting fluid confirming desired characteristics is done. Thereafter fluid is utilized for boring operations. Moreover, the used slurry is reused by treating it with an additional dose of fresh material/polymers. The amount of fresh materials/polymer to be added can be decided based on (viscosity, density and pH value) test results. The quality of recycled fluid is checked based on obtained test results. After getting satisfactory results, it is reused and finally disposed of at suitable dumping places.

Keywords Recycled polymers · Offshore and Marine piling

1 Introduction

Polymer fluid is frequently used for borehole stabilization during pile foundation construction. Polymers are available in dry granular powder/liquid form used for borehole stabilization. Support fluid performance during boring operation measures by performing density, viscosity and P_H value tests. Based on previously available researches, Schünmann (2004), Lennon et al. (2006), Jefferis et al. (2011) and Jefferis

H. S. Dewangan (✉)
AFCONS Infrastructure Ltd. Chhara Jetty Project, Chhara, Gujarat, India
e-mail: hitendrabhai.d@afcons.com

A. A. Mehta
Shri Ramdeobaba College of Engineering and Management, Nagpur, India
e-mail: mehtaaa@rknec.edu

A. M. Krishna et al. (eds.), *Deep Foundations for Infrastructure Development in India*, Lecture Notes in Civil Engineering 373, https://doi.org/10.1007/978-3-031-37256-8_19

and Lam (2013), it was observed that polymer fluids are more operational and cost-effective compared to bentonite fluid. Moreover, Wheeler (2003) and Schumann (2004) reported that polymer fluids can be also used in environmentally sensitive areas. Apart from all these benefits, there is still a lot of debate about the selection of polymer-supporting fluid. As it is well-known that borehole stability depends on types of soil, environmental aspects, and construction methodology opted during boring operation. In addition, another important parameter is to check the feasibility of recycling supporting fluid after it's used. One of the important advantages of polymer fluid over bentonite fluid is that it can be recycled easily and can be used a number of times before final disposal. For economic consideration, the used polymer fluid must be added with fresh fluid for further reuse in boring operation. The amount of fresh materials/polymer to be added can be decided based on (viscosity, density and pH value) test results. In this paper, an attempt is made to review various factors which affect supporting fluid preformation. Moreover, the procedure adopted in use and reuse of supporting fluid during boring is also discussed at length. In addition, tests conducted to ensure fluid performance is also discussed. Finally, newly available polymers, their characteristics and their application on problematic site conditions during boring operation are discussed to fulfil economic considerations.

2 Background

Natural, modified natural and synthetic polymers are used as supporting fluids all over the world since 1990. The synthetic polymer includes partially hydrolyzed poly-acrylamides (PHPA) and cellulose-modified polymers like polyanionic cellulose (PAC) Polymers. Polymers are available in dry powder form taking 30 min to an hour to fully hydrate depending on temperature and make-up water quality. Polymers are also available in water/oil emulsion or in brine emulsion. The use of support fluid requires careful consideration of the environment in which it will be applied. The selection of fluid depends on the impacts of fluid characteristics in soil and groundwater during handling, storage and disposal of the support fluid and excavated materials.

3 Methodology—Boring Operation During Pile Foundation

- Survey marking-the positioning of jack-up barge and the survey point mark as per the given coordinates
- Liner pitching-liner pitching up to sea bed level and press with the help of vibro hammer as shown in Fig. 1
- Boring of pile-with the help of rig machine/winch machine continues drilling operations carried out as shown in Fig. 2

Fig. 1 Liner lifting and pitching

Fig. 2 Liner pitching and Pile boring by Hydraulic Rig

4 Factors Considered While Selecting Suitable Supporting Fluid

- Project and site dimensions: diameter, width, length and depth of foundation
- Elements to be constructed equipment (excavation, pumping, treatment, etc.) and the length of the casing (if used)
- excavation method (static drilling or reverse circulation drilling)
- soil conditions: geotechnical profile (e.g. type of soils, permeability, cohesion and chemistry), groundwater level and chemistry
- make-up water quality

5 Activities of Supporting Fluid Application During Pile Boring

5.1 Selection of Supporting Fluid Based on Soil Strata

Selection of supporting fluid based on available soil strata is an important task during boring operation. Based on previous literature, it is well-known that for sandy and silty soil, the polymer that is preferable for comparison is bentonite.

In addition to previous literature, one field example is discussed regarding the selection of supporting fluid. The current study of Naval Base Cochin Port in the south region in the year 2021, 400 m length jetty. The available strata are dark greyish black, stiff to very stiff, elastic silt of high plasticity (−3.0 m to −24.0 m); Dark greyish, stiff to hard, fat clay of high plasticity (−24.0 m to −30.0 m); Brownish grey, hard sandy clay of medium plasticity (−30 m to −34.50 m); Brownish grey, very stiff to hard fat clay of high plasticity (−34.50 m to −41.30 m); Blackish hard and peat strata (−41.30 m to −55.0 m) first, it is decided to use polymer for the stabilization of borehole. But because of unsatisfactory results, it is further decided to use an additional additive into the polymer for stabilization of borehole. Based on observation, it is highlighted that the polymer used for flushing provides a rational solution and is advantageous compared to bentonite. Also, as recycling of polymer is possible, it fulfilled economic and environmental considerations.

5.2 Methodology Adopted During Fluid Preparation

After the finalization of supporting fluid, the following methodology is adopted for Fluid preparation.

An appropriate supply of water must be available at all times. In case of insufficient quantity, a tank or tanks for storage purposes should be dedicated to the plant. The simplest approach to wetting fine powder is to gravity feed/pour the powder slowly into a stream of water exiting a hose. If the stream is passed over a flat plate (called a splash plate), the surface area of the wetting stream is increased. The user should carefully sprinkle the powder onto the plate avoiding any buildup of material. The softer the water, the higher the viscosities achieved with lower dosages and the harder the water, the more the polymer is required to achieve the specified viscosity. If the water is coming from a natural source such as a stream or pond, the level of organics and contaminants must be considered.

5.3 Installation of Mixing Tank

The mixing tank shall be installed taking into consideration the slurry's need for constant agitation. This can be achieved by dedicating a pump and compressed air system to the mixing tank. In the case of the usage of centrifugal pumps, the speed should not exceed 60−100 rpm.

If the water flow or the slurry's agitation are insufficient versus the rate of addition of the dry polymer granulate, the formation of inert agglomerates will occur, thus resulting in a higher consumption than initially expected. The capacity of the storage tanks should take into consideration the excavation's volumes and pretended production rates preferably exceeding 1.5 times. Fluid characteristics are checked as per the recommendation given in Fig. 3

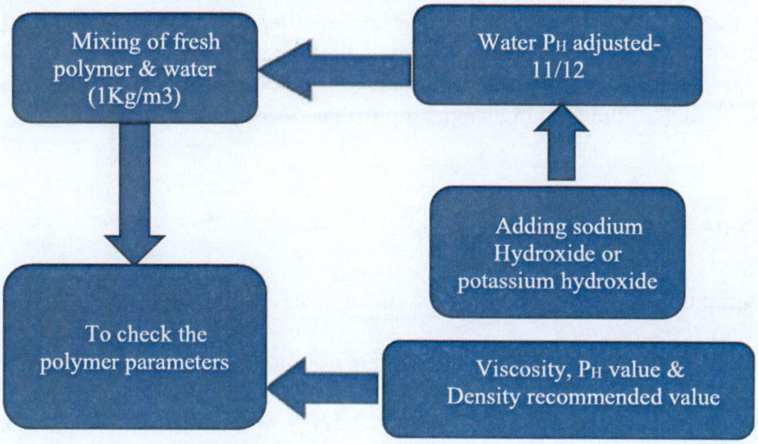

Fig. 3 Flow chart showing Fluid preparation for supporting fluid in pile boring operations

5.4 Storage and Sedimentation Tanks

Storage of slurry can be made in tanks and silos and should have the capacity for a minimum volume of 20 m^3. The sedimentation tank shall be open on top to allow easy cleaning and removal of sediments. The inclusion of two deflective vertical metal plates will assure the continuous flow of the slurry in a 'cascade' like mode, increasing the speed of sedimentation occurring between different compartments or sections. This design causes a significant increase in the sedimentation process. The inlet valve shall be directed to the top of the first deflective vertical plate. Both valves should be placed at the same height from the top of the deflective plates. Presence of salt in water affects the characteristics of polymer fluid. The salts may impact at the initial phase during hydration and have little effect on viscosity after or they may initially influence the hydration development and still persist to damage the chain through time. To avoid the ill effect of the presence of salt in water, before polymer mixing, water is to be treated with sodium hydroxide/soda ash at a dosage of 0.5 kg/cum. Casings may be partial or over the entire length of the drilled hole. The casings must be well sealed into the underlying stable ground or rock to ensure that soil does not unravel from behind the casing toe into the drilled hole during excavation. Unravelling can lead to cavities forming behind the casing which have the potential to cause the mixing of soil and water with concrete during the casing removal. These problems can also be encountered when the casing is advanced through unstable soil layers. A plug of soil within the casing may be required and/or excavation is not carried out below the toe of the casing.

Where possible, homogenization and recirculation of synthetic polymer support fluids should avoid centrifugal-type pumps where the fluid experiences high shear. Lam et al. (2010) report the results of an investigation into the effects of continued shear on the properties of polymer fluids and showed that there was significant degradation of the polymer fluid properties as a result of continuous circulation through a centrifugal pump. Alternate recirculation methods such as air bubblers or diaphragm pumps are preferred. Tanks with sloped bottoms tend to be more effective in recirculating without having stagnant portions of the tank.

5.5 Stabilization of the Borehole

The main purpose of any support fluid during excavation is the stabilization of the excavation.

Typical base cleaning methods include the use of cleaning buckets, which do not have teeth and are properly sealed so that they can 'scoop up' fines and soft debris from the very bottom of the excavation, It is important that these flocculating agents are uniformly mixed into the pile or panel otherwise they can lead to formation of a sticky sludge at the base. This sludge has been found to be difficult to remove from the base of the excavation. This can contribute to inclusions during the concreting phase. If required, the entire support fluid may have to be exchanged to avoid excessive sedimentation of suspended fines. Checks are made as per the recommendation given in Fig. 4.

The high calcium concentration and high pH of concrete will break down most polymers which assures that in most conditions there is no polymer left after a concrete pour. Any sediment that was present within the fluid will however remain in the excavation. When excavation proceeds, there is a possibility that loose soil particles of soil may fall in the drilling fluid. The presence of silt or sand may affect the density of the fluid affecting the working condition. When concrete is poured using tremie, it is required to have low-density drilling fluid, which should

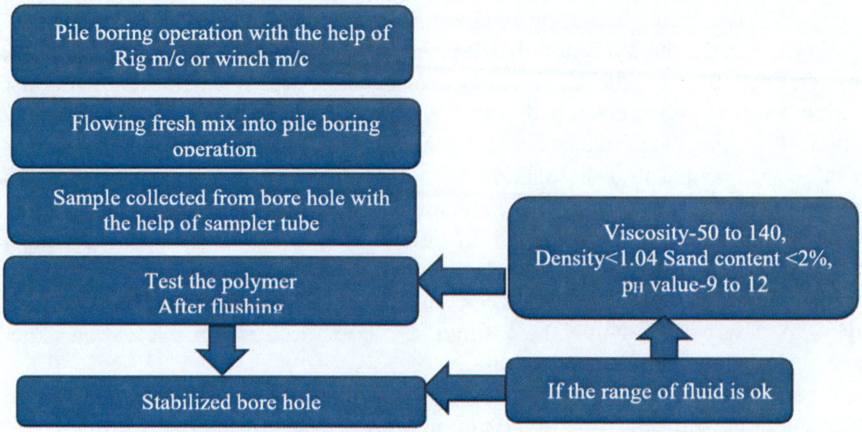

Fig. 4 Flow chart showing stabilization process for supporting fluid in pile boring operation

rise due to its lightweight. Moreover, silt if present in the suspension, may collect on the reinforcement; this may hinder the proper bonding between concrete and reinforcement.

5.6 *Slurry Regeneration*

Once the pile borehole has been completed and concreting is initiated, the used slurry will be brought back to the tanks designated for storage. To reuse the stored slurry for further pile borehole it must be regenerated. To regenerate, the reclaimed slurry will be transferred to the polymer mixing tank. Check and note the fluid characteristics of the regenerated slurry in the mixing tank. Based on the results, the amount of fresh materials to be added can be determined. Finally, initiate water flow and air agitation as required. Run the tests on the regenerated slurry before transferring it to the storage tank.

Some modern commercial polymer fluids are used for slurry (Table 1).

Table 1 Different types of polymers in the form of additive

Polymers	Occurrence	Composition/chemical reactions	Application
PolyMud	Dry granular biodegradable polymer should be mixed with water	Not exposed to high concentrations of chemical contaminants	Increased slurry viscosity and overall stabilization of the formation help to prevent fluid loss and improve the slurry's performance
MicroBond	Available in liquid form and mixed with water	Primary purpose is to strengthen the anionic and associative groups of the polymer's matrix	Used as a catalyst for the different sedimentation and cleaning processes of both the excavation or sedimentation tanks
AlfaBond	Liquid variation of the PolyMud polymer	Designed as a 'multi-role' polymer cross-linker	Main applications include fluid loss control and localized viscosity boosting
Super DS and cleaning	Water-based product that includes synthetic biodegradable polymers	Addition will cause the sudden precipitation of all solid elements suspended in the slurry to the bottom of either tank or excavation	Does not interfere with the resistance of the concrete or creates porosity

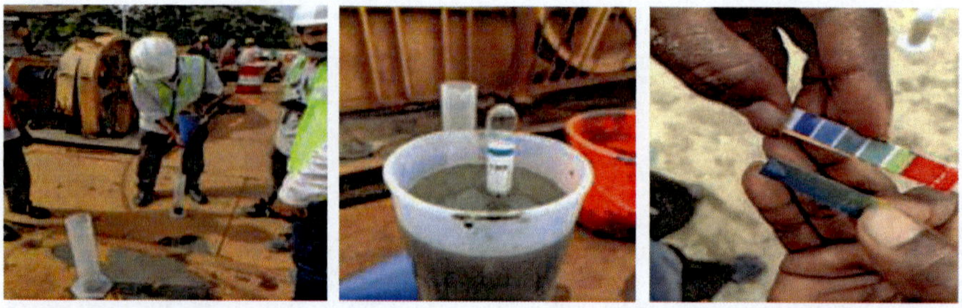

Fig. 5 Marsh Funnel and PH tests performed on supporting fluid to obtain desired recommended values during slurry regeneration

5.7 Test on Supporting Fluid

Post flushing check-polymer after flushing must be check before reused such as pH value, Viscosity, density and sand content.

Marsh Funnel is one of measures of viscosity of supporting fluid. To recycle the supporting fluid, before sending the fluid to borehole, the used fluid is treated with a fresh mix in storage tanks. Viscosity should also be checked whenever needed. The viscosity of supporting fluid should not be less than 60 s per quart (API Marsh Funnel) as per the recommendation of API. In case of higher viscosity, the performance of the polymer depends on the excavation's integrity and its consumption. The recommended % of polymer is 0.6–1.0 cum of water (Fig. 5).

5.8 Disposal

As per regulatory requirements, poly fluid polymer slurry may be broken down using household bleach (5% sodium hypochlorite) at a ratio of 1:1 of bleach and polymer used in the slurry. This is then mixed well with any recirculating pumps before disposal to landfill, or land farming.

Based on the above literature, a suitable supporting fluid is selected and it is used for the stabilization of the borehole. After confirming to following test results. Reutilization of supporting fluid—to avoid the sea pollution this bentonite it is treated with additional dosage (Fig. 6).

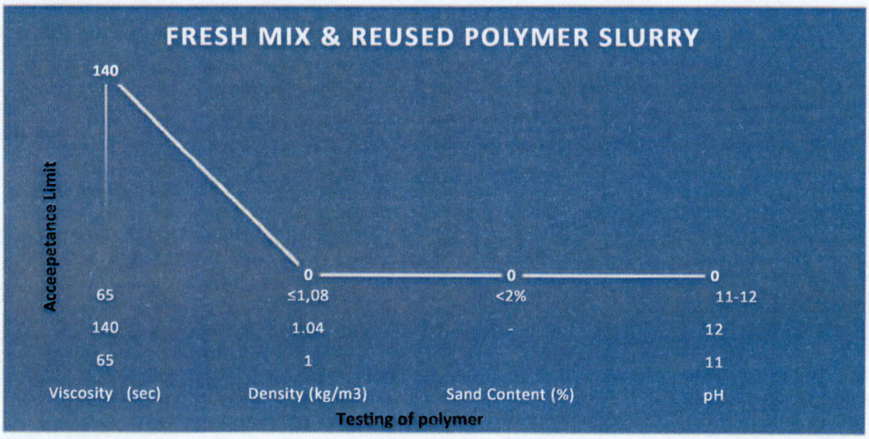

Fig. 6 Chart showing standard recommended values polymer value for maximum and minimum dosage

Table 2 Additive to be added for efficient performance of supporting fluid in extreme soil conditions

Product description	Application
Sodium Silicate	Zone stabilizer
Aluminium Silicate	Catalyst
Sodium Hydroxide	pH adjustment

5.9 *The Performance of Supporting Fluid in Problematic Soils*

In order to improve the performance of supporting fluid, the below mentioned polymers are used in extreme soil conditions like problematic granular soils such as loose sand, gravel, cobbles, and other unstable formations. Table 2 represent various additives that may be added to get the desired performance results.

6 Conclusions

Based on the previous literature and available test results, it was observed that polymer fluid can be efficiently used as supporting fluid for pile boring operations. In some extreme cases, for boring problematic soil, appropriate modern polymers may be used. The performance of supporting fluid must be checked by conducting recommendation suggested tests. Supporting fluid confirming desired results will be reasonably used for boring operations.

References

Jefferis S, Troughton V, Lam C (2011) Polymer systems for fluid supported excavations. In: Geotechnical issues in construction: short paper series and 2nd conference held on 2 November 2009, at CIRIA, London. Construction Industry Research and Information Association, pp 7–12

Jefferis SA, Carlos L (2013) Polymer support fluids: use and misuse of innovative fluids in geotechnical works. In: 18th international conference on soil mechanics and geotechnical engineering. Presses des Ponts, pp 3219–3222

Lennon DJ, Ritchie D, Parry GO, Suckling TP (2006) Piling projects constructed with vinyl polymer support fluid in Glasgow, Scotland. In: Proceedings of 10th international conferences on piling and deep foundations, deep foundations institute, Hawthorne, NJ, pp 499–506

RP, API (2009) Recommended practice for field testing water-based drilling fluids. In: API Recommendation 13B-1, ISO 10414: 2001

Schünmann D (2004) Fisherman's friend. Ground Eng 37(12):17

Wheeler P (2003) Piles unlock polymer potential. In: European Found, pp 8–9

Research, Experimental and Numerical Methods in Deep Foundations and Deep Excavation Technologies

Reliability Analysis of Load Bearing Capacity of Single Pile Placed Adjacent to Existing Buried Conduit or Utility Tunnel

Monica Malhotra and Amit Srivastava

Abstract In a congested city or urban area, the so-called "lifeline of the city", i.e., Gas, Electricity, telecommunication, as well as other utilities or facilities are supplied or transported underground through buried pipes or utility tunnels. Sometimes these buried pipes or utility tunnels are deep underground and interfere with the foundation of the new structure. It is highly probable that a pile foundation is placed near an existing underground buried pipe or utility tunnel and may influence its load bearing capacity. Hence, the estimation of reduced load bearing capacity of the pile foundation placed near the existing buried conduit or utility tunnel is a complex soil-buried structure interaction problem. It is further complicated owing to uncertainty in the estimation of geotechnical parameters of foundation soil due to measurement and testing errors, inherent soil variability, and model translation. Under such circumstances, load bearing capacity is best assessed in the probabilistic framework or through reliability analysis. The present study demonstrates how the concept of response surface method (RSM) in conjunction with Numerical Analysis and First Order Second Moment (FOSM) methods can be combined as per the reliability analysis. The study answers the following question "reliability in reduction in the load bearing capacity of a single pile placed near the existing underground pipe or utilities tunnel". It is demonstrated that the approach provides first-hand action on the reduction in the load bearing capacity and then estimates the reliability of that estimation in terms of an index called the Reliability Index (β). Such preliminary investigation can be useful for advocating the detailed analysis as well as in the decision-making for relocation of pile foundation.

Keywords Reliability analysis · Buried pipe · Numerical analysis · Response surface method

M. Malhotra (✉)
Department of Civil Engineering, PDM University, Bahadurgarh, Haryana, India
e-mail: monicamalhotra28@gmail.com

A. Srivastava
Department of Civil Engineering, Graphic Era (Deemed to be University), Dehradun, Uttarakhand, India
e-mail: amitsrivastava.ce@geu.ac.in

A. M. Krishna et al. (eds.), *Deep Foundations for Infrastructure Development in India*, Lecture Notes in Civil Engineering 373, https://doi.org/10.1007/978-3-031-37256-8_20

1 Introduction

With rapid infrastructure development, the co-existence of deep foundation structures and buried pipelines or utility tunnels for supplying cooking gas, electrical cables, water distribution, underground transport, etc., is a common problem. The consequence of this complex soil-pile-pipe interaction problem has to be studied in terms of assessing the reduction in the load bearing capacity of the deep foundation system. If there is a chance that the deep foundation of a superstructure is to be placed near the existing buried pipe or utility tunnel, it is important to assess the change in the load bearing capacity to design the foundation system, accordingly. From the published literature, it is noted that the effect of tunneling on the response of ground and pile foundation is extensively studied in the light of analytical, experimental, and numerical approaches. Although, there is a chance that the foundation has to be placed near the already existing underground utility tunnel or buried pipe and, accordingly, the foundation system has to be designed taking into consideration the % reduction in the load bearing capacity. This aspect has been studied by Malhotra et al. (2019) and it is reported that there is a considerable reduction in the load bearing capacity of the single pile when installed near the existing underground pipe or utilities tunnel. The obvious reason is the interference of the tunnel within the zone of influence of stress distribution of the pile that derives the resistance from the surrounding ground. The percent reduction in the load bearing capacity helps in designing the pile foundation accordingly if the constraint in space does not allow for the change in the location.

2 Issue of Uncertainty in Geotechnical Parameters

The complexity of the soil-pile-pipe problem gets further aggravated owing to the uncertainty in the estimation of geotechnical parameters. The uncertainty in the estimation of input geotechnical parameters brings uncertainty in the output response and it is obvious to ask for the reliability of the calculated value of the reduction in the load carrying capacity of the pile to bring confidence in the designer. Recently, considerable attention has been received to study the effect of innate random variations of soil properties on the response of geotechnical structures. Soil properties possess a wide range of uncertainties with significant variation from place to place and over short distances. Consequently, deterministic estimation of the load bearing capacity using a factor of safety approach seems more qualitative rather than quantitative as the choice of the factor of safety depends on the confidence, experience, engineering judgment, and knowledge level of the expert involved. Although not a substitute but reliability analysis is a widely used method by researchers to overcome the constraint of a deterministic approach. By treating uncertainty in a mathematical framework and providing the basis of decision-making, researchers advocated a reliability-based approach for solving geotechnical engineering problems (Baecher and Christian 2005; Phoon and Ching 2015; Li and Lo 2020).

3 Literature Background

Loganathan et al. (2001) used Boundary Element Method (BEM) to develop an analytical solution for estimating the displacement field near a tunnel and studied its influence on nearby existing single pile or pile group. Closed solutions were presented to calculate the pile displacement due to tunneling induced displacement. The results were also verified using a finite difference approach along with a parametric study using different pile diameters and pile lengths. Surjadinata et al. (2006) used a combination of boundary element (BEM) and finite element methods (FEM) for studying the tunneling effects on existing piles and proposed a solution for estimating the horizontal ground displacement along with a single pile. The results of the analysis were further verified by existing case studies and computational advantages were highlighted.

Poulos (2006) first reviewed the evolution of pile group settlement analysis and then presented the pile settlement cases using numerical methods for both cases of piles installed in sand and clay. The loss in axial capacity has been observed and the results of the analysis were compared with experimentally measured results already available in the literature. The authors also emphasized that complex geotechnical engineering problems of such nature can be analyzed in a systematic and simplified manner and also cautioned about the setbacks of theoretical understanding without practical exposure.

Soomro et al. (2015) studied the effect of tunneling on the pile group, and the associated settlement and tilting of the pile group. Finite element analysis was performed to study the influence of tunneling on pile group of 2 × 2 pattern after placing foundation in saturated stiff clay. A parametric study was done to study the effect of different covers on the diameter of tunnel ratio. It was noted that there was a reduction in the pipe capacity from 14% to 23% when the ratio was increased from 1.5 to 3.5.

Al-Omari et al. (2016) investigated the load bearing capacity of the piled raft–tunnel system using the ANSYS finite element program and verified it experimentally. The effect of pile length, number of piles, and distance between pile and tunnel were studied. A parametric study was performed by changing the pile length, number of piles, and spacing between the pile tip and crown of the tunnel. Soomro et al. (2016) further investigated the pile responses due to the presence of side-by-side twin tunnels in saturated stiff clay through three-dimensional coupled-consolidation numerical parametric study and centrifuge testing. Soomro (2021) studied the effects of twin stacked tunnels placed at different depths and having different construction sequence (TS, BS, and US) on the (2 × 2) piled raft foundation. The three-dimensional analysis was performed using hypo-plasticity soil model. It was noted that the largest settlement was caused by the US type and the largest tilting was caused by the BS type. With this background, it is noted that reliability analysis of such a complex problem, i.e., consideration of the uncertainty in the input geotechnical parameters and assessment of the load bearing capacity of pile near the existing buried conduit in the probabilistic framework is not yet studied.

4 Objective of the Present Study

Keeping in view the said aspect, the following objectives of the present study are formulated:

- To assess the reliability of % reduction in the pile capacity with due consideration of the extent of uncertainty involved in the input soil parameters
- To present a way to perform the reliability analysis using the concept of response surface methodology (RSM); combined with numerical analysis and FOSM reliability-based approach.

It is assumed that pile installation in the vicinity of the existing buried pipe does not affect the in situ geotechnical properties. The probability of adjoining structures getting affected due to pile installation is not taken into consideration assuming such disturbances are taken care of by estimating geotechnical properties after the installation is over.

5 Quantification of Uncertainty

Conventionally, geotechnical parameters are given as input in any of the analytical approaches or numerical simulations and they are considered as uniformly constant with the assumption that soil deposit is homogeneous and isotropic material. But in nature, it is never true and, to an extent, almost all soil deposits are having spatially variable properties and under extreme situations it is highly heterogeneous. The data used for geotechnical structure designs is selected from the site investigations and field as well as laboratory test results or through correlations. For this reason, data related to soil properties possess a wide range of uncertainties with significant variation from place to place and over short distances. The uncertainty in the estimation of input geotechnical parameters is expressed statistically in terms of mean and variance (or standard deviation). The coefficient of variation (CoV%) defines the extent of variation about the mean and it is dimensionless defined as the ratio of standard deviation (σ) to mean (μ) of the parameter. In case of the absence of site-specific data, one may consider the data published in the literature for assessing the extent of variation in the selected geotechnical parameters (Kulhawy and Trautmann 1996; Lacasse and Nadim 1996; Duncan 2000; Uzielli et al. 2006). Now, the effect of this uncertainty in the geotechnical parameters on the output response is studied in a probabilistic framework and most widely through reliability analysis that considers associated risk and probability of failure while measuring the performance of the geotechnical system.

6 Reliability Analysis

In general terms, reliability is defined as the probability of success of a system in a given environment of uncertainty over a period of time. Mathematically, as per Hasofer and Lind (1974), the reliability index is defined as the shortest distance between the failure surface and the origin of standard normal space (U-space). If capacity is defined as C and Demand is taken as D, the performance function is defined as M = C-D; M = 0 line defines the failure line and M > 0 define safe zone and M < 0 defines failure zone. Considering the variation in the input and output response, C and D will also be random variable, and the extent of variation can be expressed in terms of mean and variance in estimated values of C and D. If C and D are plotted in standard normal space, i.e., $C\prime = \frac{C-\mu_C}{\sigma_C}$ and $D\prime = \frac{D-\mu_D}{\sigma_D}$, M will be a straight line and the distance between the origin and failure line will be the reliability index value as depicted in the following Fig. 1. The present case can further be complicated by considering more variables, adding correlation among variables, and also defining non-linear failure surface. One has to develop a computational algorithm to solve such complicated cases with constrained minimization, Lagrangian multiplier approach, etc. The problem can also be solved using EXCEL, or writing code in mathematical software like MATLAB. A detailed discussion on the topic can be obtained in Baecher and Christian (2005).

After reliability index evaluation, it is important to judge whether the calculated value is acceptable or not. USACE (1997) made recommendations on the target reliability index values to assess the performance of the geotechnical system. A reliability index value of more than 5.0 means excellent, 5.0–4.0 relates to good performance and 4.0–3.0 for above average performance and less than 3.0 indicates poor performance. Acceptance of the reliability index level depends on technological and social acceptance along with other factors like the importance of structure, geographical location, etc. Assessment of reliability gives an answer to the most generic question "How much you are sure about your estimated value and what is the rationale behind it?" Can you provide a mathematical basis for reaching your conclusion and the logic behind it? Now-a-days engineers do not much rely on qualitative assessment, but they are looking for a more justifiable answer to the problem statement.

Estimation of the reliability index requires an explicit functional relationship between output and corresponding input geotechnical parameters. Such explicit functions are not always available, and even if available, they are too complicated and difficult to perform probabilistic calculations in a conventional way and require simulations technique like Monte Carlo Simulation (MCS) which is both computationally demanding and time-consuming. Alternatively, under such circumstances, it is possible to develop an approximate functional relationship between input parameters

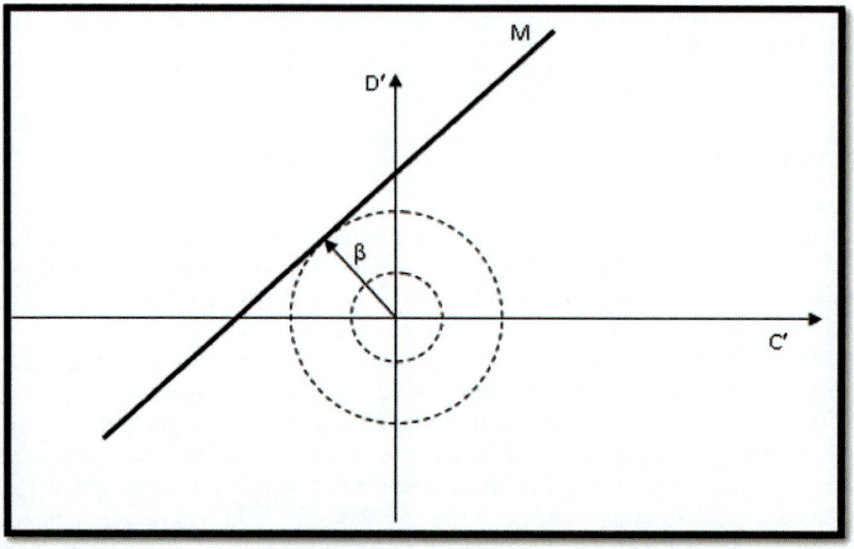

Fig. 1 Definition of reliability index for two variables as per Hasofer and Lind (1974)

and output response using the concept of response surface methodology and then use it for reliability analysis using a conventional probabilistic approach.

7 Response Surface Method (RSM)

To establish explicit functions between input and output variables, the concept of RSM is used. The RSM is commonly used in process and product optimization (Myers et al. 2016). For simple cases like uncorrelated normally distributed input parameters, using the RSM approach, explicit functions are developed through the 2^n full factorial design approach as explained, where n is the number of input variables. The objective is to first define the 2^n number of combinations for n input variables, and for each combination, the output response is obtained either through numerical simulation or through complex equations or design charts available. A linear functional relationship between input and output is then established using regression analysis available as an inbuilt option in Microsoft Excel based on the least square error approach. The values of R^2 and adjusted R^2 close to 1.0 confirm the "goodness of fit" statistical hypothesis test (Haldar and Mahadevan 2000). The approach has been well utilized in the past by researchers like Wong (1985), Humphreys and Armstrong (1993), and Tandjiria et al. (2000) for solving geotechnical engineering problems like reliability analysis of slopes, piles, etc. In the present study, the approach is used in the reliability analysis of % reduction in the load bearing capacity of a single pile installed near an existing underground pipe or utilities tunnel with the given problem statement.

8 Problem Statement

Reliability analysis of % reduction in the load bearing capacity of a single pile installed near an existing underground pipe or utilities tunnel. Firstly, an explicit functional relationship between input geotechnical parameters and output (% reduction in load bearing capacity) is established using RSM combined with numerical analysis. The linear function is then used to calculate the mean and variance of % reduction in the load carrying capacity of a single pile and is then utilized for the reliability analysis. The purpose is to know "What is the reliability, in terms of reliability index, of the estimated value of % reduction in the load carrying capacity of the pile in a given environment of uncertainty in the geotechnical parameters".

For the numerical analysis, the commercially available finite element tool PLAXIS 2D is utilized. To estimate % reduction in the load bearing capacity, load settlement curves are first obtained for the following cases, i.e., (a) in the absence of existing buried conduit or utility tunnel and (b) in the presence of existing buried conduit or utility tunnel. % reduction in the capacity is estimated by comparing the results of the two cases after estimating their ultimate load. Geotechnical parameters are considered to follow the Mohr–Coulomb failure hypothesis with non-associated flow rule. The assumed value of the coefficient of variation (CoV%) in input geotechnical parameters is taken from the published literature. The objective is to demonstrate the process of performing the reliability analysis, i.e., estimating the reliability index value of the % reduction in the load bearing capacity of the single pile installed near the existing underground pipe or utilities tunnel. The mean of geotechnical properties and their variation in terms of CoV(%) and standard deviation (σ) is presented in Table 1.

Table 1 Input parameters assumed for the reliability analysis

Geotechnical properties	Ep (KPa)	ϕ(°)	γ(kN/m^3)	Es (MPa)	ν
Mean (μ)	2,00,000	30°	17.0	20.0	0.45
COV (%)	1.00	15.0	10.0	35.0	5.0
Standard deviation (σ)	2000	4.5°	1.70	7.0	0.023

9 Numerical Modeling Procedure

For the load–settlement curves to be obtained for two scenarios, numerical analysis is performed using a finite element tool. Figure 3 depicts the steps involved in the numerical analysis. The first step is defining the geometry of the problem. It is just like plotting line diagrams of the problem on graph paper with some scale. Plaxis allows plotting the geometry with a 1:1 scale by defining the size of the graph in x- and y-direction through the initial general setting along with the setting of units. The geometry of the problem is plotted as per the problem statement shown in Fig. 2. The next step is setting the boundary conditions. The bottom of the boundary is fixed in both directions and vertical boundaries have fixity in the horizontal direction. This is the standard fixity condition. After fixing boundaries, material properties are assigned to the soil for Mohr–Coulomb model and pipe modeled as tunnel element. The rest of the operating procedures includes mesh generation, setting GWT, and generating initial conditions of effective stress. In the present study, the GWT table effect is neglected. Now, the load is applied on the top of the pile, and settlement is measured by selecting a node. The load settlement curve is then obtained by incrementally increasing the load on the pile tip. Figure 4 depicts the numerical model created as per the problem statement and the load applied on the pile tip.

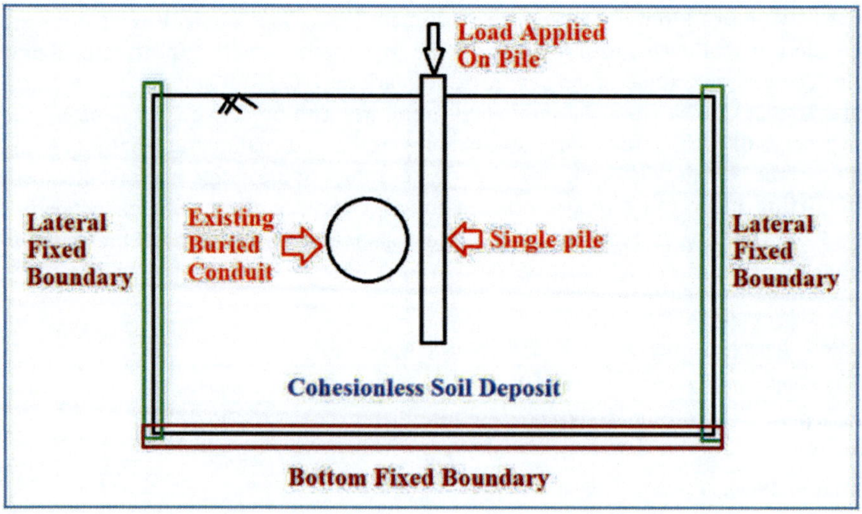

Fig. 2 Schematic of the problem statement (single pile placed in the vicinity of buried conduit or utility tunnel)

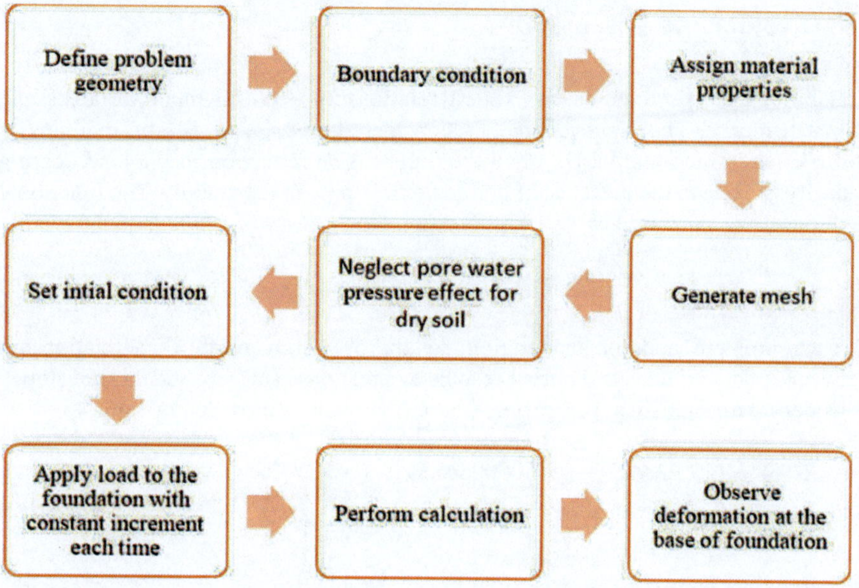

Fig. 3 Flow diagram of plaxis 2D

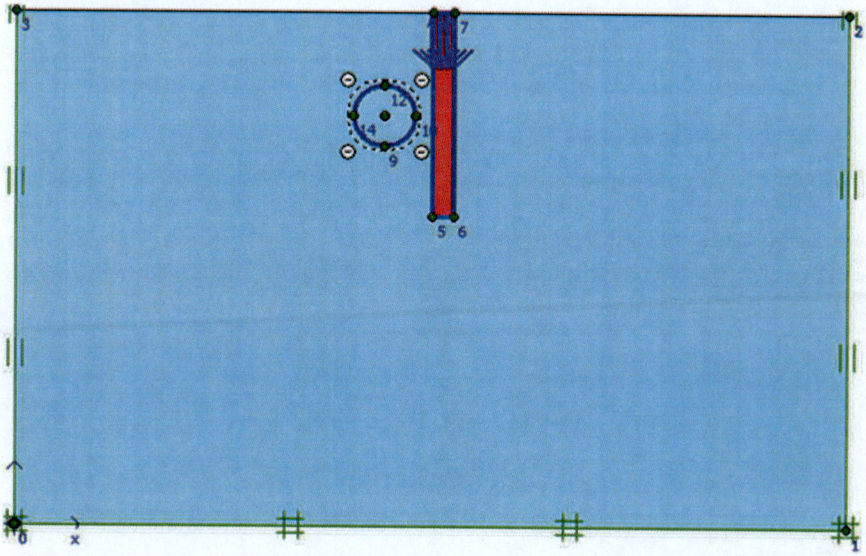

Fig. 4 Numerical model using PLAXIS 2D software (with prescribed displacement)

10 Results and Discussion

Reliability analysis requires an explicit relationship between input variables and output response. The input parameters considered are ϕ, γ, ν, Elastic modulus of soil (E_s) and pipe material (E_p) and the output is % reduction in the load bearing capacity (R) due to the presence of buried flexible pipe in the vicinity. The functional relationship is expressed as

$$R = f(\phi, \gamma, \upsilon, E_s, E_p) \tag{1}$$

For the problem under consideration, to approximately establish a relationship between input variables and output response, a response surface model is developed considering the statistical parameters of the input values provided in Table 1

$$\begin{aligned} R(\%) = {} & (49.32842) + (-0.000034)E_p + (-0.37772)\varnothing + (-1.39887)\gamma \\ & + (0.001007)E_s + (18.849)\upsilon \end{aligned} \tag{2}$$

$R^2 = 0.83$; R^2 (adjusted) $= 0.76$.

Assuming that the input geotechnical parameters are uncorrelated and normally distributed continuous random variables, the statistical parameters of output response, i.e., mean and standard deviation are obtained as $\mu_R = 15.92\%$ (Mean of expected % reduction) and $\sigma_R = 2.95\%$ (standard deviation in the expected value of % reduction in the load carrying capacity due to presence of buried pipe or tunnel) in the given amount of uncertainty in input parameters.

The demand D can be fixed as follows: with a factor of safety of 3.0, if % reduction in the load carrying capacity can be tolerated as 33%, with no variation in the demand. With these inputs, the associated reliability index (β) is evaluated as 5.88. Hence, as per USACE (1997), the corresponding reliability index is under the excellent category and one can easily tolerate such a reduction in the pile capacity in a given environment of uncertainty of geotechnical parameters.

11 Conclusions

- Consideration of variation in the input parameters and extent of variation is appropriately quantified in the probabilistic framework and the associated reliability in estimating the percent reduction in load carrying capacity helps in accepting or rejecting the challenges faced while placing the pile near the existing buried conduit or utility tunnel. A percent loss in the reduction of the load bearing capacity of the pile should be viewed in the light of uncertainties involved in the input parameters.
- The response surface method played an important role in establishing an explicit relationship between input variables and output response. The study demonstrated that the approach of utilizing the concept of RSM with numerical analysis is the most effective way to perform the reliability analysis as it requires less computational time and not a time-consuming process. The procedure demonstrates how to perform the reliability analysis combining RSM and Numerical Analysis and it can be followed in any scenario, even for a group of piles.

12 Annexure I

S.No	Ep(kPa)	ϕ	γ	E (kpa)	M	Capacity without pipe	Capacity with pipe	% Reduction
1	2.03E + 05	37.4	19.805	31,550	0.49	1340	1230	8.96
2	2.03E + 05	37.4	19.805	31,550	0.41	582.98	539.42	9.19
3	2.03E + 05	37.4	19.805	8450	0.49	534.97	518.92	4.87
4	2.03E + 05	37.4	19.805	8450	0.41	524.92	509.7	4.80
5	2.03E + 05	37.4	14.195	31,550	0.49	637.96	504.39	22.50
6	2.03E + 05	37.4	14.195	31,550	0.41	442.94	414.46	8.69
7	2.03E + 05	37.4	14.195	8450	0.49	397.82	383.25	6.18
8	2.03E + 05	37.4	14.195	8450	0.41	399.81	390.38	4.86
9	2.03E + 05	22.6	19.805	31,550	0.49	701.34	671.21	5.72
10	2.03E + 05	22.6	19.805	31,550	0.41	514.09	502.26	4.25
11	2.03E + 05	22.6	19.805	8450	0.49	515.17	505.09	3.90
12	2.03E + 05	22.6	19.805	8450	0.41	512.44	502.12	3.97
13	2.03E + 05	22.6	14.195	31,550	0.49	574.27	549.37	6.08
14	2.03E + 05	22.6	14.195	31,550	0.41	412.98	394.12	6.99
15	2.03E + 05	22.6	14.195	8450	0.49	382.98	372.04	5.47
16	2.03E + 05	22.6	14.195	8450	0.41	378.98	354.12	9.20
17	1.97E + 05	37.4	19.805	31,550	0.49	1328	1230	8.13
18	1.97E + 05	37.4	19.805	31,550	0.41	570.8	531.42	8.65

(continued)

(continued)

S.No	Ep(kPa)	ϕ	γ	E (kpa)	M	Capacity without pipe	Capacity with pipe	% Reduction
19	1.97E + 05	37.4	19.805	8450	0.49	534.07	517.92	4.90
20	1.97E + 05	37.4	19.805	8450	0.41	524.92	509.7	4.80
21	1.97E + 05	37.4	14.195	31,550	0.49	627.96	500.39	21.91
22	1.97E + 05	37.4	14.195	31,550	0.41	442.94	414.46	8.69
23	1.97E + 05	37.4	14.195	8450	0.49	397.82	387.25	5.17
24	1.97E + 05	37.4	14.195	8450	0.41	399.81	399.38	2.61
25	1.97E + 05	22.6	19.805	31,550	0.49	701.34	676.21	5.01
26	1.97E + 05	22.6	19.805	31,550	0.41	514.09	503.26	4.05
27	1.97E + 05	22.6	19.805	8450	0.49	515.17	505.09	3.90
28	1.97E + 05	22.6	19.805	8450	0.41	512.44	502.12	3.97
29	1.97E + 05	22.6	14.195	31,550	0.49	574.27	549.37	6.08
30	1.97E + 05	22.6	14.195	31,550	0.41	412.98	394.12	6.99
31	1.97E + 05	22.6	14.195	8450	0.49	382.98	372.04	5.47
32	1.97E + 05	22.6	14.195	8450	0.41	378.98	357.12	8.41

References

Baecher GB, Christian JT (2005) Reliability and statistics in geotechnical engineering. Wiley, London, UK

Duncan JM (2000) Factors of safety and reliability in geotechnical engineering. Am Soc Civ Eng ASCE J Geotech Geoenviron Eng 126(4):307–316

Haldar A, Mahadevan S (2000) Probability, reliability and statistical methods in engineering design. Wiley, New York

Hasofer AM, Lind MC (1974) An exact and invariant first order reliability. Am Soc Civ Eng ASCE J Eng Mech 100:111–121

Humphreys MP, Armstrong LW (1993) Assessing the sensitivity of numerical models using response surface methodology. In: The proceedings of the conference of probabilistic methods in geotechnical engineering, Canberra, Australia, pp 135–144

Kulhawy FH, Trautmann CH (1996) Estimation of in-situ test uncertainty. In: Shackleford CD, Nelson PP, Roth MJS (eds) Uncertainty in the geologic environment: from theory to practice, geotechnical special publication No. 58. ASCE, New York, pp 269–286

Lacasse S, Nadim F (1996) Uncertainties in characterising soil properties. In: Shackleford CD, Nelson PP, Roth MJS (eds) Uncertainty in the geologic environment: from theory to practice, geotechnical special publication No. 58. ASCE, New York, pp 49–75

Li KS, Lo S-CR (2020) Probabilistic methods in geotechnical engineering. In: Proceedings of the conference, Canberra, 10–12 February 1993, CRC Press, Taylor and Francis group, NY

Loganathan N, Poulos HG, Xu KJ (2001) Ground and pile-group responses due to tunnelling. Soils Found 41(1):57–67

Malhotra M, Sahu V, Srivastava A, Misra AK (2019) Impact of pile foundations adjacent to tunnels in sandy stratum. Sadhana (Springer). https://doi.org/10.1007/s12046-019-1169-y

Myers RH, Montgomery DC, Anderson-Cook CM (2016) Response surface method: process and product optimization. Wiley

Phoon KK, Ching J (2015) Risk and reliability in geotechnical engineering. CRC Press, Taylor and Francis Group, NY

Poulos HG (2006) Pile group settlement estimation—research to practice, ASCE geotechnical special publication. In: GeoShanghai international conference, June 6–8. https://doi.org/10.1061/40865(197)1

Al-Omari RR, Al-Azzawi AA, AlAbbas KA (2016) Behavior of piled rafts overlying a tunnel in sandy soil. Geomech Eng **10**(5). https://doi.org/10.12989/gae.2016.10.5.000

Soomro MA, Ng CWW, Liu K, Memon NA (2016) Pile responses to side-by-side twin tunnelling in stiff clay: effects of different tunnel depths relative to pile. Comput Geotech 84:101–116. https://doi.org/10.1016/j.compgeo.2016.11.011

Soomro MA, Hong Y, Ng CW, Lu H, Peng S (2015) Load transfer mechanism in pile group due to single tunnel advancement in stiff clay. Tunnelling Undergr Space Technol 45:63–72. https://doi.org/10.1016/j.tust.2014.08.001

Soomro MA (2021) 3D finite element analysis of effects of twin stacked tunnels at different depths and with different construction sequence on a piled raft. Tunnelling Undergr Space Technol 109. https://doi.org/10.1016/j.tust.2020.103759

Surjadinata J, Carter JP, Hull TS, Poulos HG (2006) Analysis of effects of tunneling on single piles. In: Bakker et al. (eds) The conference proceedings of geotechnical aspects of underground construction in soft ground. Taylor and Francis Group, London. ISBN 0415391245

Tandjiria VI, The CI, Low BK (2000) Reliability analysis of laterally loaded piles using response surface methods. Struct Saf 22:335–355

USACE 1997. Risk-based analysis in geotechnical engineering for support of planning studies, engineering and design. US army corps of engineers, department of army, Washington, DC, 1997, pp 20314-100.

Uzielli M, Lacasse S, Nadim F, Phoon KK (2006) Soil variability analysis for geotechnical practice. In: Conference: proceedings of the 2nd international workshop on characterisation and engineering properties of natural SoilsAt: Singapore, December 2006. https://doi.org/10.1201/NOE0415426916.ch3.

Wong FS (1985) Slope stability and response surface method. Am Soc Civ Eng ASCE J Geotech Geoenviron Eng 111(1):32–53

Mono-piles and Other Advanced Deep Foundation Systems

A Review on Pedestal Piles

M. Shubha, G. Srilakshmi, and S. V. Venkatesh

Abstract Choosing for the best option of pile foundations to overcome large lateral and uplift loads for the foundations in transmission line towers/offshore environment is still being a concern to achieve the required capacities but also by considering the suitability aspect in the environment. The objective of this paper is to review the methods that have been done for knowing the uplift resistance and lateral resistance of pedestal piles. Investigations by different methods such as experimental studies and numerical and analytical studies have been done by various researchers in order to evaluate the response of a belled pile. These methods are discussed in this paper and the comparison with straight shaft piles is based upon the test and analysis. The results indicated that the uplift resistance and lateral resistance were significantly increased with the adoption of belled piles than straight shaft piles.

Keywords Pile foundations · Pedestal pile · Uplift resistance · Straight shaft pile

1 Introduction

A pile is a long slender structural member manufactured from concrete, steel or wood. Piles can be precast or cast in situ and they can be either driven into the soil or holes can be drilled and concrete is filled. The main purpose of pile foundation is to transfer the heavy structural loads to a deeper stratum when the soil at a shallow depth does not possess sufficient load bearing capacity.

Belled Piles are also known as piles with an enlarged base, and these piles are widely used in the foundations of structures such as transmission towers, tall chimneys, etc., (Liu et al. 2021). These structures are often subjected to overturning loads imposed by wind, waves, etc. (Gao et al.). The belled and multi-belled piles

M. Shubha (✉) · S. V. Venkatesh
PES University, Bangalore, India
e-mail: shubham@pes.edu

G. Srilakshmi
Nagarjuna College of Engineering and Technology, Bangalore, India

A. M. Krishna et al. (eds.), *Deep Foundations for Infrastructure Development in India*, Lecture Notes in Civil Engineering 373, https://doi.org/10.1007/978-3-031-37256-8_21

can improve the base resistance and friction, and reduce the effect of negative skin friction (Moayedi and Mosallanezhad). The advantage of belled piles is that they can effectively mobilize the uplift shear resistance of the undisturbed soil above the enlarged base and thus are superior to straight-sided shafts (Quian et al. 2015).

In the soil consisting of expansive soil, the pedestal portion of the belled pile provides sufficient anchorage against uplift due to swelling/heaving of soil apart from enhancing the bearing resistance. In case of loose sandy soils, due to the driving mechanism of piles, the soil surrounding the pile gets compacted and soil derives the required bearing capacity. In case of layered strata, anchoring the base of the pile in the firm strata provides greater bearing area and the bearing capacity can be increased and also an increment in uplift capacity can be seen.

2 Bearing Capacity

2.1 Uplift Behavior of Piles

The uplift behavior of the foundation is of great importance for the construction of foundations of ports, transmission towers, and other structures that require to resist uplift loading (Liu et al. 2021). During an earthquake, the uplift force plays a vital role and acts on the piles under high-rise buildings (Tsuyoshi et al. 2011). Belled piles are widely used for the foundations of heavy structures such as transmission towers and these foundations play a vital role in the development of uplift bearing capacity (Moayedi and Mosallanezhad 2017). Moayedi and Mosallanezhad (2017) studied by installing small-scale belled and multi-belled piles in sand. Experiments were performed by installing bells at various depths. The concrete piles of 50 mm in dia and 750 mm in height were placed in a square tank of size 750 mm × 750 mm. Static tension uplift load tests were performed on eight small-scale belled piles. Initially, tests were performed on a straight pile and then followed by installing bells at various depths. The results indicated that the piles containing bells near the surface did not affect much the uplift resistance of piles. The highest uplift resistance with an increment of about 50–60% was observed in comparison with the straight shafted pile. Also, the movement of soil adjacent to the pile was measured and the results indicated that the ground deformation for a bell installed at 1/3rd depth is four times higher than the bell at a deeper depth.

Piles with enlarged bases have been widely employed in civil and coastal engineering due to their high resistance to uplift. Piles are required to exhibit sufficient uplift resistance (Yang and Qiu 2020). In order to investigate the uplift behavior of enlarged base piles, Material Point Method (MPM) was adopted by varying the pile geometries by the author. The influence factors such as bell angle, shaft/bell diameter ratio, and the contact friction angle between the soil and pile were numerically investigated and the soil and the pile were discretized into 40,442 tetrahedral elements, each element contained 4 material points. Also, MPM simulation of uplift behavior in the case of loose and dense sand was performed by varying the frictional angle of sand such as 15°, 20°, 25°, and 30° and one of the observations was, the uplift resistance increased with the increase of friction angle. According to the authors, when pedestal piles are installed in sand, dense sand gives higher uplift resistance than loose sand,which can be seen from the "Fig. 1". Also, the uplift resistance decreases with the increase of bell angle.

When belled piles are constructed in layered soils, the addition of a lower dense layer of sand beneath a loose upper layer increases the peak uplift capacity (Schafer and Madabhushi 2019). Also, especially for undrained clays, an upper loose sand layer with clay can further increase the uplift capacity under high rates of loading (Schafer and Madabhushi 2019).

When the transmission lines pass through mountainous and hilly areas, the foundations of line towers must be built at or near a sloping ground (Han et al. 2021). The belled piles can be used near the sloping ground. Han et al. (2021) made efforts to evaluate the uplift bearing capacity of belled piers adjacent to slopes. The field tests were carried out along a 220 kV transmission line with a total length of 35 km stretch, with 91% in mountains and 9% in hills. Laboratory and field tests were carried out to determine the rock properties and shear strength. Five belled piers were installed and uplift loading tests were conducted. Static loading was applied and displacements were noted. A series of numerical simulations were carried out based on field test data that was taken from full-scale belled piers installed on the project site of the transmission line. For the simulation of belled piers adjacent to slopes, 18 sets of

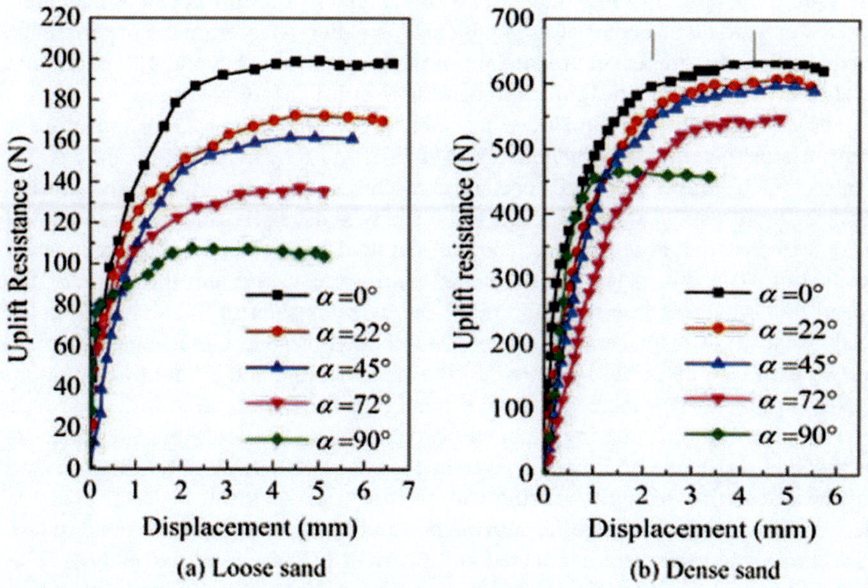

Fig. 1 Uplift load versus displacement curves

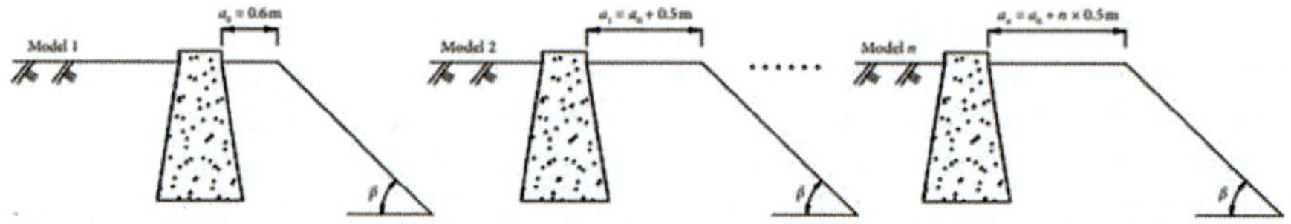

Fig. 2 Schematic diagram of pile models

slope foundation models consisting of three slope angles of 15°, 30°, and 45° and six embedment depths of 3, 4, 5, 6, 8, and10 m were established and loaded as shown in Fig. 2. Total of about 126 belled piers were used for simulation process. The results indicated that an increase in the distance from the edge of the belled pier to the slope or the embedment depth can increase the uplift resistance of piers.

The uplift resistance of a pedestal pile can be increased by providing a geocell layer at the base of the pile (Moghaddas Tafreshi et al. 2014). The tests were carried out on belled piers by incorporating geocell reinforcements in order to know the uplift capacity and also to know the upward and downward displacements. The test setup consisted of a tank (rigid box) of size 1000 mm × 1000 mm × 1000 mm.The sandy soil was compacted in the testing tank to a known density. The belled pile was placed at the center of the tank and the geocell layer was placed at the bottom of the belled pile and also above the bell of the pile. The loads were applied by a loading frame and the tests were conducted by varying the height of the geocell layer above the bell to 25,50,75 and 100 mm. The test setup is shown in Fig. 3. According to the author, the uplift capacity increases as the thickness of reinforcement increases, and the upward displacement was significantly smaller. Also, from the experiments, it can be said that increasing the height of the geocell layer beyond 100 mm is not advisable as there was no significant increase in uplift capacity was seen.

The construction of foundations in Gobi formation gravel ("Gobi gravel") for transmission towers is unavoidable (Qian et al. 2015). The most significant components of Gobi gravel are sandstone, quartz schist, quartzite, granite, and diorite, associated with small intrusive clayey and sandy soils. Forty-one full-scale belled piers were installed at seven sites, and pull-out load testing was conducted for each foundation. The dimensions of the belled piers were varied and the angle of the belled pier was kept less than 45° in all the cases as seen in Table 1. The uplift load—shaft-head displacement curves of the belled piers in Gobi gravel approximately exhibited an initial linear, a curvilinear transition, and a final linear region, and the uplift capacity increased as shown in "Fig. 4".

The ultimate uplift-bearing capacity of rock-socketed belled short piles consists of the side resistance and the pile weight (Ma et al. 2018). A simplified method for the calculation of uplift bearing capacity for rock-socketed pedestal pile was derived based on the limit equilibrium method and Balla rupture surface assumption. Field prototype tests were conducted and PLAXIS 3D was used for analysis. The comparison between numerical simulation and field test was done, and results present a good agreement between each other that can be observed from Fig. 5. According to the author in the initial stage of loading, pile, and pedestal take up the uplift

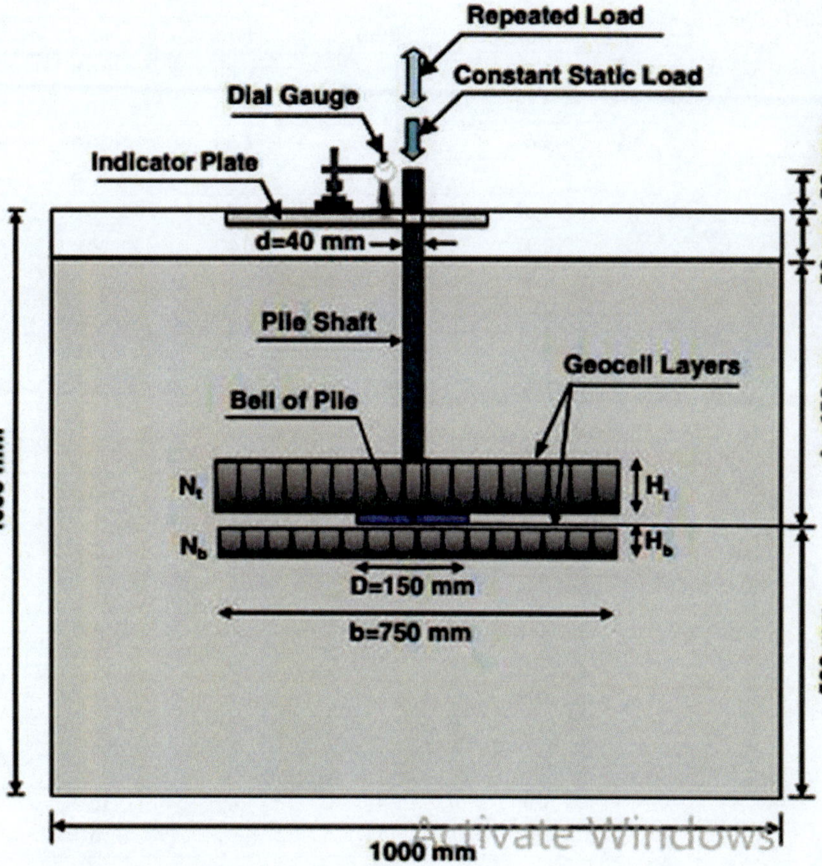

Fig. 3 Placement of Geocell and test setup

load synchronously, at the ultimate state, the uplift resistance provided by the bell accounts for about 54.9 and 34.7% of the total uplift capacity for the 6.0 m long and 7.0 m long piles, respectively. With the increase of the pile length, an increase in the ultimate uplift bearing capacity can be seen. The influence of the bell has weakened with the increase in the pile length (Ma et al. 2018).

The ratio of the bell to shaft diameter plays a significant role to provide uplift resistance (Liu et al. 2021). The uplift loading tests were performed on prototype belled piles installed in the test site, the soil consisted of quaternary holocene alluvial facies strata and tertiary sandy clay rock. The upper layer consisted of loess soil extending to a height of 25–36 m, the middle layer of sandy and gravelly soil with a thickness of about 5–6 m, and the bottom layer was sandy clay rock with a thickness of 10 m. Five sets of reinforced concrete piles were constructed by pouring concrete into boreholes with reinforcement cages. Slowly maintained load method was adopted to load the tested foundation. From the study, it was observed that the uplift capacities

Table 1 Dimensions of belled piers

D(m)	B(m)	b(m)	$\theta°$
1.72	1.01	0.80	10.0
2.65	1.64	1.20	20.0
3.64	2.29	1.60	30.0
3.73	1.41	1.20	10.0
5.29	2.04	1.60	20.0
3.93	1.49	0.80	30.0
6.54	1.81	1.60	10.0
4.53	1.24	0.80	20.0
6.82	1.89	1.20	30.0
1.72	1.01	0.80	10.0
2.65	1.64	1.20	20.0
3.64	2.29	1.60	30.0
3.73	1.41	1.20	10.0
3.93	1.49	0.80	30.0
4.53	1.24	0.80	20.0
1.72	1.01	0.80	10.0
2.65	1.64	1.20	20.0
3.64	2.29	1.60	30.0
3.73	1.41	1.20	10.0
5.29	2.04	1.60	20.0
3.93	1.49	0.80	30.0
6.54	1.81	1.60	10.0
4.53	1.24	0.80	20.0
2.05	1.20	0.80	33.7
2.60	1.23	0.84	33.0
3.55	1.20	0.80	33.7
3.20	1.40	1.05	30.3
4.75	1.60	1.05	42.5
4.08	1.80	1.25	42.5
4.80	1.85	1.30	42.5

depend on the embedment depth and base diameter. The load-displacement curve was interpreted as consisting of three segments, i.e., initial linear segment, middle curve segment, and ending linear segment. This indicates that the increase in load up to a certain extent can cause the displacement however further load increment will not affect the displacement behavior. Also, it was observed that improvement in the uplift capacity due to diameter to shaft ratio is less prominent above the value of 1.67 as seen in "Fig. 6".

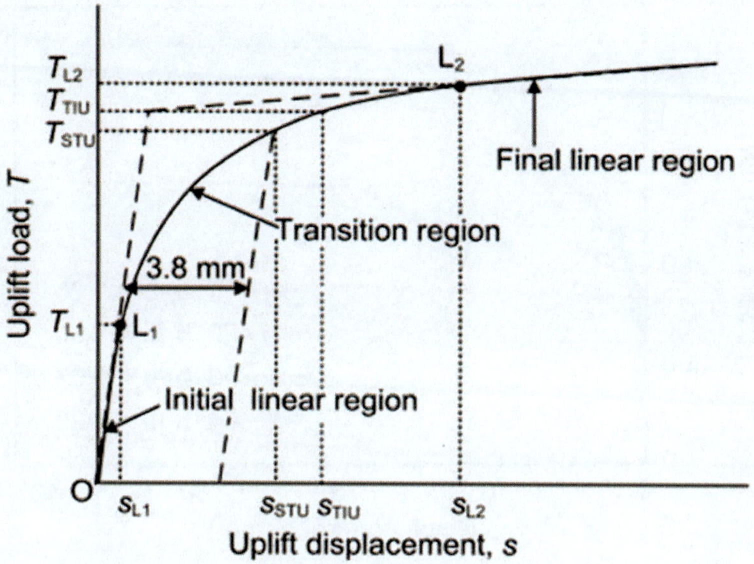

Fig. 4 Load versus displacement curve

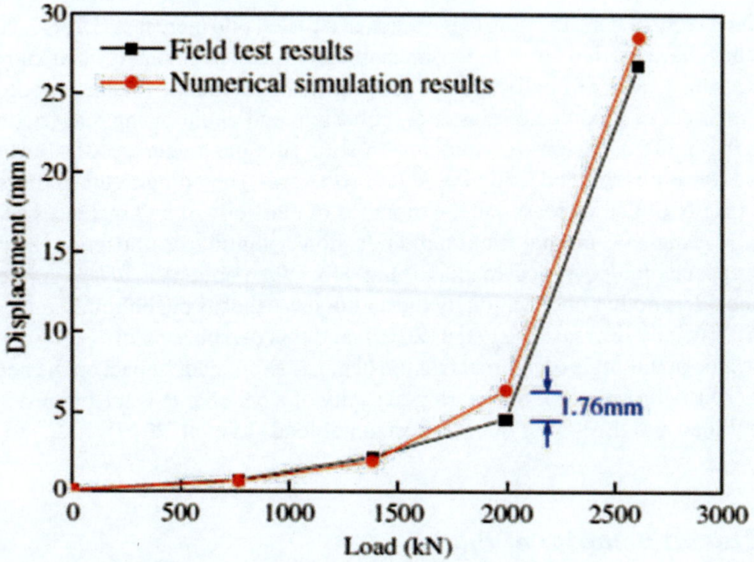

Fig. 5 Comparison between field tests and numerical results

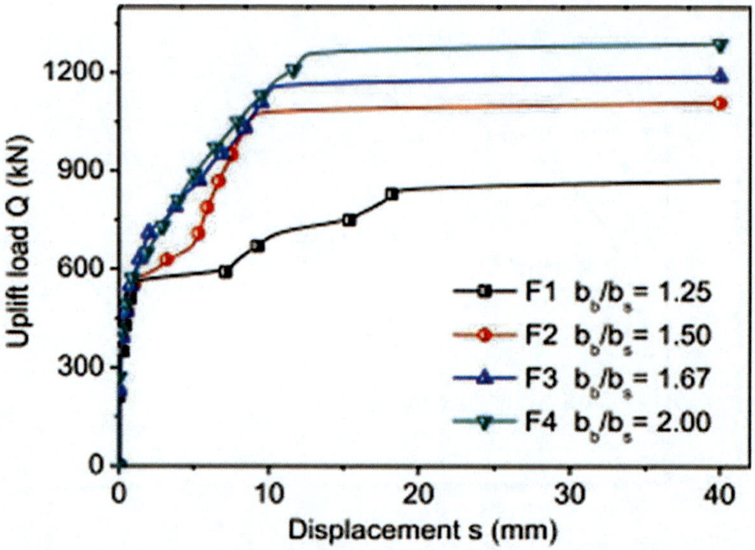

Fig. 6 Uplift load versus displacement curve

Under a vertical load, the uplift resistance of a belled pile increases (Li et al. 2019). The authors have carried out a three-dimensional finite element analysis to know the uplift bearing capacity of belled piles under vertical and oblique loading conditions. Also, the effect of modulus of elasticity, cohesion, and angle of internal friction of soil on the uplift behavior was studied. To study the parameters, a 3D FE belled pile model was constructed using ABAQUS software. According to the authors, the uplift capacity of soil increases as the modulus of elasticity of soil increases as seen in "Fig. 7". Angle of internal friction did not show a significant impact on the uplift bearing capacity under vertical loading ("Fig. 8"). When oblique load was applied on top of the pile, another observation by the authors was that as the obliquity increases, the uplift capacity decreases (Li et al. 2019), and this can be seen in "Fig. 9".

The value of the angle of internal friction has an insignificant impact on the bearing capacity of a belled pile and the bearing capacity of a belled pile is the highest under a vertical load and the lowest under a horizontal load (Li et al. 2019).

2.2 *Lateral Behavior of Piles*

Belling of the piles can effectively mobilize more soil to resist the combined uplift and lateral loads (Qian et al. 2017). The author has carried out field tests on straight sided and belled piers along transmission line towers on loess terrain which extends to a great depth on sloping ground. The results indicated that, for both straight sided and belled piles, additional embedment of the pile in the slope will increase the

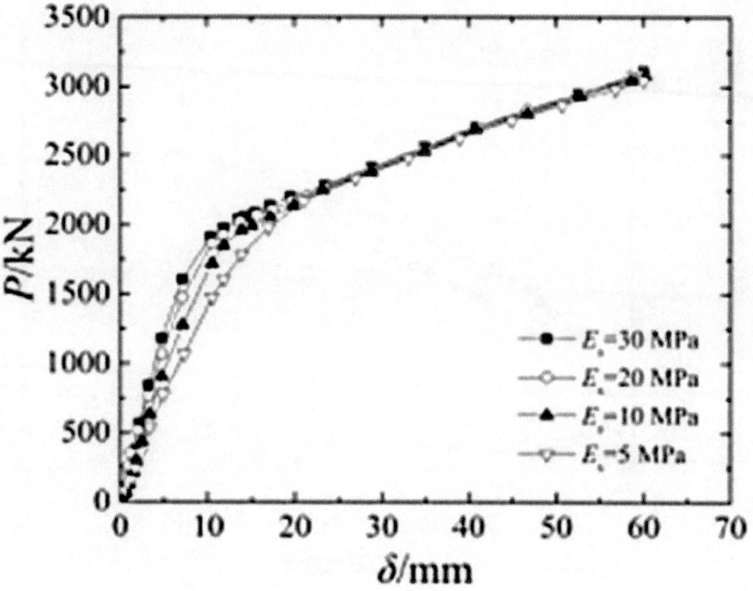

Fig. 7 Load versus displacement curve

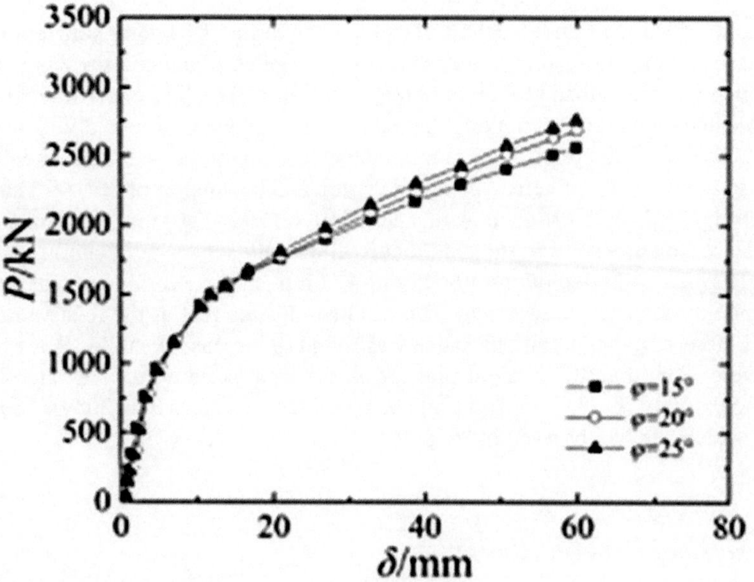

Fig. 8 Load versus displacement curve under vertical load

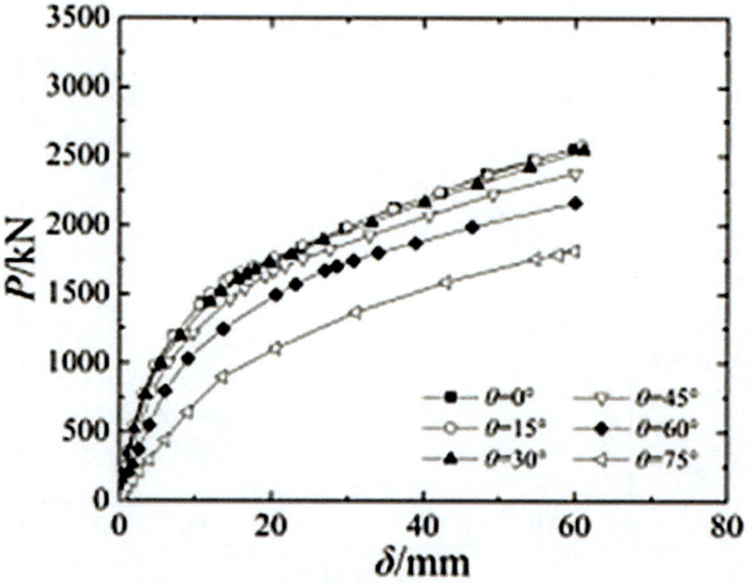

Fig. 9 Load versus displacement curve under oblique loading

uplift and lateral resistance as the resistance is governed by lateral soil failure. As applied load increases, lateral load carried by the belled pile increases which result in a higher peak load than monopile (Amaresha et al. 2020). The experimental study was conducted on aluminum model belled piles embedded in the sand with varying relative densities. The piles of 48 mm in diameter and lengths of 24, 36, and 48 cm, were used. Also, bell diameters of 72 and 96 mm and bell angles of 30°, 45°, and 60° were used. A mild steel model tank of size (1200 × 720 × 750) mm was constructed and piles were embedded in the tank that is filled with sand. Static lateral load test was performed on belled piles with varying L/d ratios. The lateral load was applied to the pile through the hand winch. The results indicated that as the relative density of sand increases, the lateral resistance was found to increase. Peak loads for belled piles were more than the straight pile for all different conditions. The lateral load carrying capacity for the pile with a bell angle of 45° was more than 30° and 60° bell angles which can be observed in "Fig. 10.

2.3 *Vertical Bearing Behavior*

The vertical bearing capacity of a belled pile is significantly greater than that of a conventional straight pile provided the debris at the bottom of the belled pile is properly removed (Gao et al. 2019). The full-scale field loading tests were conducted

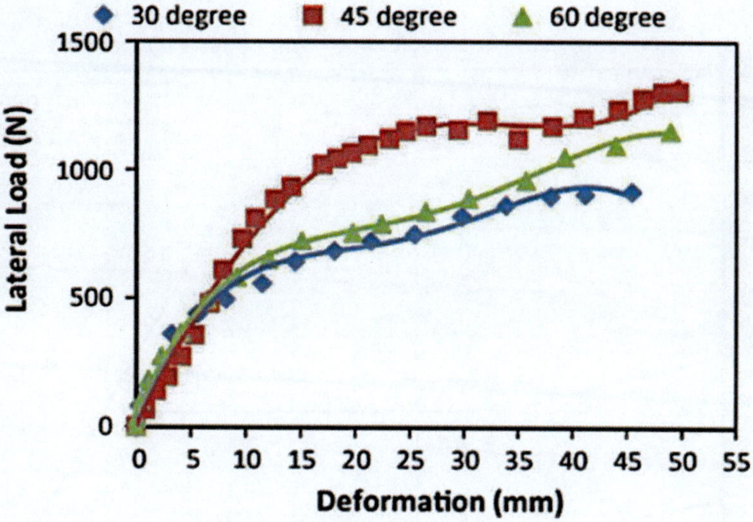

Fig. 10 Lateral load carrying capacity with varying bell angles

on 16 large diameter belled concrete piles ranging from 0.8 to 3.5 m in base diameter and 6.1 to 18.0 m in embedment length embedded in sand and gravel. Comparitive tests were carried out on belled piles and straight piles as seen in Figs. 11 and 12. A belled pile can result in higher vertical bearing capacities of about 5.5 times higher than that of straight pile. The optimum l/d ratio was found out to be 3.2 by the author and tip resistance was responsible for about 65% increase in vertical bearing capacity.

Fig. 11 Pile dimensions

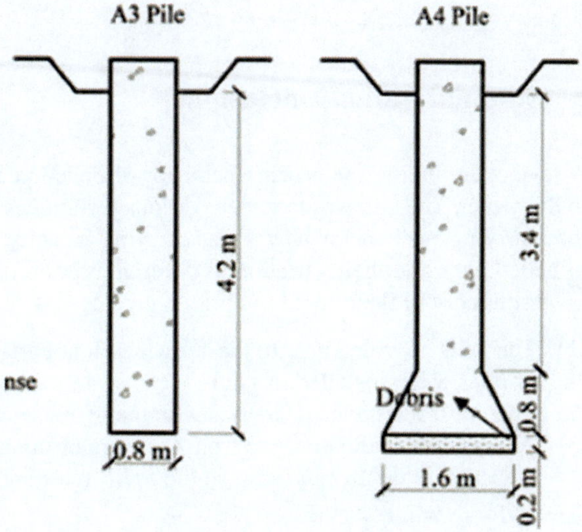

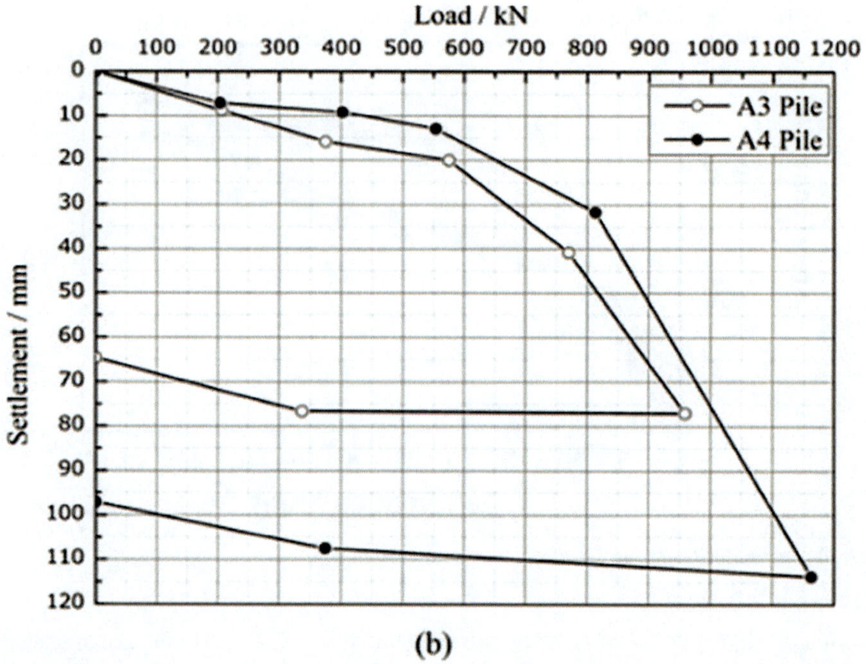

Fig. 12 Load settlement curve for straight and belled pile

The ultimate bearing capacities of belled wedge pile and traditional belled pile increase by 11.1% and 66.7%, respectively, when they are embedded in clay than in sand (Kong et al. 2017).

3 Summary and Conclusions

A review on finding the bearing behavior of belled/pedestal pile is been conducted in this paper. The aim was to review various techniques adopted to find out the uplift load carrying mechanism, lateral load carrying capacity, and vertical bearing capacity of belled piles and their variation in different types of soil.

A summary of the critical findings is given:

(1) The bells provided near to the surface will not affect much the uplift resistance of piles when installed in sand.
(2) The uplift resistance decreases with the increase in bell angle in sandy soil.
(3) An increase in the distance from the edge of the belled pier to the slope or the embedment depth can increase the uplift resistance of piers when installed on slopes.

(4) The Uplift resistance of a pedestal pile can be increased by providing a geocell layer at the base of the pile and also increasing the height of the geocell layer beyond 100 mm will not have any impact on the uplift resistance in the case of sandy soil.
(5) Uplift capacity increases with an increase in length to diameter ratio but improvement in the uplift capacity due to diameter to shaft ratio is less prominent above the value of 1.67 in case of sand and gravel.
(6) The uplift resistance of a belled pile increases as the modulus of elasticity and angle of internal friction increase. Modulus of elasticity of soil has a more significant impact than the angle of internal friction.
(7) The lateral resistance increases with an increase in the relative density of sand.
(8) The lateral load carrying capacity for the pile with a bell angle of 45° is more than any other values of bell angles when tested in sandy soil.
(9) A belled pile can result in higher vertical bearing capacities of about 5.5 times higher than that of a straight pile in sand and gravel.

References

Amaresha, Rajesh KS, Raju KVSB (2020) Lateral resistance of belled pile embedded in sand. https://doi.org/10.1007/978-981-15-6832-9_22

Gao G, Gao M, Chen Q, Yang J (2019) Field load testing study of vertical bearing behavior of a large diameter belled cast-in-place pile. KSCE J Civ Eng 23(5). https://doi.org/10.1007/s12205-019-2065-z

Han Y, Cheng J, Zheng W, Ding S (2021) Estimating the uplift bearing capacity of belled piers adjacent to sloping ground by numerical modeling based on field tests

Honda T, Hirai Y, Sato E (2011) Uplift capacity of belled and multi-belled piles in dense sand. Soils Found 51(3):483–496. https://doi.org/10.3208/sandf.51.483

Kong G, Peng H, Qin H, Wang L, Meng Y (2019) Ultimate lateral bearing capacity and group effect of belled wedge pile groups. KSCE J Civ Eng. https://doi.org/10.1007/s12205-019-0173-4

Kong GQ, Gu HW, Che P, Ren LW, Peng HF (2017) Impact of pile shaft shapes on vertical bearing capacity of belled piles. YantuLixue/Rock Soil Mech 38(2):361–367. https://doi.org/10.16285/j.rsm.2017.02.008

Li G, Zhang J, Liu J (2019) Study on the uplift bearing capacity of belled piles. IOP Conf Ser: Earth Environ Sci 376(1). https://doi.org/10.1088/1755-1315/376/1/012004

Liu G, Zhang Z, Cui Q, Peng J, Cai M (2021) Uplift behavior of belled piles subjected to static loading. Arab J Sci Eng 46(5):4369–4385. https://doi.org/10.1007/s13369-020-04779-x

Moayedi H, Mosallanezhad M (2017) Uplift resistance of belled and multi-belled piles in loose sand. Measurement 109:346–353. https://doi.org/10.1016/j.measurement.2017.06.001

Moghaddas Tafreshi SN, Javadi S, Dawson AR (2014) Influence of geocell reinforcement on uplift response of belled piles. Acta Geotech 9(3):513–528. https://doi.org/10.1007/s11440-013-0300-1

Qian Z-Z, Lu X-L, Han X, Tong R-M (2015). Interpretation of uplift load tests on belled piers in Gobi gravel. Can Geotech J 52(7):992–998. https://doi.org/10.1139/cgj-2014-0075

Qian Z-Z, Lu XL, Yang W-Z (2017) Comparative field tests on straight-sided and belled piers on sloping ground under combined uplift and lateral loads

Schafer M, Madabhushi SPG (2019) Uplift Resistance of Enlarged Base Pile Foundations. Indian Geotechnical Journal. https://doi.org/10.1007/s40098-019-00369-3
Yang YS, Qiu LC (2020) MPM simulation of uplift resistance of enlarged base piles in sand. Soils Found 60(5):1322–1330. https://doi.org/10.1016/j.sandf.2020.08.003

Trench Cutter Technology—The Multi-purpose Solution

Franz-Werner Gerressen and Alexander Blatt

Abstract The trench cutter technology (also known as hydro mill) is well-known for constructing diaphragm walls for complex retaining structures. This however is not the only purpose for which the technology can be used. The world demands much more than just retaining structures. For instance, the installation of groundwater barriers, named Cut-Off-Walls (COW), is a very common field of application. Such barriers are used for sealing dams or encapsulation of contaminated areas. Also, the system can be used to install barrettes for foundation purposes, which can be a good alternative to pile foundation systems. But there are more ideas, not only for the standard deep foundation demands. Recently, the technology could also satisfy the idea of using the trench cutter technique as bulk sampling equipment. Such kind of new ideas obviously also require the adaption of equipment and tools to facilitate corresponding demands. Some reference projects as well as developments and specific solutions will be described and a general outlook will be given, after the explanation of the installation process in general.

Keywords Diaphragm wall · Trench cutter · Cut-off-wall · Barrettes

1 Introduction

1.1 Retaining Walls

Retaining walls (Fig. 1) is one of the most common ways to use the diaphragm wall technique. These walls in general accommodate high loads and thus often provide permanent solutions for many applications, such as metro stations, underground parking areas, and shafts for e.g., sewage treatment and more.

F.-W. Gerressen (✉) · A. Blatt
BAUER Maschinen GmbH, Schrobenhausen, Germany
e-mail: franz-werner.gerressen@bauer.de

A. Blatt
e-mail: Alexander.Blatt@bauer.de

A. M. Krishna et al. (eds.), *Deep Foundations for Infrastructure Development in India*, Lecture Notes in Civil Engineering 373, https://doi.org/10.1007/978-3-031-37256-8_22

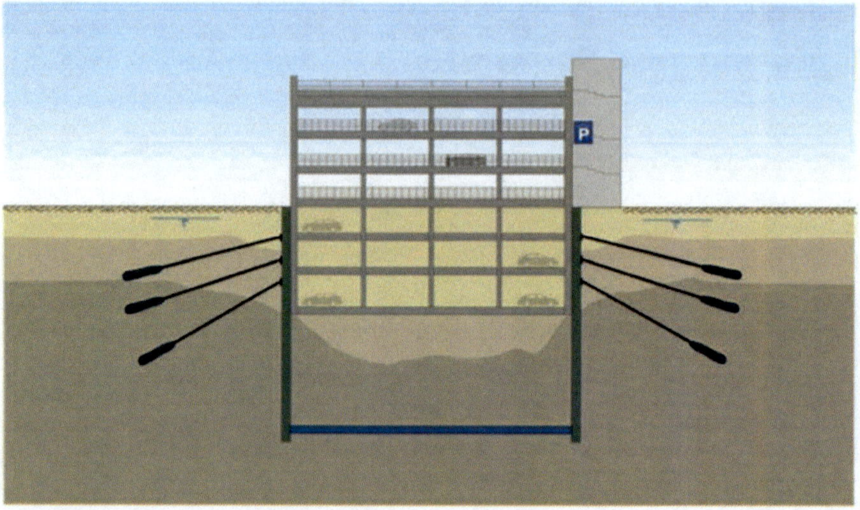

Fig. 1 Retaining wall

The advantages of the diaphragm wall technique in terms of high depth capability and flexible width possibilities solve nearly every design challenge of an underground structure. Furthermore, the possibility to work without significant ground vibration can be one answer to the rising demand of inner-city infrastructure of the future.

Over the years, the success of the method has been seen on many big projects around the world. Besides the mentioned advantages, the ability to work without restrictions in nearly every soil and/or rock environment can lead to an efficient and economical solution compared to other methods.

2 The Working Principle

2.1 Construction Sequence

The working steps follow a pre-defined sequence. The main steps are shown in Fig. 2.

Working steps:

- Site preparation
- Guide wall construction
- Trench pre-excavation
- Panel excavation
- Panel cleaning (desanding)
- Reinforcement installation (for retaining walls)
- Concreting

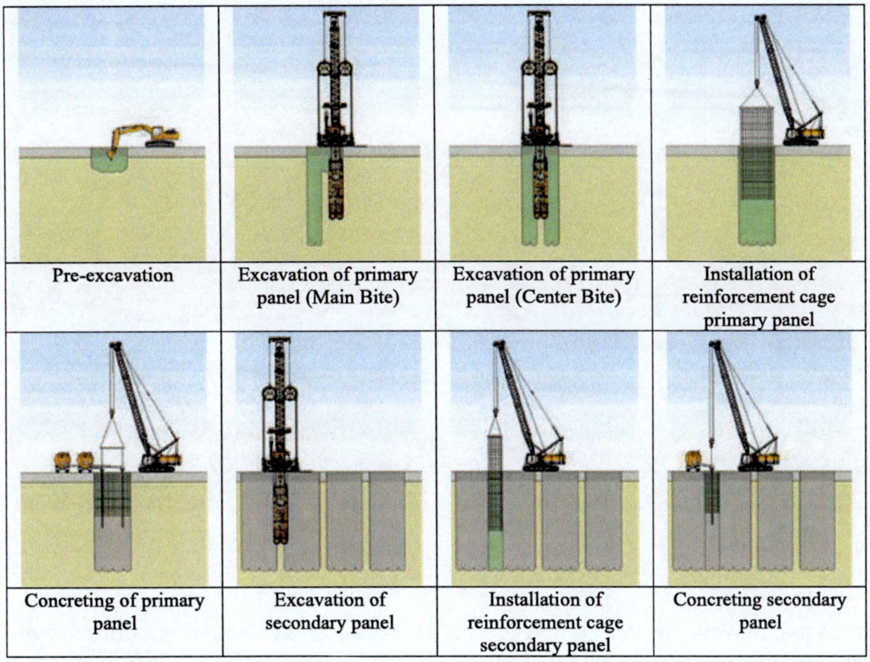

Fig. 2 Construction sequence

In confined conditions, the use of special auxiliary equipment (e.g., for installation of reinforcement cages) may become necessary. During excavation, the open trench is stabilized by supporting fluid.

2.2 Multi-purpose Solution

Retaining walls are just one possibility to use the advantages of this technique. Within the next sections, other applications are shown and described.

2.3 Cut-Off Walls

The aim of cut-off walls is to seal large areas in the ground. This, for example, can be used to rehabilitate dams or to encapsulate areas of contamination. Compared to retaining walls, cut-off walls commonly do not accommodate any reinforcement. Furthermore, the filling material has different properties than concrete. Thus, static loads cannot be transferred into the ground, but the walls are more flexible and can

Fig. 3 Cut-off wall

withstand movements to a certain degree. This is advantageous, if for example dams settle or move under water pressure.

An example of a cut-off wall is shown in Fig. 3.

2.4 Foundation (Barrettes)

An alternative to foundation piles in some cases can be a foundation with barrettes. These bearing elements can accommodate high loads and in general, have the same working process as retaining walls. The layout of these elements can be single bite panels, but also can have other geometric shapes (see Fig. 4).

An advantage of barrettes compared to bored piles can be a reduction in the number of elements and therefore reduction in construction time, especially for huge foundation projects (see Fig. 5).

Fig. 4 Examples of different shapes of load-bearing elements

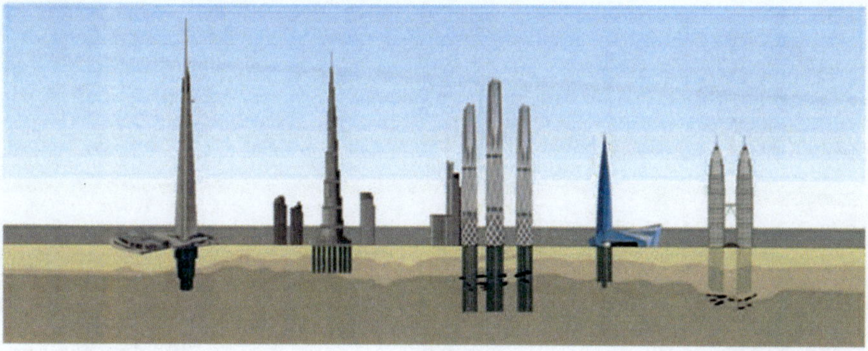

Fig. 5 Foundation

2.5 *Mining Application*

A totally different idea to use the diaphragm wall equipment is in mining applications. Here, in general, cutters with big footprints try to perforate ground and mine gold-containing, diamond-containing or other materials of value (see Fig. 6). This method applies if open-pit mining gets uneconomical or as part of a big exploration campaign.

3 Site References

3.1 *Execution of a Cut-Off Wall at the Dead Sea—Unique Conditions Require Unique Solutions (Gerressen et al. 2017)*

The Dead Sea Works Ltd., Beer Sheva, Israel operates several large-scale salt pans at the "lowest place on earth". The salt content of the Dead Sea water (Brine) is up to 33%. The mineral composition differs significantly from seawater. To ensure operation, dike heights must be adjusted accordingly. The northernmost salt pan (PAN 5) covers an area of approximately 75 km^2 with a capacity of about 150–200 million m^3. The dike, which runs parallel to the Jordanian border, has an approximate length of 18 km. Soil conditions show an alternating sequence of salt and clay layers. The salt layers are very porous, which cause sinkholes due to reverse erosion. Therefore, besides heightening, a rehabilitation of the dike was also required. This was performed in two steps. In the first step, the cracks and voids in the dike body were closed by dynamic compaction. The second step was the installation of a cut-off-wall to a depth of up to 34 m. The unique conditions required a unique approach. Therefore, a new two-phase cut-off with an internal sheet pile wall was developed and introduced as shown in Fig. 7.

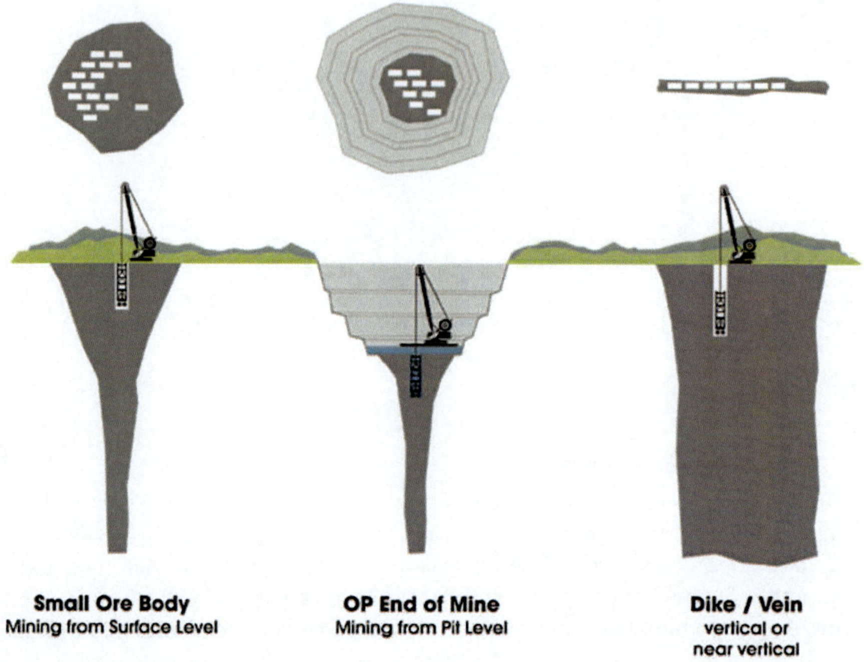

Fig. 6 Mining application

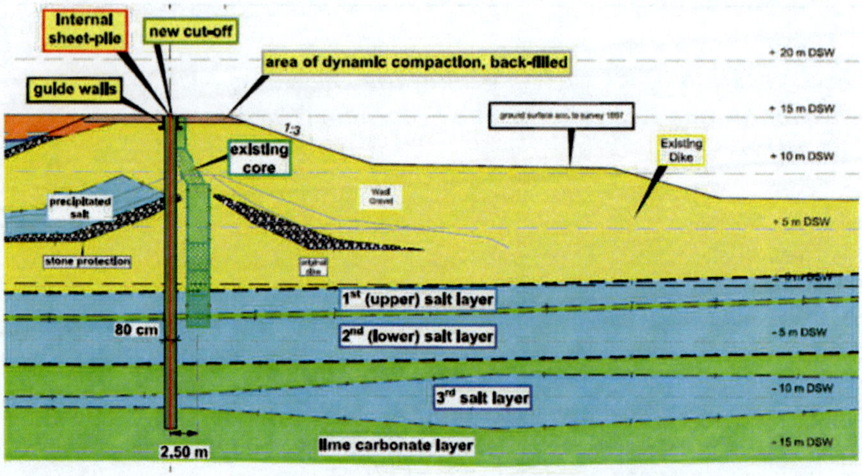

Fig. 7 Cross section of dike

Fig. 8 Cut-off wall head

Due to the high salt content of the brine, in addition to the aggressive desert climate, new rules had to be established in terms of cut-off materials and mechanical engineering. Upon completion, 550,000 m^2 of cut-off wall has been installed. Figure 8 shows the cut-off wall head with the inserted sheet piles.

3.2 *Foundation/Barrettes: Residential Tower Bullfinch Crossing, Boston USA (Fraley 2018)*

In Boston, USA, the foundation works for a residential tower was carried out (see Fig. 9) using barrettes.

For one part of the project, 16 LBEs (load baring elements/barrettes) had to be installed as shown in Fig. 10, which support the 45-story tower. The building is advertised as the tallest for-rent residential building in Boston. Further, the elevator shaft was also constructed using the diaphragm wall technique.

The depth of the elements varied between 19 and 25 m and the elements were embedded into a bedrock. Most of the elements were carried out as single bite of 2.8 m in length and 0.9 m in width. Confined conditions and only one access road created challenges during construction and could only be handled by good site logistics and equipment management. Greg Maxwell, owner of HUB Foundation, the contractor that carried out the job, described it “as a kind of structured ballet”.

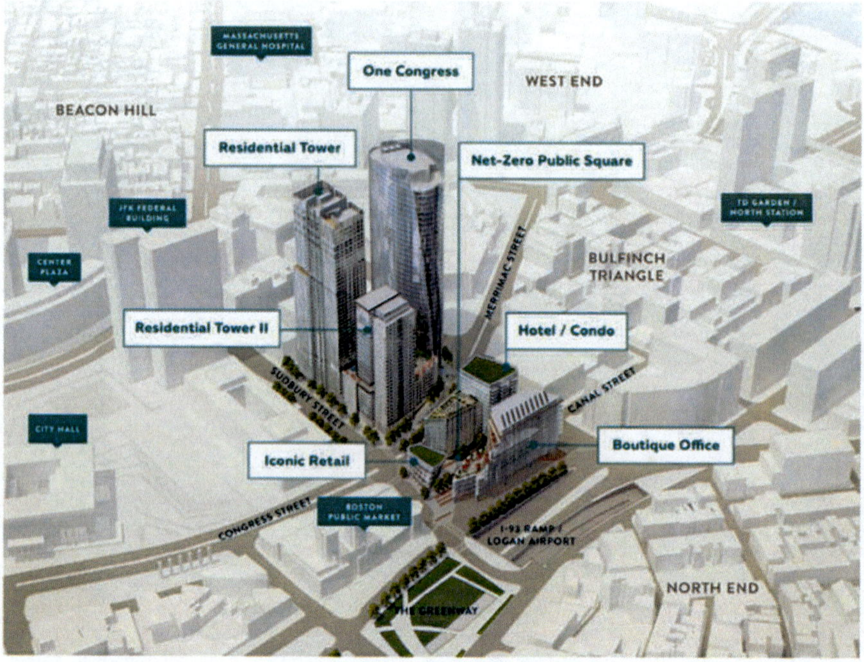

Fig. 9 Residential tower (© https://bulfinchcrossing.com/overview)

For the excavation of fill, marine deposits (Boston clay) and bedrock layers, a combination of a mechanical grab and a diaphragm wall cutter (see Fig. 11) was used. Cold weather also created challenges in terms of slurry transfer and storage.

3.3 Mining Application: FalCon Project, Sasketchewan Canada (Gerressen et al. 2020)

The following example describes the next step in depth achievement. For mine development, a reliable and detailed bulk sampling program is inevitable. For the sampling program, the use of a trench cutter system was considered. High demands for equipment were obvious to execute ten exploration trenches. Not only demands on depth capability, but also safety standards and the integrity and quality of the samples required a new technology, which was able to fulfill all specifications. Two years of planning, adaption and testing before the start of the program led to a working system.

Objective of this sampling program, which was carried out with a trench cutter system in Saskatchewan, CA, was to collect high-quality diamond-bearing Kimberlite samples. The Kimberlite zone started from around 105 m vertical depth and is

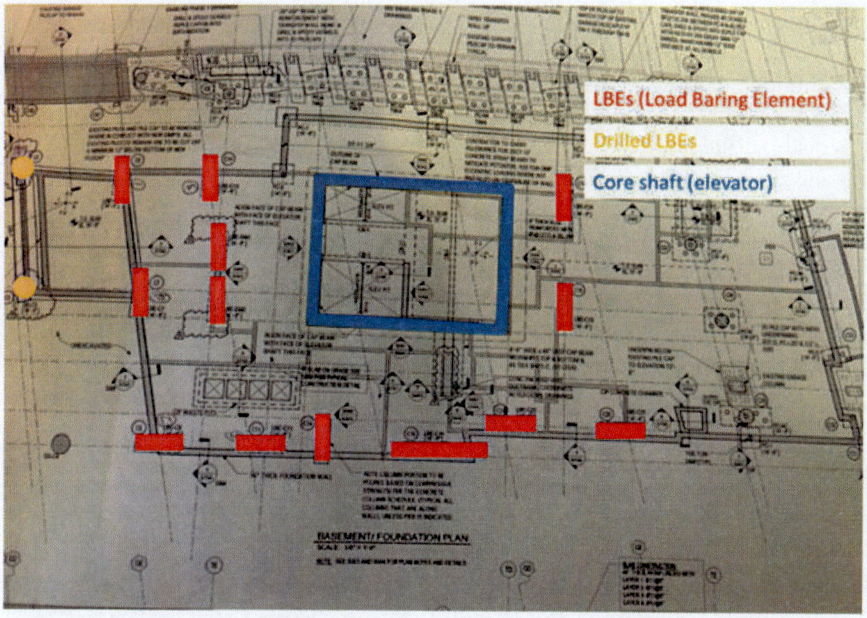

Fig. 10 Barrette layout

Fig. 11 Barrette excavation by cutter

overlain by so-called overburden material from loose sands to sticky and hard clays with cobbles and boulders (see Fig. 12: Geological situation).

During the cutting of the Kimberlite, the material was collected in big bags and stored for further processing. To sustain a high-quality analysis of the Kimberlite, it was distinguished into different zones. The zones were determined by rock parameters and geological descriptions. The thickness of these layers had high variability and the strength of the Kimberlite varied by depth and showed an average of 3,000–4,500 psi (20–30 MPa).

High attention needs to be paid, that no material got lost during the program and the most important objective was to provide high-quality samples. Several manual and remote-controlled pumps needed to work hand in hand, which required good radio communication to prevent overflows. The packing process was realized on two

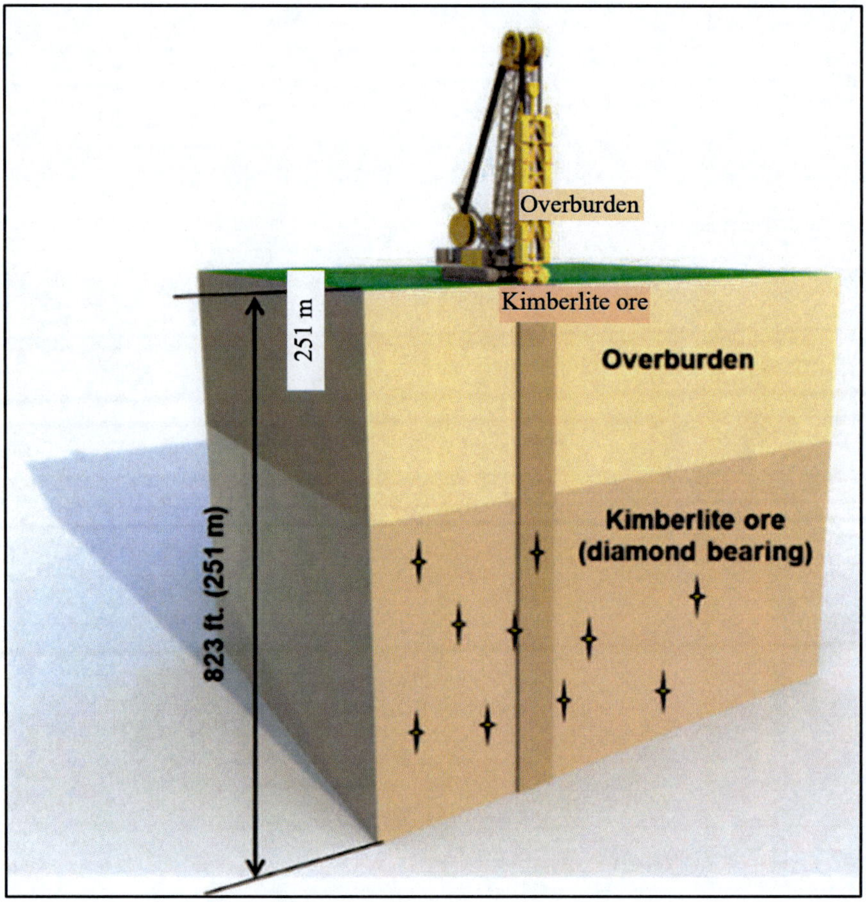

Fig. 12 Geological situation

individual and independent stations. The stations were located after 2 classifying screens, on which the material was washed. From there the material was filled into the big bags through a hopper. Once a bag was filled, a closing device was activated, and the hopper acted as a buffer volume for the arriving material. During the time the compartment was closed, a sledge guided an empty bag into the filling position. Four small telescopic handlers took care of the bag arrangement and storage in a secured area.

Aim was to produce large grain sizes to keep diamond breakage low. By a continuous optimization process, it was tried to find the ideal cutting parameters like surcharge, rotation speed and progress. Furthermore, different teeth were tested to optimize the result of chip size of the collected material.

A maximum depth of 251.1 m was reached on 25th of September 2019. It was the first time a trench cutter reached this depth for commercial purposes (see Fig. 13). In total, the project required over 2 km of cutting within 10 sampling locations. For the scope of work, 126 days of cutting were necessary. During the project duration, the crews worked 7 days per week within a 2-shift rotation. It was proven that the trench cutter successfully and repeatedly reached depths greater than 220 m. It could also be proven that the technology can efficiently fulfill high requirements even in high-pressure environments.

Fig. 13 Cutter at depth record 251 m

4 Conclusion

One standard application for diaphragm walling is the construction of retaining walls. But the technique has much more to offer, as it can serve various types of different applications.

For many years, the system has provided solutions for dam rehabilitation around the world. Such walls, to minimize seepage or to encapsulate contamination in the ground, are installed without a reinforcement cage and therefore can withstand a certain degree of movement. This is favorable in these conditions.

To transfer big loads into the ground, barrettes can be used instead of piles. These load-bearing elements act as foundation and can be favorable, if for instance a great number of piles can be replaced by a lesser number of bigger diaphragm wall elements. The elements can have different shapes and thus provide a flexible way to transfer loads into the ground.

In mining, the great depth capability, and the ability to work in nearly every rock environment creates possibilities to provide economical solutions for exploration as well as for mining of small ore bodies or for finalization of an exhausted open-pit mine.

Different jobsite examples show that the diaphragm wall is a multi-purpose solution and looking back on the developments of the past, there is still more to expect.

References

Fraley B (2018) HUB foundation breaks into Boston's Slurry wall market with key equipment. Construction equipment Guide.Com. https://www.constructionequipmentguide.com/hub-foundation-breaks-into-bostons-slurry-wall-market-with-key-equipment/41881. Accessed 13 April 2022

Gerressen F-W, Lehrer G, Scholz C (2017) 37th annual USSD conference, Anaheim, California, April 3–7, 2017; execution of a cut-off-wall at the dead sea – unique conditions require unique solution

Gerressen F-W, Schwank S, Blatt A (2020) DFI 45 annual conference on deep foundations, October 2020; diaphragm wall technique used in a different way—cutting trenches up to 823 ft. (251 m) for an exploration and bulk sampling program

Monopile Construction Technique for Mumbai Coastal Road Package 1—Mumbai

Rajesh Patel, Swarnadip Dey, D. Naveen Babu, Vipul Dobhada, and T. Vijayakumar

Abstract The most challenging scope of any bridge project in marine or intertidal zone is to construct sub-structures like well foundation, group pile, pile cap etc. Similarly, it also affects the marine environment/ecology to a great extent. Understanding the level of difficulties and aiding to provide an environment friendly solution, various options are analysed to overcome these and found monopile to be the best alternative in this context. Generally, the foundation type of any bridge constructed in river, marine or intertidal zone in India is preferred for group piles. For the first time, monopile is being accepted as the foundation of bridge in India at Mumbai Coastal Road Project. The diameters of the monopiles used in this project are 2.5 m and 3.2 m. Hence, the challenges involved in monopile construction and the experiences to overcome it are worth sharing. Here, a brief discussion on environmental and geological issues and the need of monopile as the foundation of bridge has been discussed. Furthermore, the monopile construction stages and confirmation of monopile capacity through different load tests have been elaborated.

Keywords Monopile · Reverse circulation drilling · Special bridges · Coastal road · Pile load test · Construction methodology

R. Patel (✉) · S. Dey · D. Naveen Babu · V. Dobhada
L&T Construction, Mumbai, India
e-mail: rajesh-patel@lntecc.com

D. Naveen Babu
e-mail: naveenbabud@lntecc.com

V. Dobhada
e-mail: vipuldobhada@lntecc.com

T. Vijayakumar
L&T Construction, Chennai, India

A. M. Krishna et al. (eds.), *Deep Foundations for Infrastructure Development in India*, Lecture Notes in Civil Engineering 373, https://doi.org/10.1007/978-3-031-37256-8_23

1 Introduction

Municipal Corporation of Greater Mumbai proposed to develop Mumbai Coastal Road from Princess Street flyover to Kandivali junction over a length of approximately 29 km to ease the traffic congestion in Mumbai and to create recreational spaces. This project is being implemented in 2 phases namely South and North. South phase starts at Princess Street flyover and ends at Worli end of Bandra Worli sea link (BWSL). This phase is divided into 3 Packages as mentioned below (Fig. 1):

(i). Package 4: Princess Street flyover to Priyadarshini Park (CH km 1 + 975 to CH km 5 + 900);
(ii). Package 1: Priyadarshini Park to Baroda Palace (CH km 5 + 900 to CH km 9 + 720); and
(iii). Package 2: Baroda palace to Worli end of BWSL (CH km 9 + 720 to CH km 12 + 430).

L&T Construction is constructing Package 1 of this project, which includes the construction of a 3.82 km long section over reclaimed land from Priyadarshini Park to Baroda Palace and interchanges at Amarsons Garden and Haji Ali; and Main Line Bridge. Currently, the project is into its phase of substructure construction in the marine stretch. The most challenging scope of any project is to construct substructures like well foundation, group pile, pile cap etc. for bridge or via duct in marine or intertidal zone. At the same time, it also affects the marine environment/ ecology to an extent. Understanding the level of difficulties and after aiding to an environment friendly solution, various options are analysed to overcome these and found monopile to be the best alternative in this context.

Monopile foundation is extensively preferred in offshore renewable industry by supporting the wind turbine in deep sea considering the easiness of installation. Now

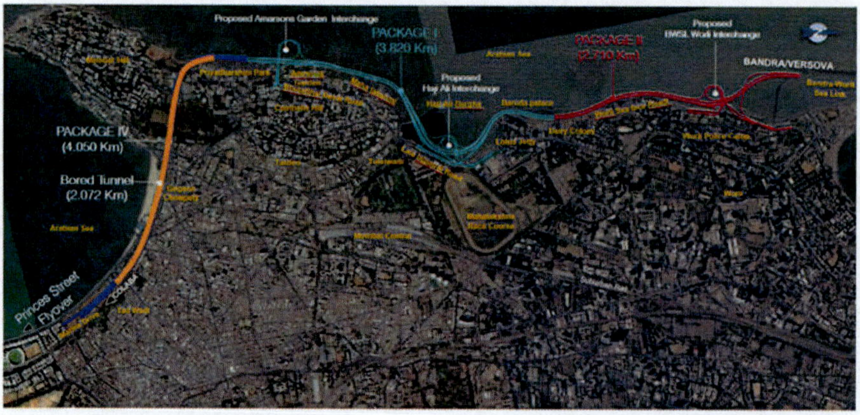

Fig. 1 Mumbai coastal road project details

a days, it is also extending its footprint to onshore region in different types of infrastructure development projects. Some of the prestigious projects where monopile foundation concept has been relied upon are.

- Incheon Bridge, South Korea-second connection between Yeongjong Island & mainland of Incheon. Monopile foundations of diameter 2.4 m & 3.0 m are used in this project which was completed in 2009.
- Dubai Metro, UAE-1st section opened in 2009 (Red line—52 km connecting Rashidiya to Jebel Ali Port) with monopile of diameters 2.2 m and 2.4 m.
- Sheikh Jaber Al Ahmead Al Sabah Causeway Project, Kuwait-Spans across Kuwait Bay between Kuwait City and Subiyah area. The diameters of the monopile are 2.5 m and 3.0 m and the construction was completed in 2018.

Generally, the preferred foundation type in India is group piles, for bridge construction in marine/intertidal zone. For the first time, monopile is accepted as the foundation in India for this Mumbai Coastal Road Project. The diameters of the monopiles being used in this project are 2.5 m and 3.2 m. Hence, the challenges involved in monopile construction and the experiences to overcome it are worth sharing.

2 Environmental Brief

One of India's key economic centres is the city of Mumbai, which is situated on the country's western coast. A tropical climate, specifically a tropical wet and dry climate, prevails in Mumbai. It alternates between a dry season from October to May and a wet season with a climax in June. The hotter season, which runs from March to May, comes after the cooler season, which lasts from December to February. The southwest monsoon season lasts from June through roughly the end of September, while the post-monsoon season lasts from October to November. For Mumbai, flooding during the monsoon is a big issue. The city is battered by the south-west monsoon rains between June and September. In May, pre-monsoon showers are common. In October, there are sporadic north-east monsoon showers.

Marine ecosystem particularly the intertidal zone is one of the most dynamic zones that is the interface between sea and terrestrial environment. The most important physical factor that influences the life and activities of organisms of the intertidal zone is the existence of waves and duration of exposure to sunlight. The intertidal area of Mumbai is also diverse and found with corals in patches indicating the water quality of the Mumbai coast. Intertidal area houses meiobenthic to macro benthos in wide diversity. Before mobilising the marine spread to construct the monopile, coral transplant has been done with proper care and procedures in front of the concerned regulatory authority.

Flora and fauna in and around Mumbai are enriched with varieties of bird species and plants. Mumbai being the major harbour since time, many invasive species of plants have made Mumbai their homeland and are found all over the Mumbai area.

Birds are found in diverse number; flamingos flock arrives in Mumbai during December to May and are found in creek water. Varieties of birds arrive in Mumbai in different seasons.

Mumbai is the only city with harbours like National Park and it houses wild animals like tigers, leopards etc. Mithi and Ulhas Rivers are major rivers running across the city. Mumbai has many lakes, and gaining the title of lake city, these lakes supply water to the city during summer and winter seasons. Therefore, the adaptation of monopile leads to minimal environmental pollution by producing less construction waste which can take care of such sensitive environmental conditions of Mumbai coast. This section is developed from the details received from Ministry of Environment, Forest and Climate Change, Government of India and Coastal Engineering Manual CEM (2011).

3 Geological Brief

The geological brief of this area is developed by taking reference from the "Greater Mumbai Disaster Management Action Plan, Risk Assessment and Response Plan, Vol-1, 2007, Maharashtra Emergency Earthquake Rehabilitation Programme, Government of Maharashtra (Greater Mumbai Disaster Management Action Plan, Risk Assessment and Response Plan 2007). The Deccan basalt flows, including their acidic and basic varieties, which were erupted between the late Cretaceous and Early Eocene period, cover the whole Greater Mumbai region. Over large areas, the basaltic flows are more or less consistent in structure and are horizontally bedded. The basalts on Mumbai Island and its environs are related with specific extrusive and intrusive mafic types. The Mumbai area also has some fossiliferous silt that is rich in fauna and primarily of tuffaceous origin.

A rigorous geological study was carried out with the help of conventional geotechnical investigation and further validated through Multichannel Analysis of Surface Waves (MASW). Geotechnical investigation was carried out before the construction of monopile which comprises of boreholes executed at each monopile location inside the diameter of it. Drilling was undertaken by hydraulic feed rotary core drilling equipment using HQ size drill and triple tube core barrel with suitable core catcher, in accordance with IRC: SP:91-2010 (2010). The investigation was undertaken as per applicable codes (IS 1892-2021 1892) and standards of practise in their latest editions as published. Furthermore, the site characterisation works around the test monopile locations were carried out through 2D MASW surveys to a depth of about 40 m by active and passive MASW methods to evaluate and confirm the presence of competent hard rock and support the geotechnical investigations in order to characterise the site and corroborate with the existing borehole data in close proximity. Based on this study the geological formation is elaborated below.

The geology of the Amarson's Garden and Baroda Palace site comprises of Basalt bedrock from the Deccan Trap Basalts, specifically the Malabar Hill Flow of the Salsette Group, and a mass horizontal basaltic and syphilitic lava flow with pillow lava fringes deposited in the late Cretaceous to early Eocene. The Basalt is characterised by a dark greyish green fine-grained groundmass when fresh, frequently crossed by white thin calcareous veins.

The geology of the Haji Ali site comprises of marine clays and sands overlying various inter-trepanned tuff deposits laid down in the late Cretaceous to early Eocene. The tuff exists as a light grey coarse grained deposit (TUFF), or fine-grained dark grey to black banded tuff (TUFF with SHALE). At depth, and around the periphery of Haji Ali Bay, a coarse grained (possible hyaloclastite) deposit is present (BRECCIA TUFF). Weathering at sea bed level is typically recorded as low (<60) RQD% (Rock Quality Designation) and modest (60%) TCR% (Total Core Recovery). Intact weathered tuff presents as pale yellowish orange to dark grey irregular gravel to cobble size fragments, often discoid where 'shale' is present. Unweathered TUFF and TUFF with SHALE material is light to dark grey, banded in fining-up sequences of centimetres to metres in scale. Unweathered BRECCIA TUFF presents as a light grey groundmass containing angular clasts of tuff with abundant white veins of hard calcareous infilling. To the north-eastern and south-eastern extents of the bay, Basalt of the Deccan Trap Formation is present as a dark greenish grey fine-grained crystalline groundmass.

As per geology, these monopiles are proposed to be rock socketed piles, which means they should be socketed inside the hard rock for a minimum depth of about 1.5 times the diameter of the pile in line with the technical specification and requirements (Fig. 2).

4 Advantages in Adopting Monopile

Faster Construction: The substructure component of any foundation of group pile is two and more piles in a group and the pile cap sink below the sea. This necessitates excavation of larger footprints below the sea bed level. Also, for the same, the contractor needs to construct a temporary coffer dam which will be sealed at bottom to avoid water seepage during the pile cap excavation and construction till the completion of pier lifts above the high-water level. Usage of large diameter piles enables to reduce the number of smaller diameter piles and thus reducing the overall total pile numbers of the project. Also, monopile eliminates construction of pile cap and related timelines thus helping to reduce the overall construction time of the foundation (Fig. 3).

Proven Innovative Technology: Though initially monopile was used for wind turbines, in the recent past the technology has been widely adopted in bridge construction due to its various technical advantages over the group of pile. Many iconic sea crossing bridges and causeways have been built using this technique globally and mentioned already.

Environment Friendly: Monopile construction is being done by providing outer steel casing till the refusal level. The pile cut off level will be kept above MSL (Mean Sea Level), from there the pier will be built till the pier cap level without

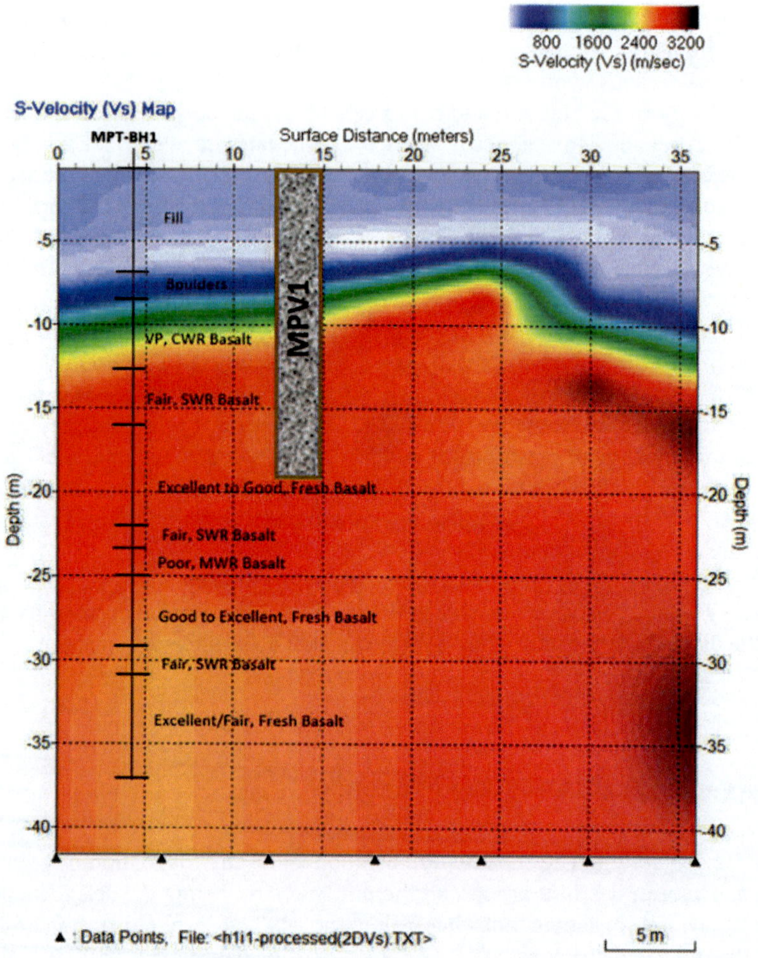

Fig. 2 Ground profile at Amarsons Garden embedded with monopile test pile—MPV1

the construction of any pile cap. Monopile construction uses about 12–15% of the footprint area of pile group and eliminates the construction of pile cap and its related sea bed excavation and cofferdam construction which reduces the impact on sea bed and its ecology (Fig. 4).

Maintenance Approachable: Throughout the design life of the bridge i.e., 100 years the structure will experience different seismic conditions (Seismic Zone III). Hence, it becomes very important to have proper access to the potential plastic hinge's location during the life of the structure and the potential plastic hinges will be formed for monopile at the junction of pier and pile cap/pile which is above the MSL. Therefore, it can be inspected routinely after each and every seismic activity to ensure structural safety (Fig. 5).

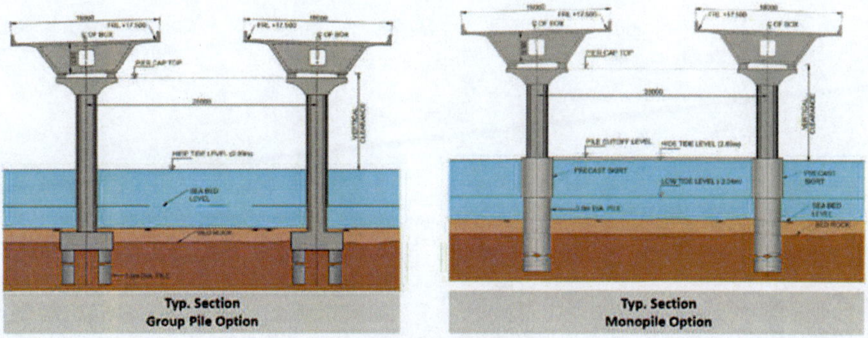

Fig. 3 Typical cross section of group pile and monopile

Fig. 4 Aesthetic view of group pile and monopile

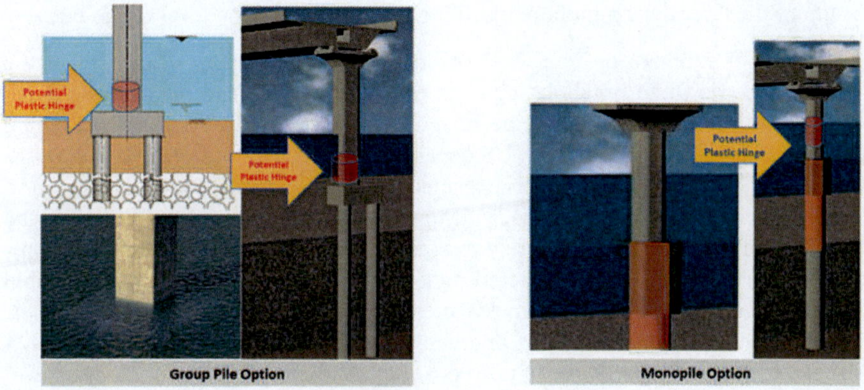

Fig. 5 Plastic hinge formation in group pile and monopile

5 Monopile Construction Technique

Stepwise different phases of work are being adopted to construct Monopiles (Fig. 6).

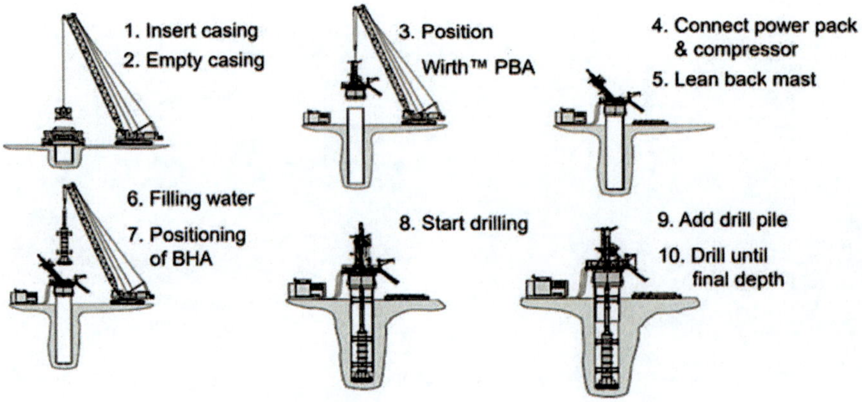

Fig. 6 Different phases of monopile drilling through RCD

5.1 Phase 1—Liner Pitching

The permanent liner for the monopiles performs several crucial purposes:

1. The liner provides formwork for the concreting above the sea bed through open water up to the defined cut off level for the pile.
2. The liner supports the ground around the pile bore during drilling and in casting of the monopile.
3. The liner provides support for the reverse circulation drilling rig, mounted on the top of the liner during monopile drilling, to the planned toe level of the pile.

In order to estimate the toe level of the liner it has been inspected from the borehole records (both borehole logs and core photography) to identify the depth below which ground is unlikely to become unstable during construction (Evaluating scour at bridges, 5th edition, FHA US department of Transportation, Circular no 18 April 2012 (2012)) where ground is expected to provide adequate vertical support to the reverse circulation drilling rig. Typically, the toe level of the liner was set to be at the base of the reported completely weathered rock, however, in certain locations, the presence of completely weathered rock has been reported beneath more stable highly or moderately weathered rock which typically possess higher RQD values. In this case, the stability of the lower completely weathered rock is less and cannot be relied upon in advance stage of pile construction, therefore a lower liner toe level has been used at the design stage to reduce the risk of unstable monopile bores.

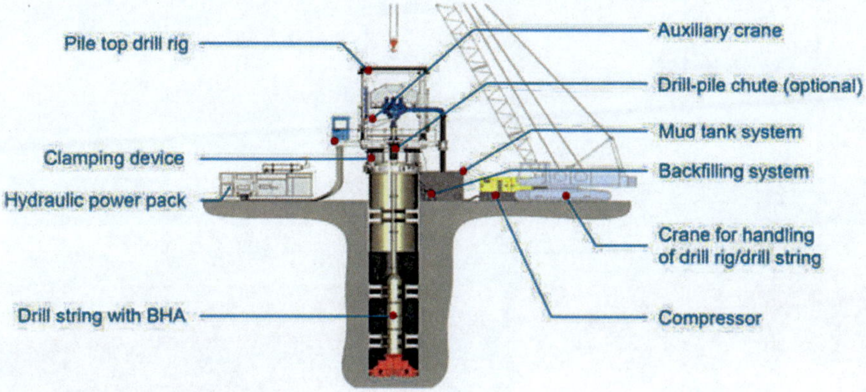

Fig. 7 Different components of top load drilling rig

5.2 *Phase II—Positioning of PBA*

Pile Top Drill Rig (PBA) provides the most efficient solution for foundation drilling in mixed ground, boulders and hard rock. Therefore, PBA Rig is adopted for this project. In order to mount the rig above the liner/cashing, crane is required.

5.3 *Phase III—Positioning of BHA*

It consists of drill string with drill pipes, drill pipe stabiliser & Bottom Hole Assembly (BHA) assembly. All stabilisers are non-rotating type (outer skids are standing still). This long and rigid BHA will stay in the centre line with high accuracy of 120 mm as it is guided inside the casing while starting the drilling process below the casing and later in the existing hole. The BHA provides the necessary weight to the drill bit for penetration. In addition to the BHA weight, the rig offers the possibility for additional thrust to the drill bit via the pull-down system of the mast (Fig. 7).

5.4 *Phase IV—Reverse Circulation Drilling*

Attached External air pipes to BHA provide compressed air and inject into the drill string above the drill bit. This reduces the density of the fluid. The water column outside the drill string is pushed down by the atmospheric pressure and the drill bit horizontally enters the pick-up tube. The drill mud, consisting of drilled cuttings, water, and air, rises inside the drill string and is discharged into the mud tank system (Fig. 8).

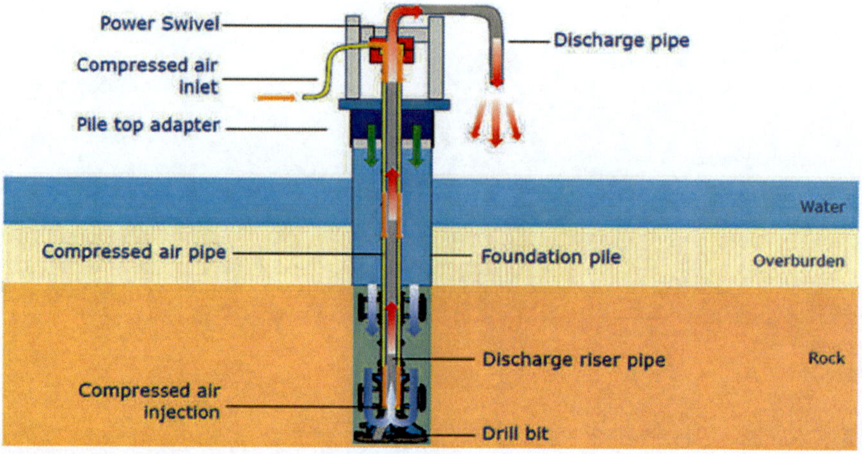

Fig. 8 Reverse circulation drilling through airlift

5.5 Phase V—Flushing and Reinforcement Cage Lowering

The RCD technology inherently allows cleaning of the pile base by using high suction and removing material from pile base. The bottom cleaning is done, with RCD until the drilling fluid coming to the surface is clear of sediment and rock cuttings, before the start of concreting and after cage lowering to ensure no soft material at pile bottom. Reinforcing steel or steel angle iron will also be welded between the top and bottom bearing plates prior to the lifting process. These temporary supports are cut out when the cage is lowered into the hole.

5.6 Phase VI—Concreting and Chipping

Concreting shall be done as per the approved mix-design for concreting works. Tremie pipe of required length with flushing head shall be lowered into the drilled pile. The Concrete placement shall start after completion of cleaning of bore hole & commencement of concrete pour shall be within 24 h. The temperature of the concrete shall be maintained below 32 °C and flow (avg. 625 mm) between 600 and 700 mm (Fig. 9).

Fig. 9 MCRP P1 Project site photographs for RCD operation (left) and CHSL Test (Right)

5.7 Phase VII—Testing of the Monopile

1. Bi-directional static axial load test and Instrumentation is being carried out after following provisions as per ASTM D8169/D8169M-18 (ASTM D8169/D8169M).
   - Two transducers have been placed to measure the jack expansion in Bi-directional Test
   - The cross hole sonic logging tests shall be conducted as per approved procedure.
   - 8 Cross hole sonic pipes shall be placed in 2.5 m Dia Pile cage & 10 no's Cross hole sonic pipes in 3.2 m Dia Pile Cage as specified in the drawing.

6 Test Requirements for the Monopile

Piles are installed upto the designed depth or refusal level to achieve estimated safe capacities. However, to ascertain that they have achieved such capacities, certain percentages (3%) of installed piles have been tested as specified in IRC:78:2014 (IRC 2014). Similarly, if doubts arise whether pile has been installed properly, then a load test may be carried out on such a pile.

As mentioned in FHWA GEC 010: AASHTO LRFD Bridge design specifications, 8th Edition 2018 (Gec 2018) pile load test is the most direct method for determining the safe loads on piles including its structural capacity with respect to support from soil/rock on which it is installed. It is considered more reliable on account of its being an in-situ test than the capacities computed by other methods, such as static formula, dynamic formulae, penetration test data or results obtained from other tests such as PDI dynamic test, etc. There are widely varying practises followed for load test results on piles, particularly, the difficulties regarding the establishment of an acceptable criterion for determining the ultimate and safe bearing capacity of piles and predicting the pile group behaviour from the test data obtained from load tests on single piles.

At present, pile load test is the most accepted method of predicting load capacity of a pile. It is probably the most reliable method of predicting the capacity of a single pile. But often from a load test on a single pile capacity of a group of piles is predicted and then the limitation of extrapolation is applied.

6.1 Types of Tests for Monopile

(1) Bi-directional vertical load test.
(2) Lateral load test.
(3) Dynamic load test.

7 Vertical Pile Load Test (Bi-Directional Pile Load Test)—Load Testing Procedure

The installation and testing of the pile is carried out under specialist supervision and in the presence of the consultant. The bi-directional static load test is carried out when the concrete strength of the pile is adequate to sustain (normally 28 days) the maximum required test load. This concrete strength is demonstrated by the crush testing of concrete cube. As per Cherian (2021) and Arda (2007), prior to commencement of the test, all hydraulic cells forming part of the assembly are pressurised to break the welded seal of the cells. Movement and pressure are recorded during this operation. The hydraulic cell is internally pressurised using a common hydraulic system (which ensures uniform and synchronised pressurisation of both hydraulic cells), creating an upward force on the shaft in upper friction and an equal, but downward force in combined lower shaft friction and/or end bearing. As mentioned, the hydraulic cell load is determined by relating the applied hydraulic pressure to load calibration. A high-pressure bourdon gauge is used to read the pressure on the pump line and a calibrated pressure transducer reads the pressure on the return line. This process is completed after achieving the required test load. The loading of the test is carried out in the stages as mentioned below (as recommended loading sequence). However, this loading sequence can be modified as per the contract specifications.

8 Lateral Pile Load Test

The Initial test pile shall be installed by the contractor as per the approved method statement & as per approved drawings. Inclinometer casings and strain gauges are installed at the time of pile casting. Locations are shown in Fig. 10. The pile surface shall be smooth at the point of loading location as well as at the LVDTs fixing location as shown in below typical photos. The testing is conducted by obtaining the reaction from the adjacent pile by means of the jack. The full load imposed by the jack shall be taken as the lateral resistance of each pile.

Refer typical photos for obtaining reaction from nearby piles or from other reaction system arrangements of excavated ground. Lateral (Horizontal) load will be

Fig. 10 MCRP P1 Project site photographs of cage preparation of test monopile (above) and reading during Bi-Directional Load Test (below)

Fig. 11 MCRP P1 Project site photographs of lateral load test set-up

transferred to test pile by means of this reaction pile and load will be transferred to the test pile by means of girder and jack system as shown in Fig. 10. LVDT shall be attached at the level indicated.

LVDT shall be held by datum bar resting on immovable supports. The support for datum bars shall be at least at the distance of 6.5 m from the edge of the pile. Hydraulic jacks, load cells, packing plates, datum bars, and Small Steel Box Girders/ members assembly shall be then erected as shown in Fig. 11. Following steps have been followed sequentially: The testing is conducted at least 28 days after the pile is installed assuming that the test pile concrete has achieved the required strength before the test. After completion of all the above arrangements, application of test load shall be started as per surge loading with standard loading schedule as per ASTM D3966 clause 8.1.5.2. According to ASTM D3966, initial pile load test shall be carried out for a maximum test load of 2 times the design load.

9 Dynamic Pile Load Test

As per ASTM D4945–12, the testing shall be conducted by fixing instrumentations that include strain sensors and accelerometers to the sides of the pile at a depth of 1.5 × pile diameter from the top of pile and concreting them to the measuring equipment. Pile shall be extended to 1 m above the cut off level after chipping the top loose concrete. The grade of concrete for extending must be the same as that of the pile. Place the helical reinforcement mesh over the extended concrete with 50 mm thick plates over mesh to avoid cracking of concrete under impact. Suitable guide arrangements for driving shall be provided and lifted hammer of 52 MT self-weight shall be dropped through a mechanically stabilised jack-lock system on the test monopile (Fig. 12).

Fig. 12 Site photographs of dynamic load test set-up

10 Conclusions

It can be concluded after having elaborated discussions about the monopile construction technique that the monopiles are comparatively faster to execute in construction field as it can save 6 months duration by avoiding pile cap construction in Mumbai coastal road project package 1. Additionally, monopile is easy for maintenance/remedial measures (if required). This type of construction is environment friendly and poses very minimal disturbance in the sensitive coastal region like Mumbai. The load test results of the monopile in this project are within the acceptable limits as proposed by the designer and successfully carried out. Thus, using this type of foundations in intertidal/Marine zone shall gain more confidence in future for the first of its kind of monopile construction concept without pile cap with pier in flyover and bridge construction sector.

Acknowledgements We are thankful to L&T Construction and entire Mumbai Coastal Road Package 1 team for making all the relevant data available for this paper. We are thankful to authors of various books/papers/reports listed below. We are also thankful to DFI-India for giving an opportunity and platform to put up this paper.

References

Arda C (2007) Vertical Load Capacity and Quality Assessments of Bored Piles. Master of Science Thesis, Civil Engineering, Yıldız Technical University
ASTM D4945—12: Standard test method for high-strain dynamic testing of deep foundations
ASTM D8169/D8169M—18: Standard test methods for deep foundations under bi-directional static axial compressive load
ASTM D3966—07: Standard test methods for deep foundations under lateral load

Cherian A (2021) Assessment of pile capacity using bidirectional static load test (BDSLT). Indian Geotech J 51:369–375
Coastal Engineering Manual CEM 2011
Evaluating scour at bridges, 5th edition, FHA US department of Transportation, Circular no 18 April 2012
FHWA GEC 010: AASHTO LRFD Bridge design specifications, 8th Edition 2018
Greater Mumbai Disaster Management Action Plan, Risk Assessment and Response Plan, Vol-1, 2007, Maharashtra Emergency Earthquake Rehabilitation Programme, Government of Maharashtra
IRC:78:2014: Standard specifications and code of practice for road bridges
IRC: SP: 91-2010: Guidelines for Road Tunnel
IS 1892:2021: Subsurface investigation for foundations—Code of practice (second revision)
Ministry of Environment, Forest and Climate Change, Government of India

Deep Foundations in Coralline and Carbonate Rocks, and Any Other Unique and Interesting Papers Related to Deep Foundation Themes

Installation of Three Up to 120 m Deep Diaphragm Wall Shafts, with Diameters Between 8 m and 35 m at Woodsmith Mine in North Yorkshire (England)

Gustav Jahnert, Norbert Hoffmann, and Michael Baltruschat

Abstract Near the port city of Whitby in the north-east of England, Anglo American (AA) is setting up the deepest mine in the UK, which will have a depth of 1,500 m. The purpose of the project is to extract and transport polyhalite, both underground, and to market it globally. As part of the construction works, BAUER Technologies Ltd, the UK subsidiary of BAUER Spezialtiefbau GmbH, constructed three different shafts using the diaphragm wall technique, with depths of up to 120 m. In order to ensure the specified vertical tolerance of maximum 200 mm, several measurement methods were applied, and the results were documented in a 3D model. Complex de-sanding plant technology and specially developed polymer products were adopted to regenerate the trench supporting bentonite slurry.

Keywords Diaphragm wall · Cutter · Polymer · Mining · Tolerances

1 Introduction

The mineral polyhalite—formula: $K_2Ca_2Mg(SO_4)_4{\cdot}2H_2O$ in eastern England as shown in Fig. 1 (Garcia Veigas et al. 2011) was formed or deposited around 260 million years ago, during the evaporation of prehistoric seas in the Permian period. Polyhalite as shown in Fig. 2 is an evaporate mineral, i.e., it was left behind when the sun heated the ancient Zechstein Sea to the point where it evaporated in its entirety (Paul 2017). After the Zechstein Sea had evaporated, mineral deposits or evaporates were left and became entombed in rock and land matter. Polyhalite's multinutrient

G. Jahnert (✉)
Bishops Stortford, England
e-mail: gustav.jahnert@bauertech.co.uk

N. Hoffmann · M. Baltruschat
Schrobenhausen, Germany
e-mail: norbert.hoffmann@bauer.de

M. Baltruschat
e-mail: michael.baltruschat@bauer.de

A. M. Krishna et al. (eds.), *Deep Foundations for Infrastructure Development in India*, Lecture Notes in Civil Engineering 373, https://doi.org/10.1007/978-3-031-37256-8_24

content makes it ideal to be used as a fertilizer in agriculture, due to its high content of potassium, sulphur, calcium, and magnesium, polyhalite offers additional benefits in comparison to contemporary products, thereby increasing its value in the agricultural industry.

In this context, AA is investing three billion dollars in the installation of a polyhalite mine in the United Kingdom to mine the largest polyhalite deposit discovered worldwide so far. Once in full operation, up to 20 million tons of polyhalite per year can be extracted and processed to the end product "POLY4", which will be sold to

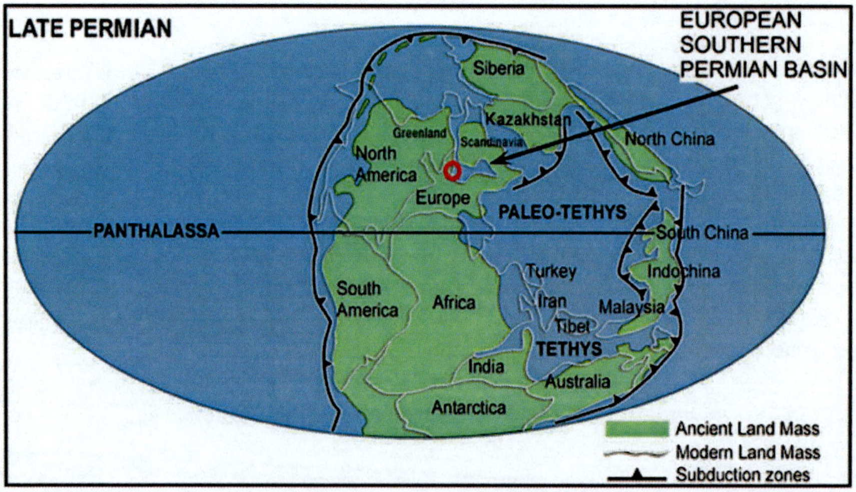

Fig. 1 Woodsmith Mine, located in the European Southern Permian Basin

Fig. 2 Specimens of polyhalite

major international clients to support the food production for the world's population on a sustainable basis in a time of rapid global population growth.

2 Project Description

In 1500 m depth, the mine will have an area of up to 775 km^2. To obtain construction approval for the operation in the middle of the North York Moors National Park, Anglo American had to develop an innovative plant concept. This concept had to enable efficient mining and extraction of the mineral on one hand, and on the other hand, ensure that there would be no long-term impact on the undisturbed natural landscape of the National Park. The approved design includes the installation of two mineshafts up to 1500 m below ground level to mine up to 70 m thick polyhalite seam as shown in Fig. 3. The shafts provide the only access to the mine and are located on the "Woodsmith Mine" site. Heavy mining equipment will be lowered after completion of the shafts to extract the mineral. The crushed rock is then lifted through two shafts to a depth of 300 m below ground level. From there it is transferred to a conveyor belt system to be exposed at the nearest cargo port (Teeside Port). The conveyor belt system is housed in a specially designed 36 km twin tunnel. Therefore, the rock is milled and compressed into pellets in a processing plant installed at the port. Afterwards, the final product POLY4 will be transported by ship to the customers around the globe.

2.1 Diaphragm Wall

Between October 2017 and December 2018, BAUER constructed diaphragm wall shafts up to 120 m in depth, with an internal diameter of up to 7.50 m and a nominal diaphragm wall thickness of 1.20 m. By means of this construction method, as shown in Fig. 4 the near surface, weathered and groundwater bearing rock strata were stabilized. Subsequently, the following works of the mine operator could safely be realized and within the time schedule. The geology in the area of the special civil engineering works is generally comprised of alternating sandstone and mudstone layers with unconfined compressive strengths ranging from 0.5 to 115 MPa. As the ground below a depth of 120 m consists mainly of stable and dry rock, conventional shaft mining methods are used for subsequent excavation down to the bottom of the shaft. In addition to the main shafts, two additional circular diaphragm walls with diameters of 35 m and a depth of 60 m each were installed. These so-called "headgear chambers" as shown in Fig. 5 enclose the main shafts and provide underground space for the conveyor systems essential to the operation of the mine. This customized design is necessary to ensure that the landscape of the park is not affected in the long-term by any industrial plant technology or emissions from the mine operations, as required by the construction permit. After completion of the construction work

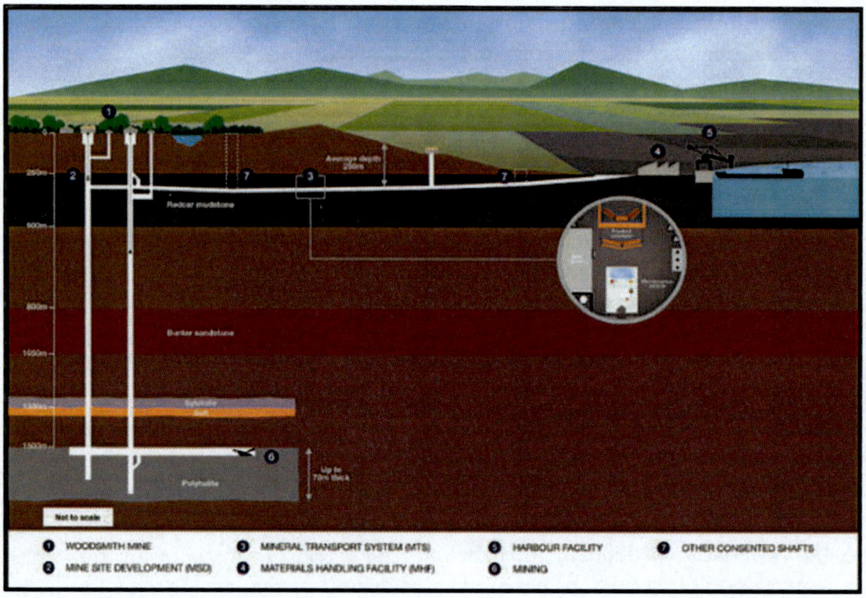

Fig. 3 Woodsmith Mine—cross section

and installation of the plant technology in the area of the Woodsmith Mine, the entire construction site will be renaturalized. All mine operations will subsequently take place underground. Only occasional flat access and administration buildings will remain on the surface of the North York Moors National Park.

2.2 *Design*

The structural analysis of the diaphragm wall shafts was compiled by us, in cooperation with the British Engineering partner Geotechnical Experts. In principle, it was discussed if the weathered rock zone could be defined by Hoek–Brown theory, however, due to the very significant degree of weathering in the upper layers as shown in Fig. 6, the design team, in accordance with the client, opted Mohr Coulomb based on K_0. The entire system was modeled in Plaxis 3D as shown in Fig. 7 (Jahnert et al. 2018). The complex geometry in combination with complicated load cases from the subsequent mine operation posed a special challenge. In addition to the high-water pressure and the influence of the over-consolidated rock-like soils, loads resulting from later operating temperature differences also had to be taken into account. In the end, a static system emerged in which the 60 m deep, outer shafts are mainly loaded by annular pressure forces, while the inner, deep shafts are primarily stressed by vertical forces, similar to an uplift protection system.

Fig. 4 Construction sequences for diaphragm wall using cutter

3 Main Equipment

Three independently operating, hydraulically driven cutter systems of the type "BC 40" were deployed to excavate the diaphragm wall panels. The BC 40 cutters have a cross-sectional area of 2.80 m × 1.20 m (Fig. 9). The cutter was mounted on MC 96 and MC 128 base carriers, and each was equipped with 120 m long Hose Drum System (HDS) as shown in Fig. 8. In order to be able to react to the highly heterogeneous soil conditions, all three above-mentioned units had different cutting wheels as shown in Fig. 9 and tooth types available. This approach ensured that the highly heterogeneous soil layers could be loosened in an individually optimized way.

During cutting operation, all cuttings are conveyed through the bentonite slurry to the de-sanding plant (Fig. 10). The de-sanding process is a particular focus for diaphragm wall projects when using cutter, compared to grab and rotary drilling sites. In order to operate the three milling machines efficiently, 3,500 m^3 of bentonite slurry was kept in stock and circulated between the cutter, de-sanding and storage tanks as shown in Fig. 10. Up to 1,000 m^3 of working slurry, from the three installed MAT BE500 and BE550 de-sanding plants, were processed every hour. In order to control the rapid increase in density of the support slurry caused by fine particles in the claystone, the so-called separation cut d50 of the de-sanding plants was reduced from 60 μm to 10 μm by using desilters and decanter centrifuges.

Fig. 5 Headgear chambers (2021)

4 Tolerances

The contractually specified tolerance for vertical deviation of individual diaphragm wall panels was 200 mm. Based on a depth of 120 m, the resulting deviation is 1:650 or 0.17%. Exceeding this limit would reduce the inner diameter of the shaft, which

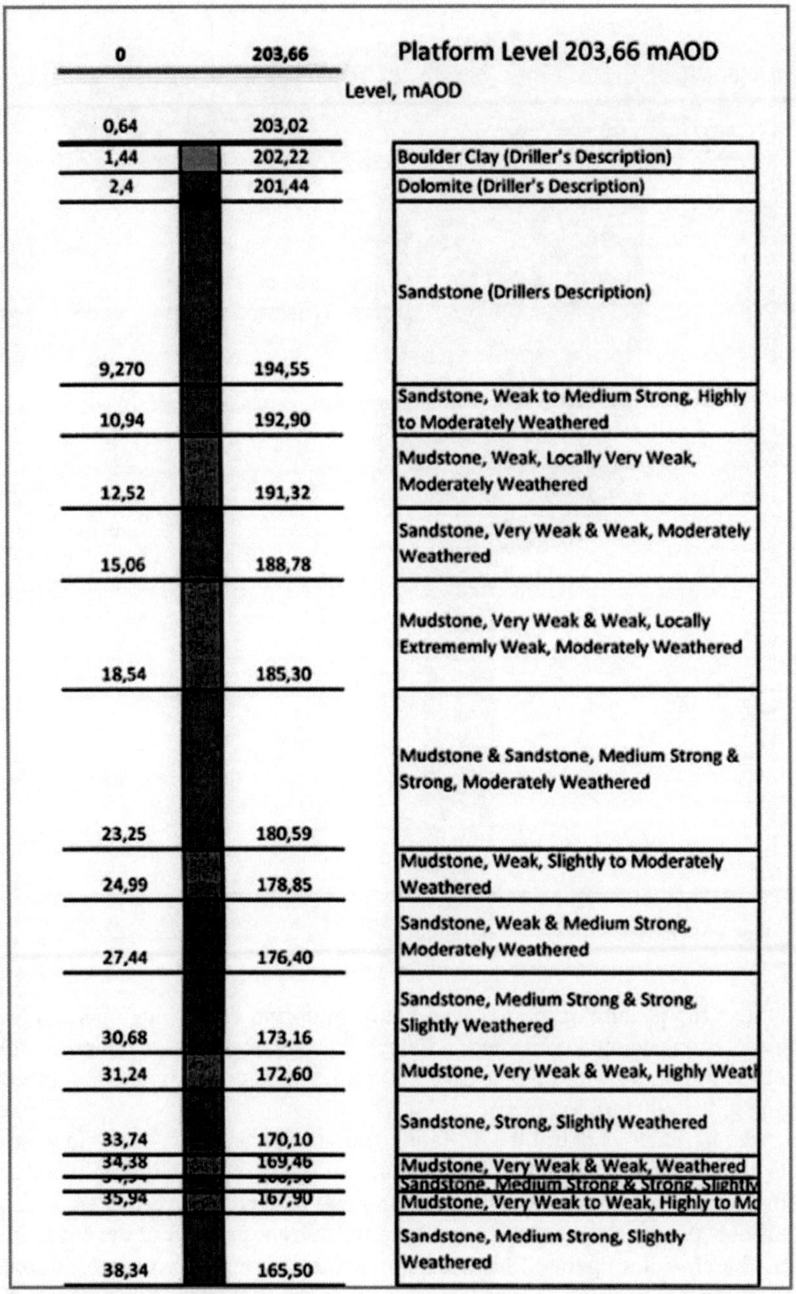

Fig. 6 Soil layers in the upper 38 m

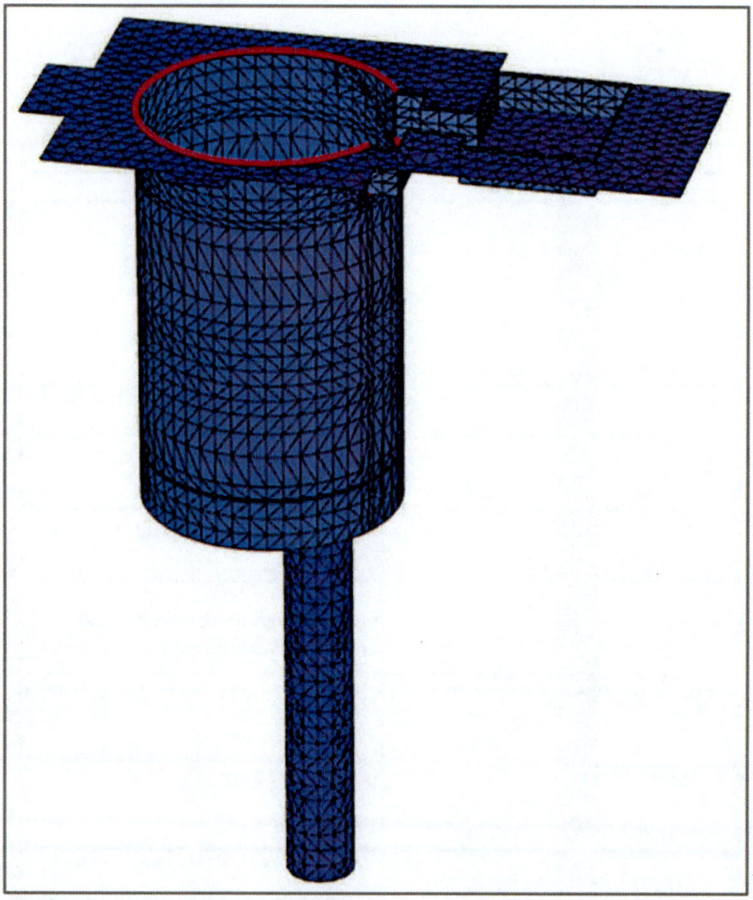

Fig. 7 Plaxis 3D-model

could affect the placing of the working platform during the mining operations and thus lead to considerable construction delays. Furthermore, the structural design is based on a minimum wall thickness of 800 mm, which can only be guaranteed by adhering to the 200 mm tolerance.

In order to reliably control the deviation from the vertical in x-axis and y-axis as well as rotation during the panel excavation, we implemented different measuring systems. In the operator's cabin, the measurement results were permanently visualized to show the operator without any doubt the current position of the cutter in the trench. Therefore, the operators have the option of rectifying the position of the cutter during excavation by selectively extending one or more of the sixteen hydraulic flaps integrated in the cutter frame. The basic measurement of the vertical deviation is continuously carried out by inclinometers integrated in the cutter frame machines. A so-called Cutter-Inclination-Survey (CIS) measurement was also carried out at

Fig. 8 Three MC base carriers during diaphragm wall construction

Fig. 9 Cutter wheels

Fig. 10 Mixing and de-sanding plant, and storage tanks for bentonite slurry

intervals of 2.5 m depth. In this system, the ropes of the cutter are measured at three points between the ground level and the mast head. By extrapolation over the panel depth, the position and rotation of the cutter can thus be determined as shown in Fig. 11. After reaching the final depth, the final dimension of each individual panel was determined by a KODEN ultrasonic survey. Afterwards, the data from all three measuring systems, i.e., inclinometer, CIS and KODEN, were combined in the software As-Built Generator (ABG). The result can be presented as a three-dimensional drawing to prove the achieved tolerances as shown in Fig. 12.

5 Spacers

The shafts constructed by means of the diaphragm wall technique consist of individual panels, each 2.80 m long and with a nominal thickness of 1.20 m. In the pilgrim step method applied, only primary panels are constructed in the first step. As shown in Fig. 13. The individual panels will be constructed using the primary/secondary panel method. At first, only primary panels get constructed. That followed the intermediate secondary panels getting constructed. The timing when the excavation of secondary panels can commence depends on the strength of the cast in-situ concrete of the previously constructed adjacent primary panels. Secondary panels will overcut

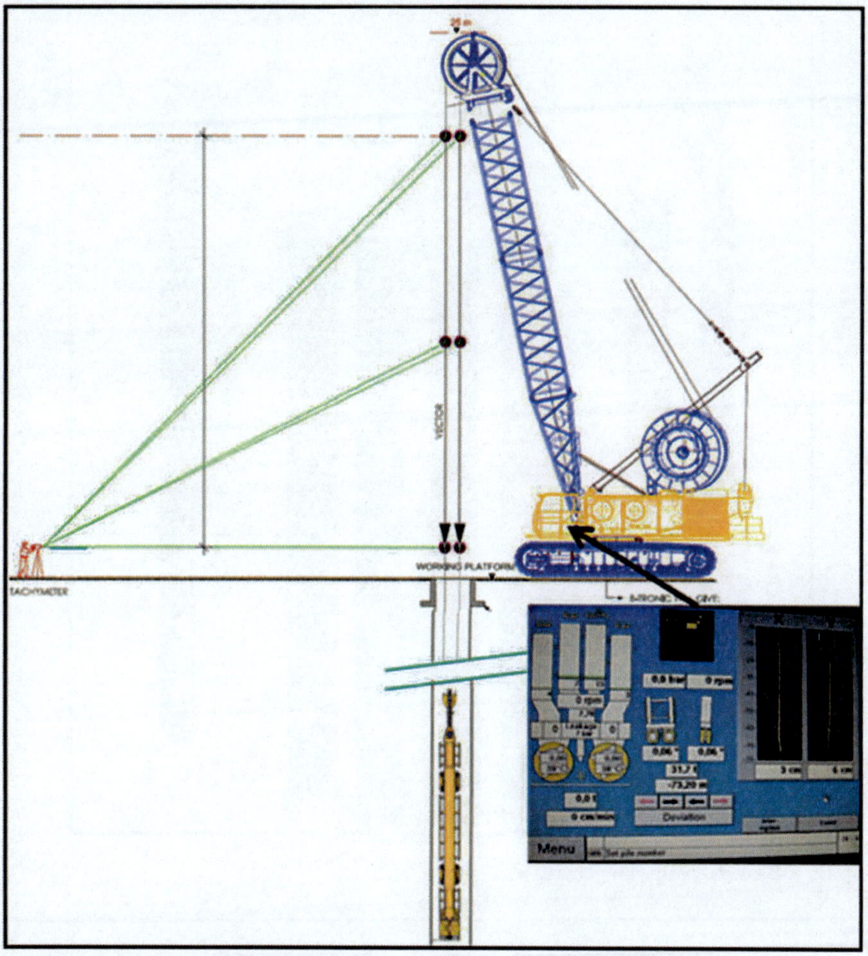

Fig. 11 Cutter-Inclination System (CIS)—schematically

on both ends into the adjacent primary panels. The overcut length at guide wall level is predefined by the panel layout to provide a continuous overcut over the whole joint length. The cutting wheels are cutting into the cast in-situ concrete of the two adjacent primary panels resulting in a roughly grooved surface ensuring a high quality of the joint between the panels.

In order to create a structural and watertight bond between the respective adjacent panels, primary panels are cut during the excavation of the secondary panels. The overcut of the secondary panels into the concrete of the two adjacent primary panels can be up to 40 cm. As the static calculation of the walls required reinforcement in all panels, prefabricated reinforcement cages had to be installed in the primary

Fig. 12. 3D-model of diaphragm wall panels completed

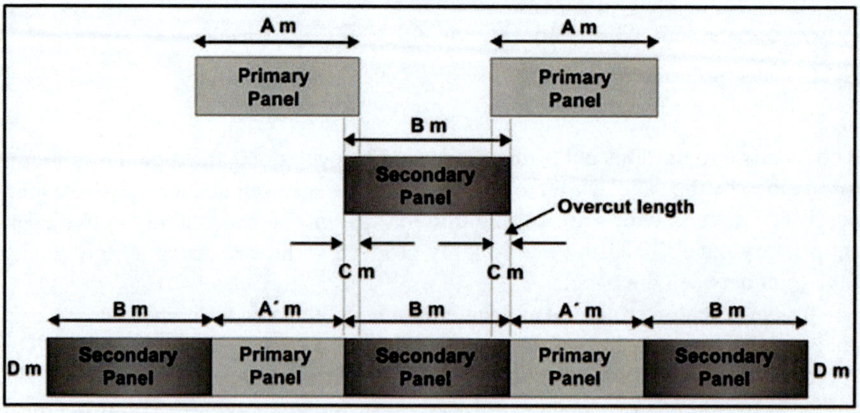

Fig. 13 Pilgrim step method

panels in an exact position to avoid any subsequent contact between the cutter and the reinforcement during the excavation of the secondary panels.

In order to secure the safe positioning or centering of the reinforcement cages in the primary panels, we in cooperation with Schoeck Bauteile GmbH—developed a system in which the spacers are made of glass-fiber reinforced plastic (GRP). The GRP spacers as shown in Fig. 14 could be solidly connected to the reinforcement cage made of steel. In comparison to steel, the glass-fiber reinforced plastic does not cause any problems when the primary panels are overcut.

Fig. 14 GRP spacers connected with the reinforcement cage

6 Polymer

In the preparation phase of the diaphragm wall construction, it could be derived from the available geotechnical data that slurry loss is to be expected at some depths due to fissures in the sandstone. While cracks with a gap width of up to 5 mm can be sealed automatically by forming a filter cake from the bentonite suspension, special measures were necessary to seal larger openings of up to 10 cm. For this purpose, a new polymer product as shown in Fig. 15 was used, which had been developed in the Group's laboratories in recent years and successfully applied on several construction sites in the Far East. The fast-swelling polymer LTA3 can be added as a powder into the slurry circulation before the cutter reached a certain fissured rock formation. The individual LTA3 particles absorb up to 300 times their own weight in water within a few minutes and become a granular, elastic structure. The resulting particles penetrate larger cracks and seal them within a very short time. The initially measured slurry loss of 250 m^3 per panel could be reduced to approximately 30 m^3 by adding LTA3.

Fig. 15 Polymer after adding water

6.1 B-tronic Activity

In order to optimize the daily excavation performance of the cutter deployed, the recording and evaluation of machine data, the cutting progress, the geology encountered, and the duration of non-productive processes are a top priority. To ensure a proper and detailed recording of the necessary parameters, the digital performance data recording system "B-Tronic Activity" was implemented. An additional tablet as shown in Fig. 16 is therefore installed in each driver's cab which is used by the operator. With a clear user interface, the tablet enables the activity of the machine to be clearly divided into different activities and recorded without gaps. The performance data of the machine itself is automatically transmitted to a central database via a SIM card inserted in the device. Together with the driver's records, this provides a comprehensive overview of the overall performance and gives important information on possible optimization measures.

Fig. 16 Tablet used by the operator

7 Health and Safety

The safety and health of all persons involved in construction is always a top priority for BAUER. By establishing a safety culture within the Group and especially on construction sites worldwide, the foundation is laid for sustainable and effective safety management.

Especially on a project running around the clock with up to 200 workers from all over the world, everyone's safety awareness must be developed—before work begins—and then continuously advanced and supported. A comprehensive and clear site induction for all new employees at the beginning ensures that the scope of work, as well as the general and site-specific safety requirements and environmental regulations, are clearly communicated and understood. Through daily construction meetings, shift briefings and toolbox talks, up-to-date information and objectives are continuously communicated to all parties involved. By means of an efficient suggestion system with regular awards and permanent feedback on the improvements implemented, it is also achieved that the entire team actively participates in the optimisation of safety, quality and performance in order to ensure a high standard in all the aspects mentioned.

7.1 Summary and Conclusions

Following the successful completion of the wall, the owner was able to start the subsequent works. Currently, the information and photos available to us indicate that the removal of the shafts, the installation of the necessary mining equipment, and the construction of the shafts down to a depth of 1500 m are successfully proceeding. For correct information, please refer to https://uk.angloamerican.com/the-woodsmith-project.

References

Garcia-Veigas J, Pueya JJ, Cendón DI, Peryt T (2011) Zechstein saline brines in Poland, evidence over overturned anoxic ocean during the Late Permian mass extinction event. Article in Chemical Geology

Jahnert G, Hoffmann N (2018) Herstellung von 120m tiefen Schlitzwandschächten, DGGT 35. Baugrundtagung, Tagungsband, Stuttgart, 273–278

Paul J (2017) Das Zechstein Meer: ein europäisches Binnenmeer vor 250 Mio. Jahren, https://www.karstwanderweg.de/sympo/13/paul/index.htm, 13. Südharz-Symposium, Bad Sachsa

Replacing Top Layer of Soft Clay with Compacted Engineering Fill to Improve the Lateral Capacity of Piles

Thomas John, Venugopal Bonasi, and Kumaran M

Abstract Liquid storage tanks are usually located in nearshore or coastal regions with soil profiles containing thick layers of compressible soils. These storage tanks are usually of large diameter up to 90 m and carry high loads, and the tanks are supported by heavy deep foundations such as piles. These structures are subjected to huge lateral loads due to wind and earthquakes, and any damage to storage tanks with highly inflammable liquids leads to disastrous conditions. So, it is important that the top layer of soil (up to the fixity depth) should be strong enough to provide the required lateral load capacity to the pile. In a project for constructing LNG terminal in Odisha containing 2nos of 89 m diameter LNG storage tanks supported with pile foundation located near to coastal region, for improving the lateral capacity of existing soil, the top 3 m was replaced with compacted engineered fill (GSB). The site was a filled-up area with dredged soil from the seabed, the top 5–10 m were soft to medium stiff clay with SPT N less than 10. The lateral design load on pile was 534 kN under free head condition. Actual lateral capacity for the strata was around 200 kN under free head condition which was less than the required capacity as per structural requirements. Therefore, for improving the lateral capacity of pile the top 3 m soil is replaced with well compacted engineered fill (GSB) in layers of 150 mm and the capacity is proved by conducting initial and routine lateral load tests. This paper describes the properties of the engineered fill used and how the lateral load capacity of pile was improved to meet the design requirement.

Keywords LNG tanks · Pile foundation · Lateral load capacity · Engineered fill · Load test

T. John (✉) · V. Bonasi · K. M
L&T Geostructure, Chennai, Tamil Nadu, India
e-mail: thomasjohn@lntecc.com

A. M. Krishna et al. (eds.), *Deep Foundations for Infrastructure Development in India*, Lecture Notes in Civil Engineering 373, https://doi.org/10.1007/978-3-031-37256-8_25

1 Introduction

A foundation of a structure will be subjected to huge lateral force mainly due to wind (if the height of the structure above foundation is more) and earthquake force. In case of liquid storage tanks which are normally constructed nearshore or coastal regions, will be subjected to huge lateral force mainly due to the wind and cyclones. Mainly for such huge structures, piles are considered as suitable foundation system to resist the huge lateral loads. Due to this huge lateral force, these pile foundations are subjected to lateral deflection, but this deflection is restricted to some definite values as per various codes & standards. So, the foundation for these structures shall be capable enough to resist lateral force. The ultimate resistance of a vertical pile to a lateral load and the deflection of the pile as the load builds up to its ultimate value are more complex matters involving the interaction between a semi-rigid structural element and the soil, which deforms partly elastic and partly plastic. Usually pile foundations find resistance to lateral loads from (a) Passive soil resistance on the face of the pile cap, (b) Shear on the base of the pile cap, and (c) soil resistance against the pile shaft upto the depth of fixity. The latter source is usually the only reliable one. So, it is important that the top layer of soil should be strong enough to give passive resistance to the pile shaft.

2 Project Scope and Stratigraphy

The project consisted of construction of Bored Cast-In-Situ pile foundation for 2 Nos of LNG tanks of 89 m diameter constructed near shore. Details of pile foundation for each LNG tank are given in Table 1.

The soil profile of the project location shows that top 20 m is very soft to soft clay. The site location is a filled-up area using dredged soil from seabed. Due to the presence of soft clay for a depth of approximately 20 m, permanent steel casing of 8 mm thk & 1200 mm inner dia is provided for working piles upto 24 m to avoid the borehole collapse while boring and pile is terminated in sand strata at 56 m at the center location and in other locations where dense sand strata are available after 60 m, the pile is terminated at deeper depth of 61 m as. Typical soil profile is shown in Figs. 1 and 2 shows the actual sub-soil strata at site near the tank location.

Table 1 Details of pile foundation

Tank no	Type of pile	Pile diameter (mm)	Pile depth (m)	Number of piles
LNG Tank-01	BCIS	1200	58	242
		1200	63	194
LNG Tank-02		1200	58	204
		1200	63	232

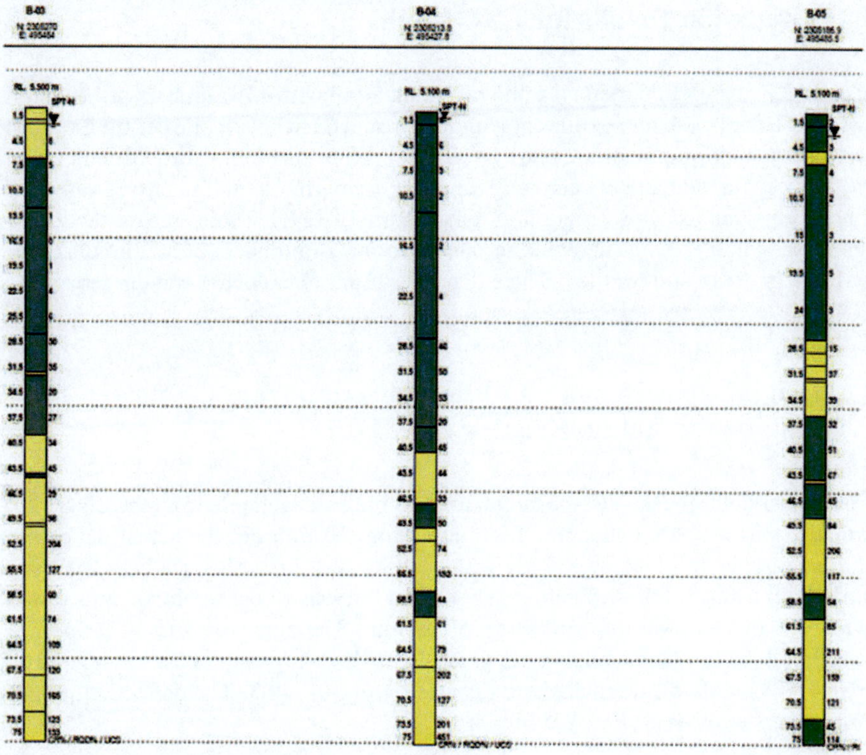

Fig. 1 Typical sub-soil profile of site

Fig. 2 Typical view of existing soil near to tank location

3 Foundation Details for LNG Tank

Each LNG tank is supported on a pile raft foundation with 436 numbers of 1200 mm diameter Bored Cast-in-situ piles of length 58 m and 63 m below pile cut-off level. As per the structural requirement, pile cut-off level is 2 m above Finished Ground Level (FGL) to avoid contact between tank base slab with ground and to ensure safety of connections due to large variation in temperature of LNG. Typical cross-section of tank is shown in Fig. 3. The working pile is having 24 numbers of 32 mm dia rebar for top 8 m depth and for the balance depth its 16 numbers of 20 mm dia rebar.

4 Design Lateral Load and Pile Deflection from L-Pile Analysis for Actual Existing Soil Data

From the structural analysis, the lateral load coming on each pile from the LNG tank structure was 534 kN under free head condition. To find out the actual deflection of pile under lateral load of 534 kN with existing soil properties, analysis has been carried out using L-pile software(2). The deflection was found to be 106 mm which is more than the allowable deflection of 75 mm. The summary of soil properties considered for the L-pile analysis is given in Table 2. Cohesion C is considered as 5xSPT-N & angle of internal friction is taken from IS 6403. The deflection profile from L-pile analysis is shown in Fig. 4.

5 Improved Soil Parameters and Initial Load Test

As the lateral capacity of the pile(1) with actual sub-soil data is less than the required capacity, some improvement is required in the top soil layer. So, to increase the lateral capacity of the pile, it is decided to replace the top 3 m of existing soil with compacted engineered fill. The engineered fill comprises of 53 mm max. size 'Crusher Run' complying with requirements of Cl.8.5.2 of IRC SP:2-2002. The grading of the fill is given in Table 3. The material is placed on a base layer of TERRAM 1000 Geotextile and is compacted in 150 mm compacted thickness layers with smooth wheeled vibrating roller of minimum 8–10 ton capacity. Each layer of material is compacted to not less than 98% of the IS heavy compaction density as per IS 2720 (Part 8). Figure 5 shows the replacement of top soil with engineered fill in site.

To confirm the lateral capacity of the pile in replaced soil strata, initial lateral load test was conducted as per the client requirement/project specification. For lateral load test, the maximum test load applied was 1335 kN which is 2.5 times the design load (534 kN). The maximum deflection observed for the applied test load was 76.06 mm which was less than the allowable deflection of 125 mm as per project specification.

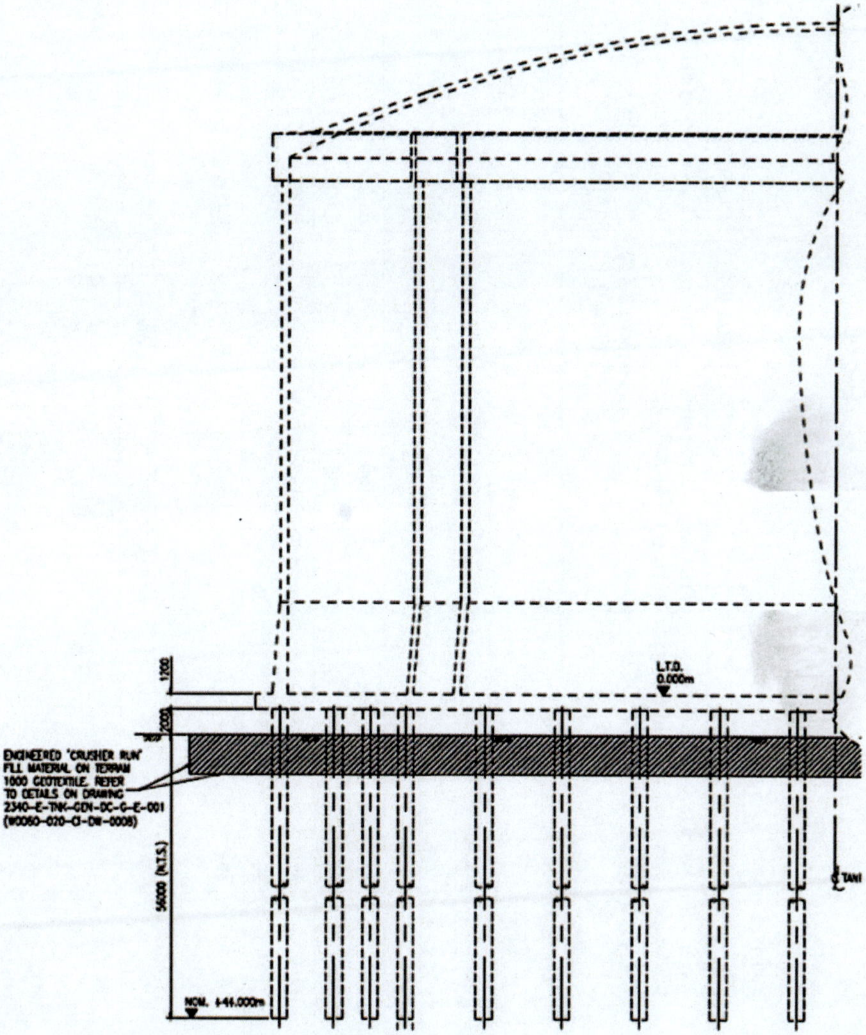

Fig. 3 Typ. Cross-section of LNG tanks along with foundation

The initial and routine load test arrangement executed at site is shown in Figs. 6 and 7. Figure 8 shows the load–deflection graph from initial pile load test results.

From the initial load test results, again L-pile analysis was carried out to find the equivalent properties of engineered fill layer. Based on this analysis the soil properties of engineered fill are given in Table 4 below. The deflection curve from L-pile analysis with improved parameters is shown in Fig. 9 (Fig. 10).

Table 2 Soil parameter considered for L-pile analysis as per existing borehole data

									Sand (IS)	Clay (IS)	
Type of soil/rock as per borelog	L-pile model	Soil layer		Layer thickness (m)	Average corrected SPT N value	Effective unit weight (kN/m³)	Friction angle (°)	Cohesion C kN/m²	Submerged	Non default strain factor E50	Non default factor k (kN/m³)
		From (m)	To (m)						Soil modulus K (kN/m³)		
C	Stiff clay with free water	2	5.0	3	3	18	–	15	–	0.020	5,538
C		5.0	11.0	6	3	18	–	15	–	0.020	5,538
C		11.0	14.0	3	11	18	–	55	–	0.010	19,800
C		14.0	32.0	18	4	18	–	20	–	0.020	7,269
S	Sand	32.0	38.0	6	42	18	36	–	6,960	–	–
C	Stiff clay with free water	38.0	41.0	3	32	18	–	160	–	0.006	57,600
C		41.0	43.0	2	46	18	–	230	–	0.005	82,800
C		43.0	50.0	7	39	18	–	195	–	0.005	70,200
S	Sand	50.0	59.0	9	32	18	36	–	4,568	–	–
C	Stiff clay with free water	59.0	63.0	4	76	18	–	380	–	0.004	1,36,800
S	Sand	63.0	72.0	9	28	18	35	–	3,992	–	–

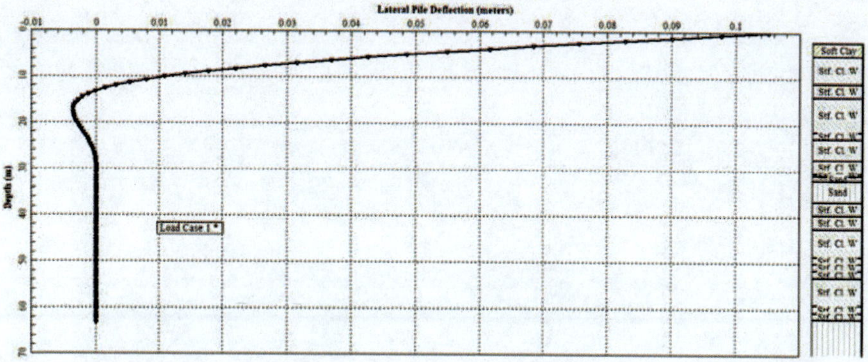

Fig. 4 L-pile output showing lateral deflection of pile for design load as per existing soil layer properties below FGL

Table 3 Aggregate grading requirement for engineered fill (GSB)

Sieve size (mm)	% Passing by weight
63	100
45	87–100
22.4	50–85
5.6	25–45
0.7	10–25
0.09	2–9

Fig. 5 Site photo for the top 3 m soil replacement by engineered fill

Fig. 6 Initial lateral load test arrangement

Fig. 7 Routine lateral load test arrangement

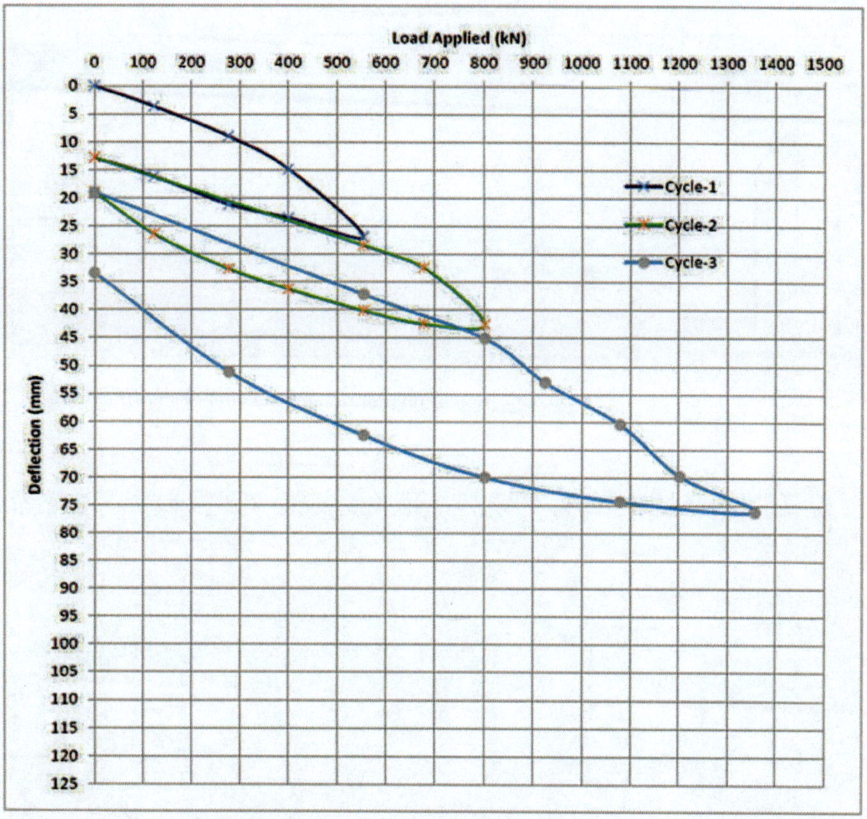

Fig. 8 Load–Deflection curve from initial pile load test results

Table 4 Properties of engineered fill from L-pile analysis

Type of replaced material	Depth of fill below FGL (m)	Effective unit weight (kN/m^3)	Angle of internal friction (degree)	Modulus of subgrade reaction (kN/m^3)
Engineered fill	3 m	23.3	38	57,015

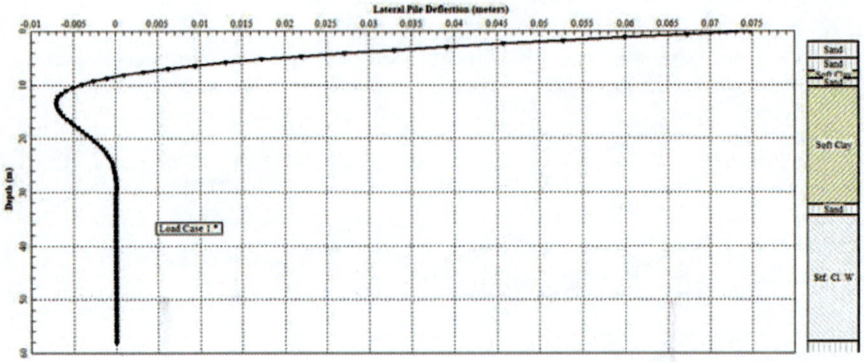

Fig. 9 L-pile output showing lateral deflection of pile for design load after engineering fill for top 3 m depth below FGL

Fig. 10 View for completed working piles for LNG tank raft

6 Conclusion

From the above data from L-pile analysis and initial lateral load test following points can be concluded,

1. For increasing the lateral capacity of piles, the top layer of existing soft or loose soil can be replaced with well compacted layer of engineered fill/crusher run/ GSB for a depth of 3 m.
2. From the L-pile analysis with existing soil properties, the lateral capacity was found to be 300 kN against the required lateral capacity of 534 kN and the deflection for 534 kN was 106 mm which was higher than the allowable limits.

3. After replacement of top 3 m with engineering fill, it was found that the lateral capacity of pile increased and the same is confirmed by conducting initial and routine pile load tests.

References

IS 2911 (Part-1, Sec-2): 2010, *Design and construction of pile foundation—Code of practice Part-1 Concrete piles, Section-2 Bored Cast-In-situ concrete piles*

Technical Manual for L-pile 2016